CROP IMPROVEMENT WITH INDUCED GENETIC VARIATION TO COPE WITH DROUGHT IN RICE AND SORGHUM

The Agency's Statute was approved on 23 October 1956 by the Conference on the Statute of the IAEA held at United Nations Headquarters, New York; it entered into force on 29 July 1957. The Headquarters of the Agency are situated in Vienna. Its principal objective is "to accelerate and enlarge the contribution of atomic energy to peace, health and prosperity throughout the world".

CROP IMPROVEMENT WITH INDUCED GENETIC VARIATION TO COPE WITH DROUGHT IN RICE AND SORGHUM

COORDINATED RESEARCH PROJECT
ORGANIZED BY THE
JOINT FAO/IAEA CENTRE OF NUCLEAR TECHNIQUES IN FOOD AND AGRICULTURE

EDITORS:
F. SARSU, S. PENNA, S. SIVASANKAR

INTERNATIONAL ATOMIC ENERGY AGENCY
VIENNA, 2024

For further information on this publication, please contact:

Plant Breeding and Genetics Section
International Atomic Energy Agency
Vienna International Centre
PO Box 100
1400 Vienna, Austria
Email: Official.Mail@iaea.org

© IAEA, 2024
Printed by the IAEA in Austria
November 2024
https://doi.org/10.61092/iaea.fngh-4elo

IAEA Library Cataloguing in Publication Data

Names: International Atomic Energy Agency.
Title: Crop improvement with induced genetic variation to cope with drought in rice and
   sorghum / International Atomic Energy Agency.
Description: Vienna : International Atomic Energy Agency, 2024. | Includes bibliographical
   references.
Identifiers: IAEAL 24-01725 | ISBN 978-92-0-135024-4 (paperback : alk. paper) |
   ISBN 978-92-0-135124-1 (pdf) |
Subjects: LCSH: | Plant breeding. | Plant breeding — Research. | Plant genetics. | Crop
   improvement.
Classification: UDC 631.528 | CRCP/IGV/004

FOREWORD

Securing a sustainable global food supply is a pressing concern. As the planet faces the profound impacts of climate change, the quest for sustainable global food security remains paramount. The fragile balance of the global ecosystem is constantly challenged, with droughts emerging as one of the most formidable adversaries. Water scarcities pose serious threats to global food staples, particularly rice and sorghum, which form the dietary foundation for billions of people.

Induced mutation has been utilized in crop breeding since the 1920s. Currently, over 3400 mutant crop varieties are documented in a database managed by the Food and Agriculture Organization of the United Nations (FAO) and IAEA. The effectiveness of breeding crop varieties can be enhanced by advancing and adapting technologies that optimize mutation density. This also involves refining the efficiency of screening large mutant populations or lines, both phenotypically and genotypically. In light of these goals, the Joint FAO/IAEA Centre of Nuclear Techniques in Food and Agriculture initiated a five year coordinated research project entitled Improving Resilience to Drought in Rice and Sorghum Through Mutation Breeding. The project brought together researchers from developed and developing countries with the aim of improving the drought resilience of rice and sorghum germplasm through induced mutation and the development and adaptation of screening techniques for sustainable food security.

This publication presents the output of the coordinated research project with contributions from nearly sixty scientists from seven countries. Part 1 contains a general introductory paper covering breeding approaches for the improvement of drought tolerance in crops. Part 2 comprises ten papers covering aspects of radiation mutagenesis and pre-field screening and field evaluation of mutants. Part 3 has papers on morphophysiological and biochemical characterization of mutants, while Part 4 includes papers on the molecular characterization of the mutants. Part 5 provides a future perspective on an integrated mutation breeding approach to develop drought resilience in crop plants.

This publication serves as a reference for scientists, plant breeders and plant biotechnologists with an interest in improving drought tolerance in rice and sorghum. It envisions the use of mutation breeding programmes with advanced biotechniques for improving crop varieties with higher yields and wider adaptability to drought conditions. These screening techniques are an essential technological advancement that lead to genetic improvement in rice and sorghum crops, allowing them to adapt to future climatic conditions.

The FAO and the IAEA are grateful to the participants of the coordinated research project for their contributions to this publication. Their efforts were key in testing and validating the protocols and providing feedback for rice and sorghum mutation breeding programmes. The IAEA officer responsible for this publication was F. Sarsu of the Joint FAO/IAEA Centre of Nuclear Techniques in Food and Agriculture.

CONTENTS

PART 1
INTRODUCTION

1. PHYSIOLOGICAL AND GENOMICS PERSPECTIVES FOR IMPROVING DROUGHT TOLERANCE IN MUTANT CROP PLANTS

S. FATMA[1], I.K. BIMPONG[1], S. PENNA[2],
S. SIVASANKAR[1]

[1]Plant Breeding and Genetics Section,
Joint FAO/IAEA Centre of Nuclear Techniques in Food and Agriculture,
International Atomic Energy Agency, Vienna

[2]Amity Center for Nuclear Biotechnology, Amity Institute of Biotechnology,
Amity University of Maharashtra Mumbai,
Mumbai, India

Abstract

Drought stress, an important environmental stress factor, has become a serious limitation for world agriculture. In several crop plants, severe yield losses to the extent of >50% have been incurred, making it necessary to understand how plants integrate drought cues with growth and development. Improving drought tolerance in crops is a top priority to increase the water use efficiency and to enhance agricultural water productivity. Speeding up the pace of breeding for developing improved varieties requires using different genomics approaches including induced mutagenesis. Diverse physio-morphological characters and secondary traits such as root architecture, leaf water potential, photosynthetic and stomatal traits, panicle water potential, osmotic adjustment, and relative water content are now used for screening of mutant population and crop germplasm. Physiological, biochemical, genomics and metabolomics approaches have become useful tools to unravel the metabolic pathways and genetic candidates that confer drought tolerance in crops. Several quantitative trait loci (QTLs) for agronomic traits associated with drought tolerance have been shown to have significant phenotypic variance, and these have been used for the development of pre-breeding material with improved drought tolerance. Further, advanced genomics resources have enabled the characterization of drought tolerance QTLs, enabling the identification and corroboration of potential candidate genes to develop transgenic plants. Considerable success is being achieved in the development of drought-tolerant mutants in some crops and, extensive research on genetic and genomic analysis is needed to establish genetic associations for use in breeding crop improvement programmes. It is also essential to combine high throughput phenomic tools in screening, and employ crop physiology, genomics and breeding approaches to develop the next-generation drought tolerant crops.

Key words: Drought stress, induced mutations, physio-morphological characters, QTLs, reverse genetics, crop improvement, drought tolerant crops

1. INTRODUCTION

Globally, climate change has made an impact on organisms and ecosystems through environmental perturbations, such as rise in temperature up to 1.5°C, higher frequency, intensity and/or amount of heavy precipitation, and incidences of droughts in several countries [1]. Since 1990, there has been an increase in land temperature and it is projected that by this century, there will be a much higher temperature rise of up to 2.6–4.8°C [2]. Hotter and drier climates will further intensify the problem of drought spells and globally this will create alterations in regional climates and shifts in agricultural systems and crop production cycles calling for new technological interventions [2, 3].

During the Green Revolution, there was a major breeding focus on the development of high-yield varieties and their adaptation to new farming practices. For example, semi-dwarf varieties in cereals (wheat and rice) were developed with better response to fertilizer and irrigation application without encountering any of the lodging losses associated with tall varieties. Despite such developments and the development of hybrid crop varieties, food security is still a major concern due to the imminent threat from environmental extremes. This, combined with the expected increase in human population that is projected to reach 10–12 billion by the turn of this century [4], calls for the development of sustainable future food resources under a progressively warmer and drier crop production environment.

The sustainability of crop production and food security is always threatened by the increasing unpredictability and severity of drought stress, caused by global climate changes. The effect of droughts on crop productivity in all regions of the world is well documented [5–8]. The last decade has seen several important reviews and publications of plant drought response and tolerance (http://www.plantstress.com/files/Recent_Reviews/index.asp) [9]. Important strategies applied to combat drought include physiological [10], genetic [11, 12], and genomics approaches [13–16]. In this chapter, we have presented different breeding approaches for the improvement of drought tolerance in crops using the information generated through physiological and genomics approaches. This paper also provides research outlines that often are necessary for screening for drought tolerance, especially by using the tools of induced mutagenesis.

2. DROUGHT STRESS

Drought majorly impacts agriculture and, up to 80% of the effects are compounded on water availability, agricultural production, food security and rural livelihoods [17, 18]. Drought stress is also a limiting factor for achieving higher plant yield through a multitude of effects on plant cellular, morpho-biochemical processes [19]. Drought stress can reduce crop yield up to 50% and, although all the plant developmental stages show sensitivity to drought stress, reproductive stages have been shown to be highly susceptible, making the ultimate impact on yield in many crop plants [20].

Plant responses to drought stress at the cellular, biochemical and physiological levels have been elegantly described by Levitt [21]. Drought stress can cause various morphological, physiological, biochemical, and molecular changes in both the below-ground and above-ground tissues of crop plants (Fig. 1). The water content being reduced, the plant experiences a diminution of leaf water potential and turgor loss. Leaf curling, partial or complete stomatal closure, decrease in cell enlargement and growth, and a decrease of internal CO_2 causing a decrease of photosynthetic activity are some of the physiological changes upon the incidence of drought [22–24].

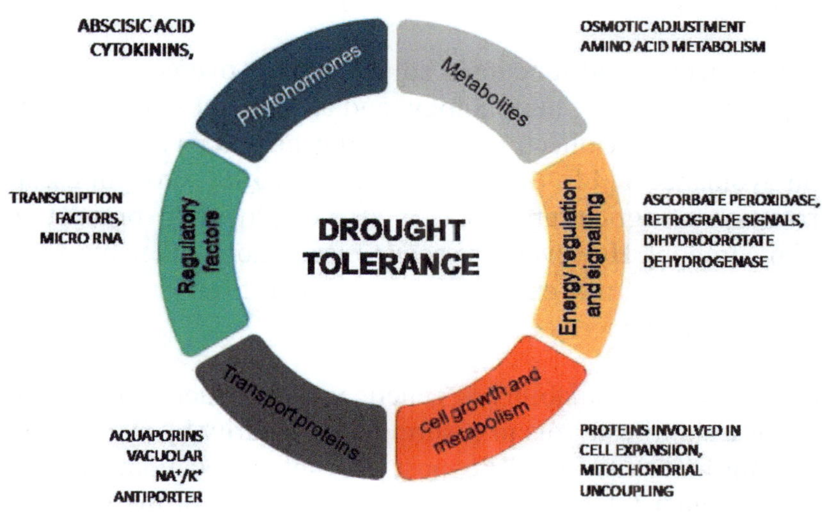

FIG. 1. Mechanisms of drought tolerance in plants.

These physio-morphological alterations lead to a reduction in leaf area and leaf development [25–27], growth rate and, consequently, thickening of the roots [27, 28], leading to an overall reduction of plant growth. If the stress is too severe, photosynthesis can stop, perturbing general metabolic activities and ultimately leading to the death of the plant [29]. In triticale, if drought occurs during the pre-anthesis stage, it shortens the time to anthesis while its occurrence post anthesis reduces the period of grain filling [30]. The major negative impact of drought stress on crop plants is the reduction in fresh and dry biomass production [31], which affects grain number and grain size in wheat [32], and grain yield in maize [33, 34].

To reduce the risk of yield losses and to develop more drought tolerant crops, we need to understand how plants integrate drought cues with growth and development. Drought tolerance is defined as the ability to sustain growth, flowering, and yields even in sub-optimally hydrated soils [31]. A common response in plants coping with drought stress is stomatal closure. This stabilizes cell turgor and permits continued cellular metabolism. However, since stomatal closure also impairs photosynthetic rate, plants must constantly adjust stomatal conductance to maintain a balance between sufficient CO_2 uptake and water loss. Plants must then permanently sense water deficit. Abscisic acid (ABA) has been described to play a crucial role in stress signalling during drought at different levels, including transcriptional changes and promoting stomatal closure [35, 36]. A drought tolerant plant can maintain cellular and physiological function, including growth and seed/fruit production, under conditions of drought stress [37]. For a plant to be able to continue its cellular and physiological functions during drought, it needs to use the available water efficiently. Water use efficiency (WUE) is the ratio between the carbon gain to the water use/loss [38]. Improved WUE is an important trait in plant breeding [39]. In the context of drought, this can be enabled by the constitutive generation of deep roots, the generation of "stay-green" cultivars, enhanced proline accumulation (via the P5CR gene), which promotes cellular protection and reactive oxygen species (ROS) scavenging, and the use of accumulated stem reserves for grain filling [40]. Drought tolerance can also be enhanced by the accumulation of late embryogenesis abundant (LEA) proteins [41]. These hydrophilic proteins offer protection against desiccation damage through various means including antioxidant activity, and membrane and protein stabilization [42].

3.   DROUGHT STRESS RESPONSES AND CROP GENETIC DIVERSITY

Plants adopt various structural and functional adjustments to overcome the effects of drought stress, such as adapting phenology, morphology, and anatomical structures and physiological and biochemical reactions [43]. Collectively, these adjustments involve drought avoidance (or 'shoot dehydration avoidance') [44], drought tolerance [45, 46], drought escape, and drought recovery [47]. Irrigation of agricultural areas is also employed to prevent substantial yield reduction imposed by drought. However, it is economically not sustainable and costly for small scale farmers and a threat to the environment as water from irrigation can trigger land degradation and soil salinization.

Genetic diversity, which is the availability of genetic variation (heritable traits in a population) of a given species, has a significant role in ensuring food security by increasing farmer income, and food production for humanity [48]. The most relevant and economically sound solution is to develop crops with higher tolerance to drought stress. Most of the research on improving drought tolerance in crops has relied on the use of domesticated accessions, which has not been successfully achieved as a result of limited genetic variation for key traits due to the domestication bottleneck [49–52]. In cereals, cultivars with higher spike fertility and higher grain number per spikelet as a result of their higher assimilate partitioning during the pre-flowering period [53] are desired and useful in breeding programmes [54] to improve drought tolerance in cereals. These key traits are limited in modern crop gene pools. In contrast, genetic diversity in the form of wild species, related species, breeding stocks and mutant lines have reportedly demonstrated greater adaptability to drought stress [55–58] and have potential resource of genes for crop breeding [59].

Several national and international gene banks are good repositories of diverse alleles that can combat the effect of drought. However, a large number of accessions makes it impossible to efficiently evaluate and identify variation in drought adaptive traits, and hence the need to use methods with higher probability of capturing beneficial allelic variation. Mutation breeding, which involves the development of new varieties by generating and utilizing genetic variability in association with enabling technologies, is a strong pillar of plant breeding as it can be used to improve both simple and complex crop traits in many crops. Next generation sequencing and high throughput genotyping platforms, which can be used to characterize allelic diversity in genetic resources and establish genetic associations for marker and candidate gene discovery, are yet to be applied at higher frequencies to mutation breeding for improved precision and breeding efficiency. Sequencing efforts, such as the 3000 Rice Genome https://onlinelibrary.wiley.com/doi/full/10.1002/ggn2.202100017 [60] and the 3000 Chickpea Genome [55, 61] have facilitated drought related studies, resulting in significant progress in the identification of related genes and gene regions, and the dissection of some of their molecular characteristics.

With better access to the genetic variability found in natural populations of wild relatives and landraces, greater opportunity exists to use mutation breeding in association with genomic technologies to identify drought adaptive traits; several of which are encoded by alleles not present in domesticated crops or which evolved individually in diverse crop lineages [62]. Methods such as Focused Identification of Germplasm Strategy (FIGS) are being used to enhance the efficiency of detecting geography specific traits in germplasm collected in crops such as faba bean [63] and wheat [64]. The approach utilizes agroecological data to generate a priori information, which is then used to identify a group of accessions possessing the desired adaptive traits.

4. PHYSIOLOGICAL AND BIOCHEMICAL RESPONSES

Drought stress is the condition where the water requirement of the plant exceeds the available water in its root zone by >50% due to inadequate water supply. Prevailing soil conditions such as excessive aluminium, sodium and chloride, can also impact water availability. Drought stress in crop growing environments is often accompanied by higher temperatures. The response of the plant to the stress is dependent on the timing, duration and intensity of the stress [65].

The nature of drought stress is highly diverse and is a complex trait collectively defined by various component traits that affect breeding efforts to enhance crop drought tolerance. Drought stress decreases plant growth and yield through effects at the morphological, physiological, biochemical and molecular levels, and plant responses at the molecular, genetic, biochemical and physiological levels define the tolerance of the crop to drought [66].

4.1. Mechanisms of drought adaptation

Plants acclimatize to drought stress by using various mechanisms at the morphological, physiological, biochemical, cellular and molecular levels [67, 68]. Drought tolerance as defined by Lewitt [21] includes dehydration avoidance and dehydration tolerance. More recently, Fang and Xiong [43] reclassify it into four mechanisms, namely, drought avoidance, drought tolerance, drought escape, and drought recovery. The different mechanisms of drought tolerance are discussed below (Table 1).

4.2. Genetic mechanisms

When drought stress occurs, plant growth slows down or totally stops due to poor morphological development of plants. Plant drought tolerance involves changes at tissue (root, leaf), physiological, biochemical and molecular levels [68, 69]. The different modes of genetic mechanisms are described below.

Drought escape (DE) is an adaptive mechanism that involves the ability of a plant to complete its full life cycle prior to the supply of water to the moisture depleted soil and to form dormant seeds before a coming drought season, which describes how the drought susceptible variety works well to avoid the drought period [70]. Plants known to be drought escapers have different traits such as early flowering time-early maturity (terminal drought) and late maturity (early season drought stress), which can be very significant for crop production. Many reports revealed that early flowering and early maturity traits offer a promising strategy for the selection of advanced drought adapted germplasm in crop plants. Shavrukov et al. [71] highlighted that early flowered/matured wheat plants can produce significantly higher grain yield than others under both stressed and non-stressed conditions.

Drought (dehydration) avoidance is the ability of plants to tolerate relatively high tissue water potential despite a shortage of soil moisture, i.e. setting up barriers against water loss or establishing deep roots [70, 72]. Drought avoidance in plants is a consequence of improved water uptake under water stress:

(1) Performed by maintenance of turgor through roots that lead to increase in root development in the soil.

(2) Involves reduced photosynthesis through stomatal control of transpiration and by reduction of water loss through reduced epiderm.

(3)   Involves various morphological features of plants such as leaf characteristics, leaf angle, leaf rolling, etc., which reduce water loss through reducing transpiration to help the plant to avoid drought, i.e. reduced surface by smaller and thicker leaves.

(4)   Includes deep rooting, conservative use of water to ensure complete grain filling and life cycle modification to match rainfall.

Plant root architecture is an important trait to avoid dehydration by increasing water uptake under drought conditions. Early vigour is another significant character which allows crops to limit the loss of water as a result of evaporation [73].

Drought (dehydration) tolerance, building up the capacity to survive extreme desiccation, is the ability to withstand water stress with low tissue water potential. Plants maintain good tissue water content involving tolerance characters which support better photosynthesis under drought conditions. Biochemical and physiological changes, such as the maintenance of turgor osmotic adjustment, stable membrane, protein and chlorophyll and better membrane repair and cell elasticity are some of the desirable attributes. Dehydration tolerance requires the plant's ability to partially dehydrate but remain viable and grow again when rehydration resumes.

TABLE 1. MECHANISMS AND TRAITS TO IMPROVE DROUGHT RESILIENCE

Mechanisms of drought tolerance	Means of drought tolerance	Component traits
Drought escape (accelerating the life cycle)	Completing life cycle before the onset of drought	Early flowering Early maturity Plant phenology High leaf N_2 level High photosynthetic capacity
Dehydration avoidance	Reduced transportation	Stomatal susceptibility Leaf rolling
	Osmatic adjustment	Stomatal closure Stem waxiness High leaf sugar content High transportation efficiency
	Increased water uptake	Deeper tap root
Dehydration tolerance	Recovery after stress	Seedling survival
	Stay green	Low leaf senescence at grain filling
	Plant growth	Early vigour under stress
	Osmatic adjustment	High proline
		High stomatal conductance
		Desiccant tolerant enzymes

5. MUTATION BREEDING STRATEGIES FOR IMPROVING DROUGHT TOLERANCE

Drought is the one of the most devastating abiotic stress factors affecting crop production and is projected to worsen with climate change [70]. It severely limits plant growth and development as well as agricultural characteristics resulting in reduction of crop yields [69]. Improving drought tolerance in crops is among the top priorities for most countries to increase the efficiency of water use and to enhance agricultural water productivity under rainfed conditions. Although varieties tolerant to biotic and abiotic stresses have been developed by conventional breeding programmes, speeding up the pace of breeding is essential for developing improved varieties.

Breeders create new gene combinations and useful variability in crop plants. Cross-breeding is dependent on genetic diversity and if the required traits are not present in the primary gene pool, the breeder may not apply crossing to develop new cultivars that lengthen the breeding process. In addition to the required variation, many undesirable traits are also introduced which need to be eliminated by additional rounds of crossing [74]. Selection and breeding of similar lines with desired traits for specific environments also cause narrowing of the genetic base in breeder's lines. However, when natural genetic variability within a species is narrow, it can be enlarged through the induction of mutations. Induced mutations are defined as inheritable changes in the DNA, not derived from genetic recombination. Since spontaneous mutation rates in higher plants are too low (10^{-5} to 10^{-8}), artificially induced mutations using chemical and physical mutagens have been used [75], or by over-expressing or silencing specific genes [76]. For forward genetics screening of drought tolerance, induced mutants are advantageous because they can be freely distributed and tested under field conditions. Once mutation is generated, the next steps include screening and selection of mutants with desired traits [76].

The induced mutation approach has been widely applied in crop improvement; thus far, more than 3402 varieties in 228 crop species across 73 countries have been released for planting throughout the world (Fig. 2) [77]. Among the mutants, 7% (237) are abiotic stress tolerant, and less than half (105) are improved for drought tolerance.

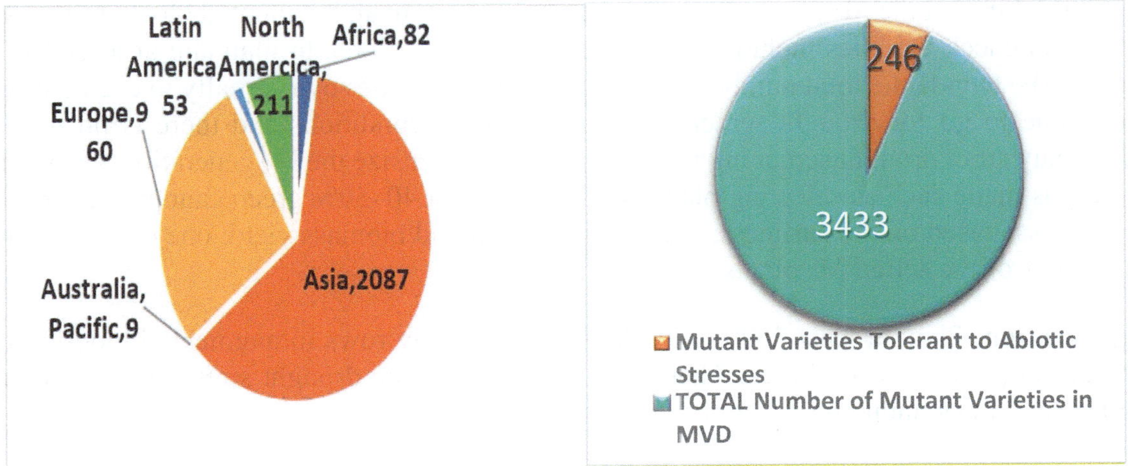

FIG. 2. Number of registered mutant varieties in different continents; and proportion of abiotic stress tolerant mutant varieties [77].

One significant example is cv. Calrose 76, described as the first semi-dwarf table rice cultivar, released in USA in 1976. Calrose 76 had a 15% yield advantage over conventional tall cultivars [78]. Another success in rice is the induced mutant cv. Nagina 22 (N22), which is a deep-rooted, drought and heat tolerant variety released by India [79]. The mutant rice variety Zhefu 802, with high yield potential, wide adaptability, high resistance to rice blast and tolerance to cold was induced by gamma rays, and was the most extensively rice variety planted on one million hectares between 1986 and 1994 [80, 81]. A significant achievement from Indonesia for mutation breeding was the official release of three mutant varieties by the Ministry of Agriculture in 2013–2014. These grain sorghum mutant varieties are drought tolerant, of semi-dwarf stature, early maturing and high yielding, and are recommended for dry season cultivation, especially in drought prone areas of the eastern part of Indonesia [82].

5.1. Mutation breeding approach to develop drought resilient varieties

In Section 5.2, the mutation breeding approach of inducing mutations and their selection for drought tolerance are described. Figure 3 illustrates the flow chart of several steps in the development of mutant population for evaluation. There are certain considerations about the size of population for handling in M_1 and other generations, and the standard practices to exercise selection for drought responsive traits in crops like rice and sorghum.

5.2. Handling mutated populations/lines

The size of the M_1 generation is rather small in comparison with the following generations. For each treatment dose, a few thousands of seeds can be adequate to obtain 10 000–30 000 seeds for the M_2 generation.

M_0 generation. Irradiate 2000 homogeneous seeds for each dose and each variety having 10–12% seed moisture and 90% germination. Irradiate at least 2000 homogeneous seeds for each dose and each variety.

M_1 generation. Irradiated seeds grown in blocks, until maturity, maintain at least 1500 plants in the M_1 generation. Self all the plants and keep them as a single plant. Look for any dominant mutations in comparison with parent and tag them separately.

M_2 generation. All the M_1 plant progenies need to be sown in an augmented block design with control and check varieties included in each block; it is advisable to maintain at least 10 000 plants in M_2. Divide the total number of M_2 plants per row to be sown equally in each block. It is advisable to go for drought screening in natural field conditions and there is no need to exercise any strict drought screening in M_2. In order to downsize the M_2 generation, agronomic selection is made such as selecting plants showing at least 40–50% green canopy, better yield and biomass, based on the earliness, seed weight, grain and biomass yield; one can then select top 25–30% of the selfed M_2 plants.

M_3 generation. Plant at least 2500 individual plant progeny rows in augmented block design with control plants in each block. There is a need to simulate drought stress, as per the crop specific guidelines, inducing drought stress:

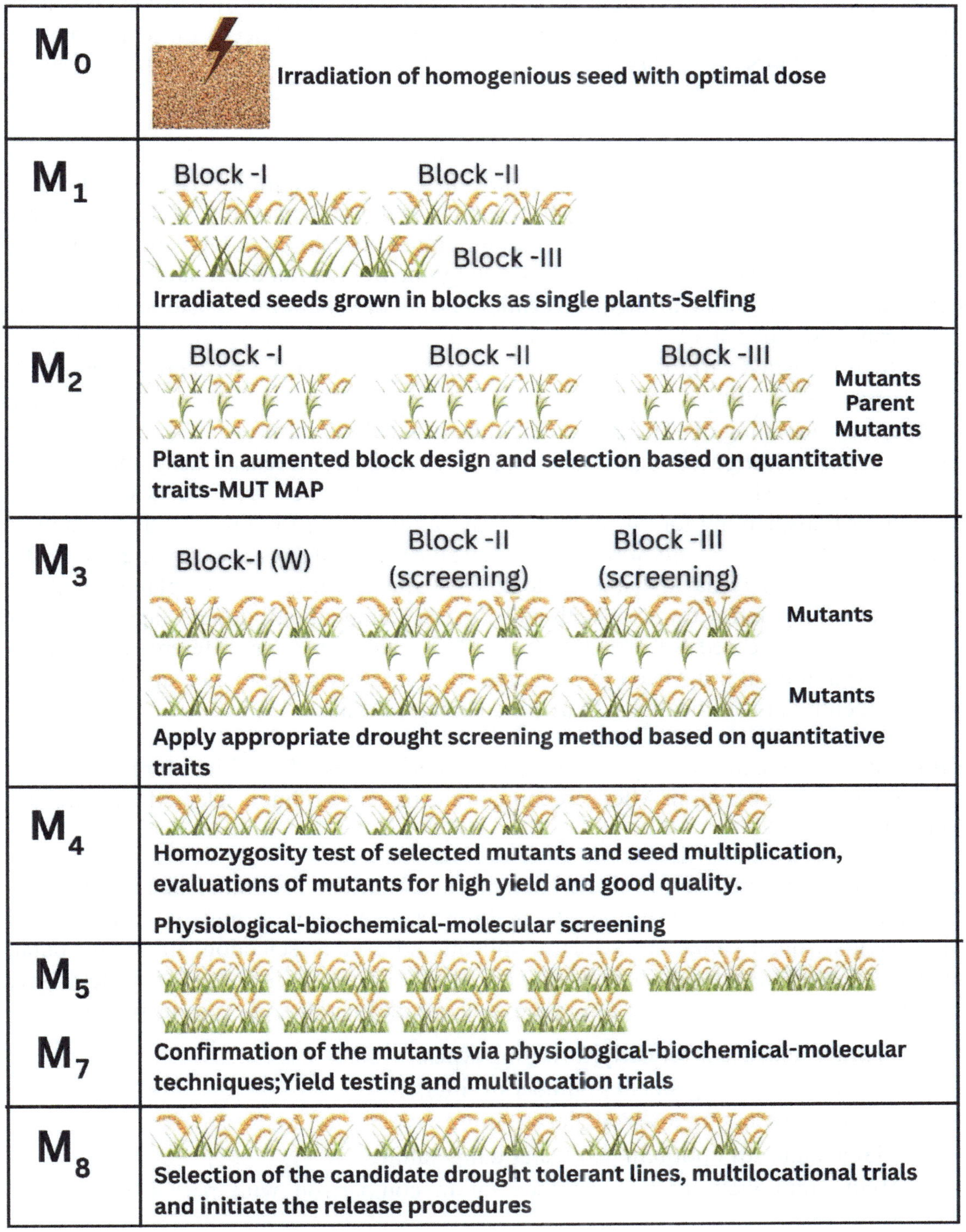

M_0	Irradiation of homogenious seed with optimal dose
M_1	Block -I Block -II Block -III Irradiated seeds grown in blocks as single plants-Selfing
M_2	Block -I Block -II Block -III Mutants / Parent / Mutants Plant in aumented block design and selection based on quantitative traits-MUT MAP
M_3	Block-I (W) Block -II (screening) Block -III (screening) Mutants Mutants Apply appropriate drought screening method based on quantitative traits
M_4	Homozygosity test of selected mutants and seed multiplication, evaluations of mutants for high yield and good quality. Physiological-biochemical-molecular screening
M_5 M_7	Confirmation of the mutants via physiological-biochemical-molecular techniques; Yield testing and multilocation trials
M_8	Selection of the candidate drought tolerant lines, multilocational trials and initiate the release procedures

FIG. 3. Mutation breeding approach to develop a drought resilient variety.

The entire field is divided into three blocks and an equal number of progeny rows planted in each block, along with parent/check varieties; screening can be carried out as explained below.

For sorghum. Impose drought treatments as follows: (1) well watered; (2) withdrawal of irrigation after 40 DAS (panicle emergence stage); and (3) withdrawal of irrigation after 65 DAS (post-flowering stage).

If the susceptible checks show severe leaf rolling with less chance to recover upon watering, and the soil tensiometer shows a reading of -50 kPa, then the field needs to be irrigated once (10–20 mm) at 60 DAS and 85 DAS.

For rice

(1) *Reproductive stage drought screening in lowland situations.* The field is drained about 23–25 days after transplanting, or 44–45 days after sowing. If all the susceptible checks show severe leaf rolling with minimum probability to recover upon watering, and the trial as a whole also shows rolling, the field is flood-irrigated.

(2) *Reproductive stage drought screening in upland situations.* Irrigation in stress trials can be stopped at 42–45 days after sowing. For irrigation, if severe stress appears, the field is visited early in the morning at 10.00 am and, if the susceptible checks show severe leaf rolling with less chance to recover upon watering and the soil tensiometer shows a reading of -50 kPa, the field is irrigated (40 mm).

(3) *Vegetative stage drought screening.* For lowland screens, transplant 14-day-old seedlings. Allow for establishment for 10 days and then drain out water at 10 days after transplanting for lowland or 18 days after sowing for direct seeded upland.

(4) *Seedling stage drought screening.* Initiate the stress 10 days after sowing and continue the stress for 15 days, that is, up to 25 days after sowing. Selected plants which show leaf rolling and waxy stem with better green leaf area until dough stage and maturity need to be scored for drought stress (0–9 scale scoring as reference). In addition, morphological traits need to be scored. It is also advisable to take up the above field screening in two locations, to avoid unexpected rains.

Selection. Based on the morphological traits and drought scores, 25% of the M_3 plants are selfed and advanced to M_4 generation. After screening of the mutant M_3 lines, selected lines are grown as M_4 generation.

M_4 generation. Confirmation of selected mutants according to physiological and biochemical parameters, homozygosity test of mutants, preliminary evaluation of mutants to identify high yielding and good quality lines and seed multiplication. Based on selection, 30% of the putative mutant progenies are selected and can be forwarded to M_5 generation. Overall, in order to screen large, mutagenized population lines, the following are the best drought responsive traits:

— Identify and select for early flowering types.
— Mutants possessing smaller number of tillers.
— Mutants possessing low spikelet sterility.
— Maintenance of optimum leaf water temperature.
— High grain yield in the test location in comparison with parents/known check varieties.

M_5–M_n generation. M_4 plants will be planted in an RBD design with replicated trial, with checks under water stressed and irrigated plots, the true breeding nature of the mutant lines will be checked again. Validation of drought resilient mutant lines using specific markers associated with drought responsive traits, such as stay green, root traits, earliness, sugar, protein metabolism, etc., are carried out selecting eight–ten mutant lines and testing for advanced yield under drought conditions.

M_7 generation. Yield testing and multilocation trials of advanced mutant lines and selected drought tolerant mutant lines.

M₈ generation. Selection of the candidate drought tolerant advanced mutant lines and initiation of the variety release procedures.

6. GENERAL SELECTION STRATEGIES FOR BREEDING DROUGHT RESILIENCE

During the past century, theories of Mendelian genetics have facilitated a better understanding of the genetic bases of simply inherited (qualitative) traits. Such traits are often controlled by one or few genes with perceptible effects and discrete variation. As a result, considerable breeding progress has been made for these traits. However, most agriculturally important traits, including tolerance to drought stress exhibit complex inheritance. Such traits are often conditioned by the effects of several (and in some cases many) genetic loci which interact with each other and with the environment. The quantification of stress tolerance poses serious difficulties. Direct selection in the field is difficult because uncontrollable environmental factors adversely affect the precision and repeatability of such trials. In order to identify sources of drought tolerance, it is necessary to develop screening methods that are simple and reproducible under the target environment conditions. Several field and laboratory screening methods are available to screen the crops for drought tolerance. In this context, it is important to design a screening technique that could be used for selection of stress tolerance using phenotypic measurements in field nurseries or greenhouse facilities [83].

Plant breeders directly select drought tolerant germplasm using empirical breeding methods considering yield and yield components, while breeders emphasize the improvement of yield in analytical breeding approaches using morphological, physiological and biochemical traits associated with yield [84]. Due to the complexity of drought tolerance traits, several studies highlighted that empirical breeding has limited success in the development of drought tolerance in crop improvement [84, 85]. Plant breeders have changed their approach in some experiments to analytical breeding based on the selection of secondary traits associated with plant performance under targeted drought conditions. The improvement of drought tolerant and high yielding varieties involves initial emphasis on physiological mechanisms and then secondary traits like osmotic adjustment, biochemical components, leaf and root architecture, relative water content, etc. [86]. Four approaches are mainly applied: (i) Breed for high yields under optimal conditions; (ii) breed for maximum yield by selection in field in target drought prone areas; (iii) incorporate selected physiological and morphological traits confirming drought; (iv) identify key traits for drought tolerance at specific growth stages in the high yielding background. A study on drought tolerance in maize was investigated to determine the proficiency of indirect selection for grain yield under drought conditions through secondary traits and genome wide selection (GWS) [87]. The experiment showed that secondary traits are reliable indicators of drought tolerance and GWS is superior to indirect selection to increase genetic gains under drought conditions.

The nature of the crop is very important to consider in any approach. For example, rice is exceptional due to cultivation across diverse ecosystems in upland and lowland areas. Moreover, some investigations revealed that direct selection for yield under drought will be effective under both lowland and upland drought stresses [86]. Another important finding is that some lowland adapted drought tolerant varieties demonstrated medium to late maturity and enabled anaerobic growing environments, whereas upland drought tolerant varieties are characterized by early flowering and root architecture [86].

The complexity of the drought stress pattern needs be resolved with a holistic approach integrating physiological assays of resistance traits and molecular genetic tools, together with conventional breeding and agronomical practices that lead to better conservation and utilization of soil moisture as well as matching crop genotypes with the environment [70].

Approaches to develop drought tolerant material include the following:

(1) A systematic characterization of the drought environments is needed where the crops are grown to enable adequate targeting of drought tolerance traits using climatic information, GIS tools, water balance and crop simulation models. Since any variety can be adapted to several environments, the target environment needs to be defined to develop the target population to reduce genotype environment interaction [73]. Though field phenotyping is critical for grain yield and yield components, managed stress environment facilities (rainout shelter, green house, growth chamber) have advantages for secondary traits since the environment is controlled, i.e. light, humidity, temperature, etc. [88]. Controlled environments can provide alternative methods to explore breeding materials/varieties for drought resistance. However, growing and/or irrigating plants in pots can create stress on plants [73].

(2) In terms of phenology, the appropriate crop duration is a compromise of various factors, including season length, yield potential and the timing of when drought stress occurs. For example, using an escape mechanism, the development of short duration genotypes can help mitigate the effects of terminal drought.

(3) Pathways and physiological/biochemical mechanisms involved in drought tolerance strategies, such as the role of sugars, potassium and calcium nutrition status, scavenging of reactive oxygen species, cell wall biosynthesis and modification need to be thoroughly investigated using physiological methods and molecular techniques [89] in mutant population and crop germplasm.

(4) An ideotype approach is needed for incorporating the relevant stress resistance traits into major crops of interest, which requires a better knowledge of the physiological mechanisms involved in stress resistance and their genetic control. The genetic enhancement of root system architecture to make them more effective in water extraction is a high priority research effort for rainfed crops.

(5) The strategies for drought tolerant crop improvement need more focus on maximum extraction of available soil moisture and its efficient use in crop establishment, growth, and maximum biomass and seed yield, and not only on crop survival. An improved knowledge of probable soil moisture availability is necessary (neutron probe, etc.) to further exploit the drought escape option.

(6) With regard to drought avoidance, adaptation of root architecture, e.g. a large and/or deep root system, can be useful in greater extraction of available soil moisture. Smaller leaf area could reduce the transpirational water loss under water deficit conditions.

(7) Secondary traits include germination and early vigour, leaf traits, leaf area maintenance, root and shoot growth rate and development plasticity. Early growth vigour is an important factor in stress resistance as it permits establishment of a root system more effective in extracting water during later stress periods. Tillering characters in cereals could be useful for drought stresses.

(8) Delayed senescence or stay-green is considered a useful trait for plant adaptation to post-flowering drought stress, particularly in environments in which the crop depends largely on stored soil moisture for grain filling.

(9) In terms of transpiration efficiency (TE), depending on the crop, there is a great scope for the genetic improvement of the efficiency of crop water use under dryland conditions. Research has also shown that TE and carbon isotope discrimination in leaf

($\otimes$) are well correlated in several crop species, suggesting a possibility of using $\otimes$ as a rapid, non-destructive tool for selection of TE.

(10) Improving drought tolerant genotypes using QTLs and molecular breeding techniques, combined with physiological characterization and conventional breeding, can significantly improve the ability of crops to withstand stress in defined target environments. The most appropriate populations for mapping and marker assisted transfer of QTLs are probably inbred backcross (IBC) populations.

6.1. Drought responsive traits

Phenotyping of plants with variable parameters for screening of tolerant and sensitive breeding germplasm would allow the identification to select the promising lines. The identification of morpho-physiological traits, which is reflective of mechanisms and processes that confer tolerance, can be a high priority activity in drought research [70]. The traits need to show heritability and have a link with yield under drought stress. The morphological, physiological, biochemical and root related traits under drought stress associated with some promising stress tolerant germplasm are listed in Table 2.

TABLE 2. LIST OF MORPHOLOGICAL AND ROOT FEATURES, PHYSIOLOGICAL, BIOCHEMICAL AND PHENOLOGICAL TRAITS UNDER DROUGHT STRESS

Sl. No	Category	Traits
	Morphological and root features	Plant height; stem width; number of leaves; biomass weight; grain weight; leaf number–area; leaf rolling; No. of internodes; internodal length; panicle length–width; harvest index (%); deeper thicker roots; greater root volume; root fresh/dry weight; root pulling resistance; root anatomical features; greater root penetration ability.
	Physiological and biochemical	WUE; osmatic adjustment; carbon isotope discrimination; stomatal density and conductance; canopy temperature; leaf relative water content; AA (proline and glycine); betaine polyamines and organic acids; protective proteins; oxidative stress associated enzymes H2O2; malondialdehyde; ascorbate peroxides.
	Phenological	Days to flower and maturity; seedling vigour; early to maturity; anthesis; photosensitivity.

The above morphological, physiological and biochemical traits are crucial to enhance drought stress tolerance of plants that are effective for the expression of several mechanisms, such as germination and early vigour, leaf area maintenance, root and shoot growth rate, all of which are very important traits to improve drought stress tolerance [69]. Efficient crop breeding programmes for drought tolerance are supported by precise phenotyping with physio-morphological traits and molecular approaches [90]. Accordingly, the quantification of yield related physio-morphological characters under stress conditions increases to select and identify drought tolerant germplasm to develop a variety. According to the inheritance of these traits, the prevailing additive or dominance variation is particularly essential to decide the breeding methods required to develop a drought tolerant variety [80].

6.2. Secondary traits related to drought tolerance

Improving crops using secondary traits such as root architecture, leaf water potential, stomatal traits, panicle water potential, osmotic adjustment and relative water content depends on either direct selection or molecular approaches, which require their genetic correlation with yield under drought [90]. Physiological/biochemical responses can facilitate measurement of drought tolerance if a correlation exists between specific metabolites and the trait [67]. These physiological traits could be used as the confirmation tools after selection in the M_3 stage based on phenotyping in the field at the M_3 stage in mutation breeding programmes. Some of the secondary traits are described below.

6.2.1.   Canopy temperature depression

Canopy temperature depression (CTD) is measured by thermal imaging, which is the difference in temperature between the canopy surface and the surrounding air. CTD is a highly integrating trait resulting from the effects of several biochemical and morpho-physiological features acting at the root, stomata, leaf and canopy levels. At the field level, genotypes with a cooler canopy temperature under drought stress, or a higher CTD, use more of the available water in the soil to avoid excessive dehydration. CTD can be a good predictor of yield.

6.2.2.   Stomatal conductance

When plants sense water stress, the first adaptation mechanism is the narrowing or closing of stomata to prevent water loss. Plants quickly close their stomata to minimize water loss by transpiration under drought conditions. As a result of this, there is a decrease in the plant photosynthesis rate as the carbon dioxide uptake is also reduced. Stomatal regulation is a crucial component that directs plant development and survival. There is a positive correlation between stomatal conductance and yield. Indirect selection for yield under drought conditions can be performed since natural oxygen isotope composition (leaf and grain ^{18}O) is also correlated with stomatal conductance. Stomatal conductance measured on a fully expanded young leaf of three different plants measures the water vapour flux from the leaf surface to the atmosphere. Leaf water potential indicates the whole plant water status and expresses the energy needed to pull the water out of the leaf.

6.2.3.   Carbon isotope discrimination

Carbon isotope discrimination ($\Delta^{13}C$–CID) is used to measure the ratio of stable carbon isotopes ($^{13}C/^{12}C$) in the plant dry matter compared to the ratio in the atmosphere [16]. That method is a surrogate for water use efficiency to select drought tolerant genotypes under drought conditions. CID is negatively associated with WUE over the period of dry mass accumulation, and hence it is a good predictor of stomatal conductance and WUE under drought stress and an attractive breeding target for improving yield and WUE.

6.2.4.   Chlorophyll content and chlorophyll concentration

The chlorophyll fluorescence (Fv/Fm) ratio is a critical indicator of direct or indirect effects of abiotic stress on photosynthesis [91]. A hand-held chlorophyll meter (SPAD-502 Chlorophyll meter) is used on a fully expanded young leaf of a randomly chosen plant taken once at mid-pod filling 45 days after planting.

6.2.5. *Canopy temperature and canopy biomass components*

The canopy leaf temperature is taken on a fully expanded leaf at mid-pod filling in the morning and afternoon using an infrared thermometer. Canopy biomass components (mid-pod filling, photosynthetic efficiency, CO_2 assimilation rate) are essential for crop growth monitoring as the yield results from the accumulation and transportation of translocates between different organs.

7. NON-SENESCENCE STAY-GREEN

Stay-green is an important trait for drought tolerance in several crops, e.g. sorghum, rice and maize specifically when terminal drought is a recurrent problem. The stay-green trait is the heritable delayed foliar senescence. It is observed after post-flowering during the grain filling phase of the plant in the course of drought exposure to give resilience to premature senescence, stalk rot and lodging [13, 92, 93]. Stay-green is the breakdown of leaf chlorophyll, reduced photosynthesis and the general reduction in cellular capacity for various life functions. Normal senescence is accelerated when drought occurs during the late developmental stage, while constitutive non-senescence is more effective towards plant production when drought stress occurs than with non-stress conditions.

8. ROOT SYSTEM ARCHITECTURE

The root system architecture (RSA) constitutes the formation, spatial and temporal configuration of a plant root system. Plant drought tolerance is directly affected by root morphological characteristics. Roots exhibit morphological plasticity to soil conditions and selection for faster growing and deeper roots could enhance water harvest. RSA is an important characteristic for adaptability to a drought environment.

9. FLOWERING TIME

Flowering time is a critical factor to optimize adaptation in environments altered in water availability and distribution during the growing season.

10. PLANT HEIGHT

Plant height is one of the important indicators of growth based parameters. Vegetative growth can be assessed as plant height to distinguish drought tolerant and sensitive genotypes.

11. PHENOLOGICAL TRAITS

Plants growth period from sowing to flowering–maturity days is frequently applied to determine earliness and is a very important indicator for drought avoidance.

12. EARLY VIGOUR

Early vigour trait is used to improve water use efficiency of crops:

— Early vigour optimizes WUE and limits the loss of water due to direct evaporation from the soil surface. It saves on soil water available for later developmental stages.
— Vigorous canopy development may cause early depletion of soil moisture.
— Depends on the environmental characteristics and target population.

13.   LEAF BASED TRAITS

Drought stress causes a series of leaf damage, such as reduced leaf area and transpirations, leaf rolling, yellowing of leaves (chlorosis), tissue death, wilting and trough leaf rolling, which are significant indicators of dehydration avoidance mechanism to evaluate drought tolerance in crops [69]. Leaf based traits are being used for drought assessment in crops; for example, in rice according to the assessment of the Standard Evaluation System for Rice of the International Rice Research Institute (SES IRRI) in 2002 of drought affected leaf shapes, changes are recorded (starts to fold, V shaped leaves, U shaped leaves, curled leaves, leaf roll tight). Leaf rolling enhances stomatal closure by increasing leaf resistance to water loss and reduces leaf temperature and loss of water by decreasing incident radiation. Leaf rolling is a good indicator of drought tolerance.

14.   LEAF WATER POTENTIAL

Leaf water potential (LWP) can be used as an easy and fast way to screen genotypes for drought avoidance. Higher LWP can be achieved by maintaining higher turgor at a given level of moisture stress. Stomatal conductance (SC) and leaf rolling are reliable physiological indicators of drought tolerance (associated with LWP).

15.   BIOCHEMICAL TRAITS

Abscisic acid. Abscisic acid (ABA) is a natural growth hormone, and is produced under abiotic stress conditions, including drought. ABA is a vital component of the mechanisms allowing the plant to match the water demand with the water supply. An increase in ABA concentration is a universal response observed in plants subjected to drought and other abiotic stresses. ABA modulates the expression of a large number of genes whose products protect the cell from the harmful effects of dehydration. It has been shown to affect many of the traits that influence the water balance of the plant through both dehydration avoidance and dehydration tolerance. Upon drought stress imposition, ABA increases, which may affect stomatal closure leading to reduced transpiration rate. The levels of proline and malondialdehyde (MDA) and chlorophyl are also indirect traits of drought tolerance in plants [94].

Electrolyte leakage. Tissue damage can be measured by membrane damage based on electrolyte leakage, which is a good marker of drought tolerance in several crops.

Proline. Plants have adaptive response to defeat the effect of the stresses of which the accumulation of free amino acids such as proline in higher levels is observed during flowering stage whereas very low levels are seen in vegetative growth stage [70].

Methylglyoxal (MG). Methylglyoxal (MG) is a cytotoxic molecule and damages DNA and protein [95]. MG homeostasis plays an essential role in promoting plant growth whenever a stress is perceived. It leads to buildup of toxicity responses to drought stress. MG levels increase under various stress conditions.

16.   MOLECULAR BREEDING STRATEGIES

Abiotic stresses cause severe yield losses in food crop production and hence, improvement in stress tolerance of crops is considered a major breeding goal. Sustainable crop production requires the development of tolerant varieties with desirable traits. Molecular breeding for tolerance is facilitated by the identification of unique genetic source of tolerance traits, knowledge of the genetics of inheritance/resistance, mapping and cloning of resistance quantitative trait loci (QTLs) and identification of candidate genes. The advent of next generation sequencing technologies (NGSs) and the availability of draft genome sequences of many agronomically important crops has opened a new possibility to develop extensive genomic resources for mapping and cloning of QTLs. Advances in genomics have also led to the identification, functional characterization and introgression of genes associated with drought adaptation in many crops to drought stress. Several molecular breeding applications such as QTL mapping, marker assisted backcrossing and association mapping studies, genomic selection, transgenics and genome editing are being carried out to develop new resistant cultivars (Fig. 4) [96–98]. To sum up, the rapid development of crop genomics and transcriptomics has provided great scope for the genetic improvement of crops for drought tolerance with yield stability [99].

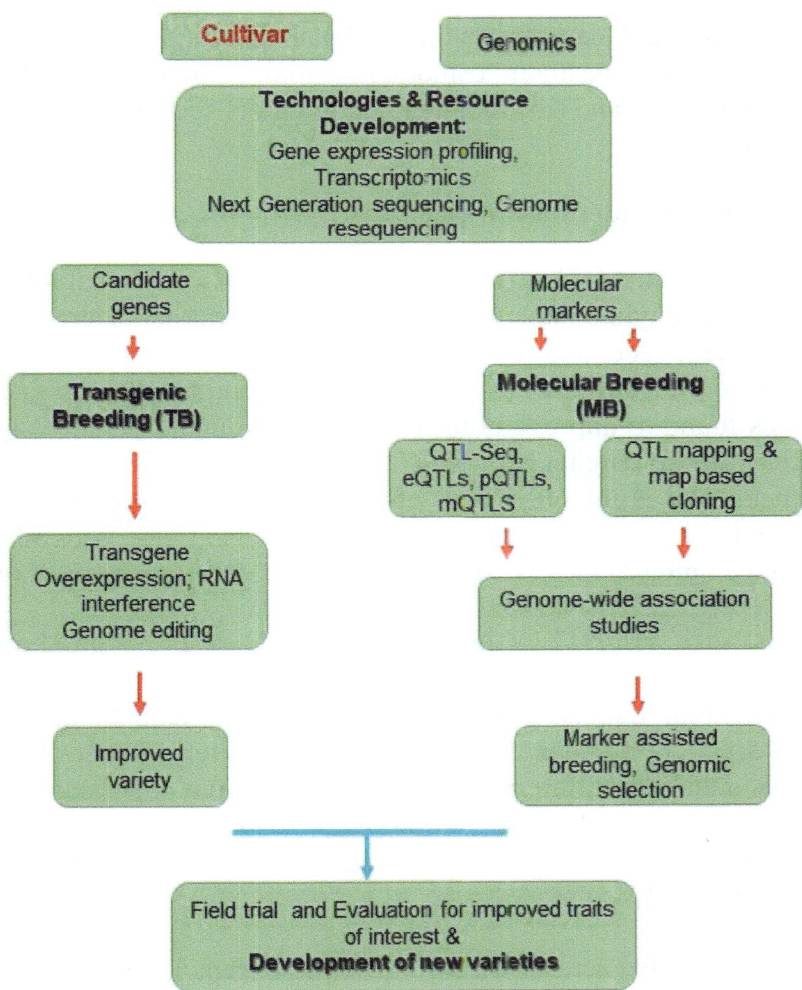

FIG. 4. Integration of transgenic and molecular breeding tools for drought tolerance.

16.1.   QTL mapping in plants

Conventional breeding for drought resistance in crops is limited due to the quantitative nature of drought resistant traits. Several studies on QTL mapping have been conducted to improve the inheritance of quantitative traits as the most important agronomic traits, including drought stress, are quantitative in nature and are controlled by more than one gene (polygenes). One of the effective alternatives to breed high yielding drought resistant crops is using genetic mapping of QTLs to identify genomic regions controlling a trait under drought and subsequent utilization of markers linked to the QTLs in marker assisted breeding of desirable alleles. Chromosomal positions (loci) of genes that control quantitative traits can be identified by the use of specific molecular markers (i.e. SNPs, SSR, RFLP) and statistical software (i.e. MAPMAKER/QTL, QTL Cartographer, QGene, MUTMAP, PLABQTL). These identified loci are termed 'QTLs' and may be a single or cluster of genes that affect traits of interest. Several different traits, such as biotic and abiotic stress tolerance including drought in crop plants, have been reported from QTL mapping studies. Many of these QTLs/genes associated with drought tolerance in some major crops can be accessed at https://www.plantstress.com.

Several of the identifiable stable QTLs for agronomic traits associated with drought tolerance in major crops have shown significant phenotypic variance, ranging between 20 and 45% due to the variable nature of drought stress. Most of these major QTLs were identified for traits such as: (i) yield and yield related traits, including biomass, grain filling, grain number, spike density, 1000 grain weight, grain yield, days to heading and plant height; (ii) physiological responses, such as leaf rolling, osmotic adjustment, osmotic potential, relative water content, chlorophyll contents, canopy temperature; and (iii) root architecture traits such as root length and number, root angle and network area.

16.2.   QTLs for drought tolerance in rice

In rice, most of the yield related QTLs have been reported from studies that focused on different components associated with yield rather than grain yield under drought stress from research on drought resistant local land races and ecotypes [100–102]. The first reported QTL for rice grain yield under upland reproductive stage drought stress conditions (*qDTY12.1*) located on chromosome 12 was found to improve grain yield under drought tolerance significantly with explainable genetic variance ranging between 43 and 51%. This QTL had consistent effects in multiple environments and is the only locus utilized to develop drought resistant lines with increased grain yield for highly diverse upland and lowland rice ecosystems [103–105]. However, map based cloning of the QTL to identify causal gene(s) is not yet available.

Several other QTLs, such as *qDTY1.1* on chromosome 1, *qDTY2.3* on chromosome 2 and *qDTY3.2* on chromosome 3 have been identified and play major roles in enhancing grain yields and yield related traits (e.g. harvest index, variation in canopy temperature during flowering and seedling shoot dry weight under stress and drought recovery) under drought in different genetic backgrounds and from different donors [106–110]. Five other stable QTLs (*QSnp1b, QGyp2a, QSnp3a, QSf8, and QSnp11)* from a set of 20–30 QTLs for drought resistance related traits between a drought sensitive parent (Lemont) and resistant parent known as Teqing have also been reported in rice [112].

Sixteen other QTLs distributed across rice chromosomes except 5, 7 and 8 have been reported to enhance grain yield under drought [86]. It is interesting to note that QTLs for grain yield under drought in managed stress environments (i.e. controlled greenhouse conditions) may not be translatable to target field environments where the timing and severity of drought may vary over the years. Yi et al. [113] performed a genome-wide association study on the drought resistance index of traits and identified nine QTLs associated with drought related traits. The study also predicted ten candidate genes associated with various metabolic activities related to drought stress.

16.3. QTLs for drought tolerance in maize

Several QTLs associated with drought related traits have been identified in maize; for instance, Nikolic et al. [114] detected 45 QTLs for yield and yield components and mapped them on nine out of ten maize chromosomes with phenotypic variation ranging from 0.1 to 45%. Twenty two QTLs for drought related traits, such osmotic potential, leaf surface area and stay-green (a desirable trait for crop production), have been mapped on various chromosomes of maize [115]. A total of 18 stable QTLs for various traits, including grain yield, ear setting percentage and anthesis–silking interval under seven environments under well watered and drought regimes have been mapped using a 6 K SNP assay and 756 SNP (single nucleotide polymorphism) markers [116]. Analysis of the QTL-by-environment interaction in the same studies revealed 28 environment dependent QTLs associated with the drought conditions and 22 associated with the well watered or non-drought conditions for traits such as grain yield, ear setting percentage and anthesis–silking interval. Over 55% of the QTLs identified under the non-drought conditions and all those identified under the drought conditions were observed to be located in the environment dependent QTL regions. Other QTLs have also been detected and mapped to QTLs on other chromosomes that influenced adventitious root formation in waterlogged conditions, grain yield and root architecture under different water regimes [117, 115].

16.4. QTLs for drought tolerance in wheat

In wheat, a large number of QTLs have been reported for several traits related to drought tolerance, such as coleoptile length, water soluble carbohydrates, root system, grain yield, and related traits, with some QTLs contributing as much as over 20% for phenotypic variation of the individual traits [118]. In hexaploid bred wheat, some QTLs have been reported for root architecture, including deep root ratio, root dry weight, root length, root number and root anatomical characters. Fine mapping for major chromosomes on 3B for durum wheat, which has the potential to affect grain yield across a wide range of soil moisture regimes, is in progress [115]. Four major markers (*Xwmc11, Xgwm314, Xwmc296* and *Xgwm400*) reported to be linked with important agro-physiological traits can be deployed in MAS for wheat improvement to find major controlling genes for drought mechanisms [119]. One major QTL explaining up to 20–60% phenotypic variation in three physiological traits (i.e. stem reserve mobilization, water soluble carbohydrates and chlorophyll content) that lead to the enhancement of yield under drought has been demonstrated as the potential candidate to be used in MAS of cereal crops.

16.5. QTLs for drought tolerance in barley

Several studies have been conducted to identify QTLs linked with yield and yield related and physiological traits under drought stress in barley [120–122]. It has been observed through comparative genomics that many QTLs in wheat associated with drought tolerance also co-localized with QTLs identified in barley under drought condition [123]. Several drought tolerances associated QTLs such as QDT.TxFr.2H on chromosome 2H and QDT.TxFr.5H on chromosome 5H, are linked with proline content and QTLs for moisture content have been reported [124, 121].

16.6.   QTLs for drought tolerance in sorghum and pearl millet

The drought adaptation mechanism at the post-flowering stage in sorghum has been found to be associated with the stay-green trait [125, 126]. Three QTLs with phenotypic variation ranging between 17 and 21% linked with CO_2 assimilation and transpiration have been reported to be co-localized with yield related traits such as biological yield and leaf area under drought conditions [127]. Four other QTLs for nodal root angle linked to the drought adaptation mechanism in sorghum under drought conditions have been detected in addition to two other QTLs for shoot dry weight, three for root dry weight, and three for leaf area [128].

Research has been conducted in pearl millet to map the QTLs associated with stover and grain yields under drought conditions so as to maintain yield under terminal water stress [129, 130]. Some QTLs associated with grain yield with phenotypic variation of 32% and for low transpiration rate under drought stress have also been reported under terminal water stress [131, 132].

16.7.   Meta-QTLs and their associated candidate genes

The meta-QTL (MQTLs) analysis approach has become an integral part of QTL studies, whereby non-dependent data sets obtained from different mapping populations in various environments for traits of interest are analysed. This approach has been used to authenticate drought responsive regions in rice, for QTLs related to yield using 15 independent mapping populations with a consensus map of 531 genetic markers and map length of 1821 cM. Fourteen MQTLs on seven chromosomes which enhance rice yield under drought by 4–28% were obtained which can be utilized by MAS in rice [133]. The report further indicated that one QTL, (*qDTY12.1*), is very common in 85% of the cultivars, whereas the other QTLs (i.e. *qDTY1.1, qDTY1.2, qDTY3.2, qDTY4.2* and *qDTY8.1*) were present in >50% of the cultivars.

Meta-QTL analysis in wheat has been performed to study drought responsive regions involving over 500 QTLs for agronomic as well as physiological traits under conditions of drought. As many as 19 MQTLs for drought tolerance spread over 13 chromosomes were reported. Each MQTL corresponded to two–eight individual QTLs and had a narrow confidence interval of 5.8 cM. The small physical and genetic intervals of MQTLs make them suitable to be deployed for MAS. The candidate genes associated with MQTLs can be cloned to study the underlying molecular mechanism responsible for the regulation of yield under drought stress.

16.8.   Marker assisted backcrossing

The low nature of heritability and instability of yield and related agronomic traits for crops under drought conditions makes a conventional breeding approach for drought difficult and unattractive, hence the use of molecular breeding approaches such as MAS to develop crops which are more adaptable to drought. Significant genetic variation for traits associated with drought tolerance has been found in the germplasm of many crop species, making it useful to deploy MAS involving the available QTLs for drought related traits for the development of pre-bred material with improved tolerance to drought stress [134, 135]. Several agronomically important QTLs/genes associated with drought tolerance traits are representative of only a small part of phenotypic variation, making them attractive to be deployed directly for MAS in crop breeding programmes. However, marker assisted backcrossing (MABC) has only been rarely attempted for the improvement of drought tolerance in crops on a large scale. For instance, at the International Rice Research Institute (IRRI), several drought QTLs identified for yield under drought stress [110, 136] have successfully been introgressed into high yielding rice varieties susceptible to drought using MABC [108, 137, 138].

In wheat, MABC has also been used to introgress QTLs for several drought related traits including grain yield, and its related traits into some elite wheat cultivars at CIMMYT [119, 139]. Efforts are under way to introgress several different QTLs for various traits into different genetic backgrounds of wheat and for their potential to be released as new improved cultivars. MABC is an efficient breeding tool in terms of time, cost and precision for the selection of target traits [135]. Several QTLs and meta-QTLs for drought responses and associated physiological and grain yield traits have contributed to understanding the genetic architecture of drought tolerance [135].

16.9. Genetic modification for drought tolerance

Drought resistant plants can be cultivated in extreme environments, thus maintaining and widening the area for plant cultivation and food production. Multiple genes and their signalling pathways under abiotic stresses have been identified that regulate key plant drought stress responses [97, 136,]. The presence of several genes has been reported in water and ion transport, amino acid metabolism, signalling pathways, cell energetics, plant defence, and phytohormone biosynthesis [140, 141]. Different signalling pathways, including calcium-sensing, protein phosphorylation and/or dephosphorylation, protein degradation, phospholipid metabolism, etc., and stress hormones such as abscisic acid (ABA) and ethylene have been shown to regulate several stress related genes, molecular chaperones and defence enzymes [142, 143]. The applications of genetic engineering are being extensively explored to increase plant resistance to drought by engineering regulatory genes [144]. Several abiotic, stress related genes have been transferred to cereal crops using genetic engineering approaches. The transformed crops demonstrate improved physiological and morpho-biochemical characteristics compared to non-transgenic crops [145, 146]. The challenge is now to realize cultivation of drought tolerant plants under realistic field conditions, as most of the transgenic plants are evaluated in greenhouses rather than performing field testing focusing on their tolerance responses and yield [147]. Transcriptional factors and micro RNAs regulate a chain of downstream genes for stress adaptation and hence they are a suitable target for regulation of genes to engineer robust crops resistant to drought [144, 148].

16.10. Genome editing

Genome editing (GE) has been used for improving the various plant agronomic traits in rice, tomato, Arabidopsis, wheat, barley, cotton, sugarcane, soyabean and pepper [149]. The technology has been evolving rapidly with simpler and targeted base editing tools. The plant GE techniques, based on on-site specific endonucleases, are categorized into meganucleases, zinc finger nucleases (ZFNs), transcription activator-like effector nucleases (TALENs), and clustered, regularly interspaced, short palindromic repeats (CRISPR)/CRISPR-associated protein 9 (Cas9). The CRISPR/Cas9 system has been used predominantly for genome editing of different traits, including traits for abiotic stress tolerance.

Drought affects the growth of plants at various morphological, physiological, and biochemical levels [150] and its negative impact is expected to intensify due to climate change. The identification and cloning of several stress related candidate genes has been shown to provide long term resistance against abiotic stresses. In this regard, several genetic engineering strategies are in place to manipulate stress related genes. Genome editing has been successfully employed for improving the drought tolerance of different cereal crops (Table 4). The OPEN

STOMATA 2 (OST2) gene encodes for an H+-ATPase and is involved in creating proton gradients in plant cells. Precise modification of this gene via CRISPR/Cas9 has been reported to modulate stomatal closing in response to water deficient conditions, thus conferring drought stress tolerance [151]. The loss of function sapk2 mutant rice plants produced using CRISPR/Cas exhibited more sensitivity against drought stress that was linked to the modulation of the expression of several genes that acts downstream of the SAPK2, including OsOREB1, OsRab21, OsRab16b, OsLEA3, OsbZIP23, OsSLAC1 and OsSLAC7 genes [152]. ARGOS8 is yet another drought stress responsive gene modulated using genome editing. It is a negative regulator of the ethylene signalling pathway, and its increased expression is known to confer drought stress resistance in plants [153].

TABLE 4. SUCCESSFUL EXAMPLES OF CRISPR/CAS BASED GENOME EDITING FOR IMPROVING DROUGHT TOLERANCE IN PLANTS
(modified from Kaur et al. 2022) [154]

Plant	Gene	Responsive mechanism	Reference
Arabidopsis	*OST2*	a H+-ATPase	[150]
Maize	*ARGOS8*	Negative regulator of ethylene response	[153]
Wheat	*TaDREB2*	Dehydration responsive gene	[155]
Wheat	*TaERF3*	Ethylene responsive factor 3	[55]
Rice	*miR535*	Regulation of abiotic stress responsive gene expression	[156]
Rice	*SAPK2*	ABA signalling	[157]
Rice	*DST*	Zinc finger transcription factor	[158]
Rice	*OsSAP*	*Oryza sativa* senescence associated protein	[159]

The replacement of the promoter sequence of ARGOS8 using CRISPR/Cas9 with the GOS2 promoter resulted in the increased ubiquitous expression of this gene in maize, thus enhancing their resistance towards drought stress [153]. Further, it has been realized that genome editing can be successfully used for improving the different traits of diploid plants; however, its implementation on the polyploid and complex genomes is a major challenge. The CRISPR/Cas9 has been used to edit the TaDREB2 gene that encodes for a dehydration responsive element binding protein in the protoplasts of wheat to generate drought stress resistant plants, and this approach opens new opportunities to improve the stress resistance of polyploid plants using genome editing [155]. Moreover, editing of the trehalase gene that plays a key role in trehalose catabolism using CRISPR/Cas improved the drought tolerance in *Arabidopsis thaliana* [160]. Trehalase enzyme catalyses the only reaction of trehalose catabolism in plants and results in the hydrolysis of trehalose into D-glucose molecules. The plants with edited substrate-binding domain of trehalase enzyme showed drought tolerant phenotypic traits. Besides, the alteration of DST and miR535 genes using CRISPR/Cas leads to improved drought tolerance in addition to salinity stress acclimatization in rice [156, 161]. The studies have successfully demonstrated that genome editing technique is very useful for the identification and characterization of important drought stress related genes and for targeted modification for improving drought tolerance in crop plants [162].

17.   CONCLUSIONS AND FUTURE PERSPECTIVES

The main approach for breeding for drought prone environments is to: (i) improve yield potential and, depending on the type of drought, select for the appropriate combination of maturity to avoid stress during the reproductive stage; and (ii) select for tolerance to drought stress during the reproductive period, and avoid plant types that use a lot of water prior to flowering (i.e. produce large amounts of dry matter (DM) and run out of water at the critical stage of flowering. In upland rice, as in other aerobic crops, there may also be opportunities to increase the amount of water transpired through more vigorous root systems.

Remarkable progress has been made not only improve genotyping and phenotyping methods but also the statistical tools in order to increase speed and accuracy of conducting QTL analysis. Recent developments in genomics and phenomics has helped in developing an understanding of the genetic architecture that is responsible for providing adaptation that has enabled a more accurate and detailed characterization of the QTLs that regulate a particular trait. However, improved QTL meta-analyses, better estimation of QTL effects and improved crop modelling will enable a more effective exploitation of the QTLs that regulate a particular trait. Some major and stable QTLs and MQTLs along with candidate genes have been identified for drought tolerance in important crop plants and have been found to be associated with yield and other important agronomic and physiological traits. There is therefore an urgent need to exploit this genetic and genomics resource of drought tolerance as well as genomic selection and other strategies like epi-QTLs, e-QTLs, cloning, to design programmes for breeding crops tolerant to drought stress.

REFERENCES TO PART 1

[1]   UNITED STATES ENVIRONMENTAL PROTECTION AGENCY, Washington, DC (2023); https://www.epa.gov/climate-indicators/weather-climate (accessed 14/06/2022).

[2]   LEISNER, C.P., Climate change impacts on food security — focus on perennial cropping systems and nutritional value, Plant Sci. **293** (2020) 1–7.

[3]   TRIPATHI, A., et al., Paradigms of climate change impacts on some major food sources of the world: A review on current knowledge and future prospects, Agric. Ecosyst. Environ. **216** (2016) 356–373.

[4]   UNITED NATIONS, World Population Prospects 2019: Highlights, ST/ESA/SER.A/423, UN, New York (2019).

[5]   HOSEINLOU, S.H., et al., Nitrogen use efficiency under water deficit condition in spring barley, Int. J. Agron. Plant Product. **4** (2013) 3681–3687.

[6]   MOHAMMADI, R., Breeding for increased drought tolerance in wheat: A review, Crop Past. Sci. **69** (2018) 223–241.

[7]   ROY, S., VERMA, B.C., BANERJEE, A., Genetic diversity for drought and low-phosphorus tolerance in rice (*Oryza sativa* L.) varieties and donors adapted to rainfed drought-prone ecologies, Sci. Rep. **11** (2021) 1–9.

[8]   ISEKI, K., et al., Diversity of drought tolerance in the genus Vigna, Front. Plant Sci. **9** (2018) 1–18.

[9]   http://www.plantstress.com/files/Recent_Reviews/index.asp

[10]   SINCLAIR, T.R., Challenges in breeding for yield increase for drought, Trends Plant Sci. **16** (2011) 289–293.

[11]   MESSINA, C.D., et al., Yield–trait performance landscapes: from theory to application in breeding maize for drought tolerance, J. Exp. Bot. **62** (2011) 855–868.

[12]   HAMMER, G.L., COOPER, M., REYNOLDS, M.P., Plant production in water-limited environments, J. Exp. Bot. **72** (2021) 5097–5101.

[13]   CATTIVELLI, L., et al., Drought tolerance improvement in crop plants: An integrated view from breeding to genomics, Field Crops Res. **105** (2008) 1–14.

[14]   FLEURY, D., et al., Genetic and genomic tools to improve drought tolerance in wheat, J. Exp. Bot. **61** (2010) 3211-22.

[15]   VARSHNEY, R.K., Agricultural biotechnology for crop improvement in a variable climate: hope or hype? Trends Plant Sci. **16** (2011) 363–371.

[16]   BALDONI, E., Improving drought tolerance: Can comparative transcriptomics support strategic rice breeding? Plant Stress **3** (2022) 1–9.

[17]   ASHRAF, M., et al., Quantifying climate-induced drought risk to livelihood and mitigation actions in Balochistan, Nat. Hazards **109** (2021) 2127–2151.

[18]   FOOD AND AGRICULTURE ORGANIZATION OF THE UNITED NATIONS, Drought and Agriculture, FAO, Rome (2017) https://www.fao.org/land-water/water/drought/droughtandag/en/ (accessed 14/06/2022).

[19]   OSAKABE, Y., et al., Response of plants to water stress, Front. Plant Sci. **5** 86 (2014) 1–8.

[20]   DIETZ, K.-J., ZÖRB, C., GEILFUS, C.-M., Drought and crop yield, Plant Biol. **23** (2021) 881–893.

[21]   LEVITT, J., Responses of Plants to Environmental Stresses, Academic Press, New York and London (1972).

[22]   RIDLEY, E.J., TODD, G.W., Anatomical variations in the wheat leaf following internal water stress, Bot. Gaz. **127** (1966) 235–238.

[23]   PANDEY, V., SHUKLA, A., Acclimation and tolerance strategies of rice under drought stress, Rice Sci. **22** (2015) 147–161.

[24]   KAPOOR, D., et al., The impact of drought in plant metabolism: How to exploit tolerance mechanisms to increase crop production, Appl. Sci. **10** (2020) 1–19.

[25]   RIZZA, F., et al., Use of a water stress index to identify barley genotypes adapted to rainfed and irrigated conditions, Crop Sci. **44** (2004) 2127–2137.

[26]   NEZHADAHMADI, A., PRODHAN, Z.H., FARUQ, G., Drought tolerance in wheat, Sci. World J. 610721 (2013) 1–13.

[27]   KADAM, S.R., FAKRUDIN, B., Marker assisted pyramiding of root volume QTLs to improve drought tolerance in rabi sorghum, Res. Crop. **18** (2017) 683–692.

[28]   BUDAK, H., et al., From genetics to functional genomics: improvement in drought signaling and tolerance in wheat, Front. Plant Sci. **6** 1012 (2015) 1–13.

[29]   JALEEL, C.A., et al., Drought stress in plants: A review on morphological characteristics and pigments composition, Int. J. Agric. Biol. **11** (2009) 100–105.

[30]   ESTRADA-CAMPUZANO, G., MIRALLES, D.J., SLAFER, G.A., Genotypic variability and response to water stress of pre-and post-anthesis phases in triticale, Eur. J. Agron. **28** (2008) 171–177.

[31]   FAROOQ, M. et al., Plant drought stress: Effects, mechanisms and management, Agron. Sustain. Dev. (2009) 153–188.

[32]   DICKIN, E., WRIGHT, D., The effects of winter waterlogging and summer drought on the growth and yield of winter wheat (*Triticum aestivum* L.), Eur. J. Agron. **28** (2008) 234–244.

[33]   KAMARA, A.Y., MENKIR, A., BADU-APRAKU, B., IBIKUNLE, O., The influence of drought stress on growth, yield and yield components of selected maize genotypes, J. Agric. Sci. **141** (2003) 43–50.

[34]   MONNEVEUX, P., SANCHEZ, C., BECK, D., EDMEADES, G.O., Drought tolerance improvement in tropical maize source populations: Evidence of progress, Crop Sci. **46** (2006) 180–191.

[35]   SHINOZAKI, K., YAMAGUCHI-SHINOZAKI, K., Gene networks involved in drought stress response and tolerance, J. Exp. Bot. **58** (2007) 221–227.

[36]   SREENIVASULU, N., et al., Contrapuntal role of ABA: Does it mediate stress tolerance or plant growth retardation under long-term drought stress? Gene **506** (2012) 265–273.

[37]   CHAVES, M.M., MAROCO, J.P., PEREIRA, J.S., Understanding plant responses to drought—From genes to the whole plant, Funct. Plant Biol. **30** (2003) 239–264.

[38]   LEAKEY, A.D.B., et al., Water use efficiency as a constraint and target for improving the resilience and productivity of C3 and C4 crops, Annu. Rev. Plant Biol. **70** (2019) 781–808.

[39]   PRASCH, C.M., SONNEWALD, U., Signaling events in plants: Stress factors in combination change the picture, Environ. Exp. Bot. **114** (2015) 4–14.

[40]   BLUM, A., Stress, strain, signaling, and adaptation — Not just a matter of definition, J. Exp. Bot. **67** (2016) 562–565.

[41]   HERNÁNDEZ-SÁNCHEZ, I.E., et al., Leafing through literature: Late embryogenesis abundant proteins coming of age-achievements and perspectives, J. Exp. Bot. **73** (2022) 6525–6546.

[42]   YANG, X. et al. Response mechanism of plants to drought stress, Horticulturae **7** (2021) 1–36.

[43]   FANG, Y., XIONG, L., General mechanisms of drought response and their application in drought resistance improvement in plants, Cell. Mol. Life Sci. **72** (2015) 673–689.

[44]   WIMALASEKERA, R., "Breeding crop plants for drought tolerance", Water Stress and Crop Plants: A Sustainable Approach (AHMAD, P., Ed.), Wiley, Oxford (2016).

[45]   LUO, L.J., Breeding for water-saving and drought-resistance rice (WDR) in China, J. Exp. Bot. **61** (2010) 3509–3517.

[46]   PANDA, D., MISHRA, S.S., BEHERA, P.K., Drought tolerance in rice: Focus on recent mechanisms and approaches, Rice Sci. **28** (2021) 119–132.

[47]   COHEN, I., et al., Meta-analysis of drought and heat stress combination impact on crop yield and yield components, Physiol. Plant. **171** (2021) 66–76.

[48]   BHANDARI, H.R., et al., Assessment of Genetic Diversity in Crop Plants: An Overview, Adv. Plants Agric. Res. **7** (2017) 279-286.

[49]   MUNNS, R., JAMES, R.A., Screening methods for salinity tolerance: A case study with tetraploid wheat, Plant Soil **253** (2003) 201–218.

[50]   TORRES, R.O., et al., Screening of rice Genebank germplasm for yield and selection of new drought tolerance donors, Field Crops Res. **147** (2013) 12–22.

[51]   MONKHAM, T., et al., Genotypic variation in grain yield and flowering pattern in terminal and intermittent drought screening methods in rainfed lowland rice, Field Crops Res. **175** (2015) 26–36.

[52]   ZHU, Q., et al., Multilocus analysis of nucleotide variation of *Oryza sativa* and its wild relatives: Severe bottleneck during domestication of rice, Mol. Biol. Evol. **24** (2007) 875– 888.

[53]   ROYO, C., ABAZA, M., BLANCO, R., GARCIA DEL MORAL, L.F., Triticale grain growth and morphometry as affected by drought stress, late sowing and simulated drought stress, Aust. J. Plant Physiol. **27** (2000) 1051–1059.

[54]   TSHIKUNDE, N.M., MASHILO, J., SHIMELIS, H. ODINDO, A., Agronomic and physiological traits, and associated quantitative trait loci (qtl) affecting yield response in wheat (*Triticum aestivum* L.): A review, Front. Plant Sci. **10** (2019) 1–18.

[55]   VARSHNEY, R.K., ROORKIWAL, M., SUN, S., A global reference for chickpea genetic variation based on the sequencing of 3,366 genomes, Nature **526** (2021a) 68–74.

[56] BIMPONG, I.K. et al. Determination of genetic variability for physiological traits related to drought tolerance in African rice (*Oryza glaberrima*), J. Plant Breed. Crop Sci. **3** (2011) 60–67.

[57] BAPELA, T., SHIMELIS, H., TSILO, T.J., MATHEW, I., Genetic improvement of wheat for drought tolerance: Progress, challenges and opportunities, Plants **11** (2022) 1–32.

[58] ZHANG, Q., et al., Identification of drought tolerant mechanisms in a drought-tolerant maize mutant based on physiological, biochemical and transcriptomic analyses, BMC Plant Biol. **20** (2020) 1–14.

[59] PALMGREN, M.G., et al., Are we ready for back-to-nature crop breeding? Trends Plant Sci. **20** (2015) 155–164.

[60] WANG, W., et al., Genomic variation in 3,010 diverse accessions of Asian cultivated rice, Nature **557** (2018) 43–49.

[61] VARSHNEY, R.K., ROORKIWAL, M., SUN, S., A global reference for chickpea genetic variation based on the sequencing of 3,366 genomes, Nature **526** (2021b) 68– 74.

[62] VARSHNEY, R.K., THUDI, M., ROORKIWAL, M., Resequencing of 429 chickpea accessions from 45 countries provides insights into genome diversity, domestication and agronomic traits, Nat. Genet. **51** (2019) 857–864.

[63] KHAZAEI, H. et al., The FIGS (focused identification of germplasm strategy) approach identifies traits related to drought adaptation in *Vicia faba* genetic resources, PloS One **8** (2013) 1–10.

[64] REYNOLDS, M., et al., "Exploring genetic resources to increase adaptation of wheat to climate change", Advances in Wheat Genetics: From Genome to Field (OGIHARA, Y., TAKUMI, S., HANDA, H., Eds), Springer, Tokyo (2015) 355–368.

[65] SIVASANKAR, S., "Developing drought- and heat-tolerant varieties of grain legumes", Achieving Sustainable Cultivation of Grain Legumes, Vol. 1, Advances in Breeding and Cultivation Techniques (SIVASANKAR, S., et al., Eds), Burleigh Dodds Science Publishing, Cambridge, UK (2018) 1–21.

[66] HARB, A., KRISHNAN, A., AMBAVARAM. M.M., PEREIRA, A., Molecular and physiological analysis of drought stress in Arabidopsis reveals early responses leading to acclimation in plant growth, Plant Physiol. **154** (2010) 1254–1271.

[67] SIVASANKAR, S., et al., Eds, Achieving Sustainable Cultivation of Grain Legumes, Burleigh Dodds Science Publishing, Cambridge, UK (2018) 1–27.

[68] OMPRAKASH, et al., Resistance/tolerance mechanism under water deficit (drought) condition in plants, Int. J. Curr. Microbiol. Appl. Sci. **6** (2017) 66–78.

[69] BADIGANNAVAR, A., et al., Physiological, genetic and molecular basis of drought resilience in sorghum (*Sorghum bicolor* (L.) Moench), Ind. J. Plant Physiol. **23** (2018) 670–688.

[70] PAMIRELLI, R., SRINIVASA RAO, M., "Breeding for drought resistance", Plant Breeding: Current and Future Views (ABDURAKHMONOV, I., Ed.), IntechOpen, London (2021).

[71] SHAVRUKOV, Y., et al., Early flowering as a drought escape mechanism in plants: How can it aid wheat production? Front. Plant Sci. **8** (2017) 1–8.

[72] YADAV, S., SHARMA, K.D., Molecular and morpho physiological analysis of drought stress in plants molecular and morphophysiological analysis of drought stress in plants, Plant Growth (2016).

[73] RAUF, S., et al., "Breeding strategies to enhance drought tolerance in crops", Advances in Plant Breeding Strategies: Agronomic, Abiotic and Biotic Stress Traits (AL-KHAYRI, J.M., JAIN, S.M., JOHNSON, D.V., Eds), Springer International Publishing, Cham, Switzerland (2016) 397–445.

[74] DRIEDONKS, N., RIEU, I., VRIEZEN, W.H., Breeding for plant heat tolerance at vegetative and reproductive stages, Plant Reprod. **29** (2016) 67–79.

[75] MANJAYA, J.G., "Contribution and impact of mutant varieties on food security" Advanced Crop Improvement (RAINA, A., WANI, M.R., LASKAR, R.A., TOMLEKOVA, N., KHAN, S., Eds), Vol. 1, Springer International Publishing, Cham, Switzerland (2023) 47–73.

[76] Manual on Mutation Breeding, 3rd edn (SPENCER-LOPES, M.M., FORSTER, B.P., JANKULOSKI, L., Eds), Food and Agriculture Organization of the United Nations, Rome (2018).

[77] FAO/IAEA Mutant Variety Database, IAEA, Vienna (2023) https://nucleus.iaea.org/sites/mvd/SitePages/Home.aspx [accessed August 6, 2023].

[78] RUTGER, J.N., "The induced sd1 mutant and other useful mutant genes in modern rice varieties", Induced Plant Mutations in the Genomics Era (SHU, Q., Ed.), Food and Agriculture Organization of the United Nations, Rome (2009) 44–47.

[79] POLI, Y., et al., Characterization of a Nagina 22 rice mutant for heat tolerance and mapping of yield traits, Rice (New York) **6** (36) (2013) 1–9.

[80] SARSU, F., BIMPONG, I.K., JANKULOSKI, L., Contribution of induced mutation in crops to global food security, ACI Avances EnCiencias E Ingenierías **12** (2021) 1–10.

[81] SUPRASANNA, P., SHIRANI BIDABADI, S., JAIN, S.M., "Mutation breeding to promote sustainable agriculture and food security in the era of climate change", Mutation Breeding for Sustainable Food Production and Climate Resilience (PENNA, S., JAIN, S.M., Eds), Springer, Singapore (2023) 1–23.

[82] SOERANTO, H., SIHONO, INDRIATAMA, W.M., Sorghum improvement program by using mutation breeding in Indonesia, IOP Conf. Series, Earth Environ. Sci. **484** 012003 (2020) 1–7.

[83] SARSU, F., PENNA, S., NIKALJE, G.C., "Strategies for screening induced mutants for stress tolerance", Mutation Breeding for Sustainable Food Production and Climate Resilience (PENNA, S., JAIN, S.M., Eds), Springer, Singapore (2023) 151–176.

[84] COOPER, M., MESSINA, C.D., Breeding crops for drought-affected environments and improved climate resilience, The Plant Cell **35** (2023) 162–186.

[85] CECCARELLI, S., GRANDO, S., BAUM, M., UDUPA, S.M., "Breeding for drought resistance in a changing climate", Challenges and Strategies of Dryland Agriculture (RAO, S.C., RYAN, J., Eds), Wiley, New York (2004) 167–190.

[86] KUMAR, A., BASU, S., RAMEGOWDA, V., PEREIRA, A., "Mechanisms of drought tolerance in rice", Achieving Sustainable Cultivation of Rice (SASAKI, T., Ed.), Burleigh Dodds Science Publishing, Cambridge, UK (2017).

[87] ZIYOMO, C., BERBARDO, R., Drought tolerance in maize: Indirect selection through secondary traits versus genome wide selection, Crop Sci. **53** (2013) 1269–1275.

[88] TUBEROSA, R., Phenotyping for drought tolerance of crops in the genomics era, Front. Physiol. **3** (2012) 1–26.

[89] KAUR, G., ASTHIR, B., Molecular responses to drought stress in plants, Biol. Plant. **61** (2017) 201–209.

[90] MIR, R.R., et al., Integrated genomics, physiology and breeding approaches for improving drought tolerance in crops, Theor Appl. Genet. **125** (2012) 625–645.

[91] KALAJI, H.M., Chlorophyll fluorescence as a tool to monitor physiological status of plants under abiotic stress conditions, Acta Physiol. Plant. **38** (2016) 1–11.

[92] THOMAS, H., OUGHAM, H., The stay-green trait, J. Exp. Bot. **14** (2014) 3889–3900.

[93] KAMAL, N.M., et al., Stay-green trait: A prospective approach for yield potential, and drought and heat stress adaptation in globally important cereals, Int. J. Mol. Sci. **20** (2019) 1–26.

[94] ZU, X., A new method for evaluating the drought tolerance of upland rice cultivars, Crop J. **5** (2017) 488–498.

[95] KHORASGANI, O.A., PESSARAKLI, Manipulation of plant methylglyoxal metabolic and signaling pathways for improving tolerance to drought stress, J. Plant Nutr. **42** (2019) 1268–1275.

[96] SHAW, R.K., et al., Molecular breeding strategy and challenges towards improvement of downy mildew resistance in cauliflower (*Brassica oleracea* var. botrytis L.), Front. Plant Sci. **12** (2021) 1–20.

[97] SUPRASANNA, P., Plant abiotic stress tolerance: Insights into resilience build-up, J. Biosci. **45** (2020) 1–8.

[98] SATPUTE, G.K., et al., "Breeding and molecular approaches for evolving drought-tolerant soybeans", Plant Stress Biology (GIRI, B., SHARMA, M.P., Eds), Springer, Singapore (2020).

[99] GOVIND, G., "Plant abiotic stress tolerance on the transcriptomics atlas", Advancements in Developing Abiotic Stress Resilient Plants (KHAN, M., REDDY, P.S., GUPTA, R., Eds), CRC Press, Boca Raton (2022) 196–236.

[100] IKEDA, M., MIURA, K., AYA, K., MATSUOKA, M., Genes offering the potential for designing yield-related traits in rice, Curr. Opin. Plant Biol. **16** (2013) 213–220.

[101] SUN, L., et al., Identification of quantitative trait loci for grain size and the contributions of major grain-size QTLs to grain weight in rice, Mol. Breed. **31** (2013) 451–461.

[102] YAN, W., BAO, J. (Eds), Rice–Germplasm. Genetics and Improvement, Rijeka, InTech (2014).

[103] BERNIER, J. et al. Characterization of the effect of a QTL for drought resistance in rice, *qtl12.1*, over a range of environments in the Philippines and eastern India, Euphytica **166** (2009) 207–217.

[104] BOOPATHI, N.M., et al., Evaluation and bulked segregant analysis of major yield QTL *qtl12.1* introgressed into indigenous elite line for low water availability under water stress, Rice Sci. **20** (2013) 25–30.

[105] MISHRA, K.K., et al., *qDTY12.1*: A locus with a consistent effect on grain yield under drought in rice, BMC Genet. **14** (2013) 1–10.

[106] VIKRAM, P., et al., *qDTY₁.₁*: A major QTL for rice grain yield under reproductive-stage drought stress with a consistent effect in multiple elite genetic backgrounds, BMC Genet. **12** (2011) 1–15.

[107] GHIMIRE, K.H., et al., Identification and mapping of a QTL (*qDTY1.1*) with a consistent effect on grain yield under drought, Field Crops Res. **131** (2012) 88–96.

[108] DIXIT, S., et al., Increased drought tolerance and wider adaptability of *qDTY* 12.1 conferred by its interaction with *qDTY* 2.3 and *qDTY* 3.2, Mol. Breed. **30** (2012) 1767– 1779.

[109] TANG, S.Q., et al., QTL mapping of grain weight in rice and the validation of the QTL *qTGW3.2*, Gene. **527** (2013) 201–206.

[110] YADAW, R.B., et al., A QTL for high grain yield under lowland drought in the background of popular rice variety Sabitri from Nepal, Field Crops Res. **144** (2013) 281–287.

[111] SAIKUMAR, S., et al., Major QTL for enhancing rice grain yield under lowland reproductive drought stress identified using an *O. sativa/O. glaberrima* introgression line, Front. Crop Res. **163** (2014) 119–131.

[112] ZHANG, M., et al., Mapping and validation of quantitative trait loci associated with concentrations of 16 elements in unmilled rice grain, Theor. Appl. Genet. **127** (2014) 137–165.

[113] YI, Y., QTL mapping and analysis for drought tolerance in rice by genome-wide association study, Front. Plant Sci. **14** (2023) 1–11.

[114] NIKOLIC, A., Identification of QTLs for drought tolerance in maize. II Yield and yield components, Genetika **45** (2013) 341–50.

[115] ANDLEEB, T., "QTL mapping for drought stress tolerance in plants", Salt and Drought Stress Tolerance in Plants, Signalling and Communication in Plants (Hasanuzzaman, M., Tanveer, M., Eds), Springer Nature Switzerland, Basel (2020).

[116] HU, Y., et al., Genome assembly and population genomic analysis provide insights into the evolution of modern sweet corn, Nat. Comm. **12** 1227 (2021) 1–13.

[117] MANO, Y., MURAKI, M., FUJIMORI, M., TAKAMIZO, T., Varietal difference and genetic analysis of adventitious root formation at the soil surface during flooding in maize and teosinte seedlings, Jpn J. Crop Sci. **74** (2005) 41–46.

[118] GUPTA, P.K., BALYAN, H.S., GAHLAUT, V., KULWAL, P.L., Phenotyping, genetic dissection, and breeding for drought and heat tolerance in common wheat: Status and prospects, Plant Breed. Rev. **36** (2012) 85–147.

[119] PINTO, R.S., Heat and drought adaptive QTL in a wheat population designed to minimize confounding agronomic effects, Theor. Appl. Genet. **121** (2010) 1001–1021.

[120] VON KORFF, M., WANG, H., LÉON, J., PILLEN, K. AB-QTL analysis in spring barley: III. Identification of exotic alleles for the improvement of malting quality in spring barley (*H. vulgare* ssp. spontaneum), Mol. Breed. **21** (2010) 81–93.

[121] FAN, Y., Using QTL mapping to investigate the relationships between abiotic stress tolerance (drought and salinity) and agronomic and physiological traits, BMC Genom. **16** (2015) 1–11.

[122] MORA, F., SNP-based QTL mapping of 15 complex traits in barley under rain-fed and well-watered conditions by a mixed modelling approach, Front. Plant Sci. **7** (2016) 1–11.

[123] PEIGHAMBARI, S.A., et al., QTL analysis for agronomic traits in a barley doubled haploids population grown in Iran, Plant Sci. **169** (2005) 1008–1013.

[124] SAYED, M.A., et al., AB-QTL analysis reveals new alleles associated to proline accumulation and leaf wilting under drought stress conditions in barley (*Hordeum vulgare* L.), BMC Genet. **13** (2012) 1–12.

[125] ABREHA, K.B., ENYEW, M., CARLSSON, A.S., Sorghum in dryland: morphological, physiological, and molecular responses of sorghum under drought stress, Planta **255** (2022) 1–23.

[126] XU, Y., Global view of QTL: Rice as a Model. Quantitative Genetics, Genomics, and Plant Breeding, CABI Publishing, Wallingford (2002) 109–134.

[127] KAPANIGOWDA, M.H., et al., Quantitative trait locus mapping of the transpiration ratio related to preflowering drought tolerance in sorghum (*Sorghum bicolor*), Funct. Plant Biol. **41** (2014) 1049–1065.

[128] MACE, E.S., et al., QTL for nodal root angle in sorghum (*Sorghum bicolor* L. Moench) co-locate with QTL for traits associated with drought adaptation, Theor. Appl. Gen. **124** (2012) 97–109.

[129] YADAV, R.S., et al., Genomic regions associated with grain yield and aspects of post-flowering drought tolerance in pearl millet across stress environments and tester background, Euphytica **136** (2004) 265–277.

[130] BIDINGER, F.R., et al., Field evaluation of drought tolerance QTL effects on phenotype and adaptation in pearl millet (*Pennisetum glaucum* (L.) R. Br.) topcross hybrids, Field Crops Res. **94** (2005) 14–32.

[131] YADAV, R.S., SEHGAL, D., VADEZ, V., Using genetic mapping and genomics approaches in understanding and improving drought tolerance in pearl millet, J. Exp. Bot. **62** (2011) 397–408.

[132] KHOLOVÁ, J., et al., Water saving traits co-map with a major terminal drought tolerance quantitative trait locus in pearl millet (*Pennisetum glaucum* (L.) R. Br.), Mol. Breed. **30** (2012) 1337–1353.

[133] SWAMY, B.P., Meta-analysis of grain yield QTL identified during agricultural drought in grasses showed consensus, BMC Genom. **12** (2011) 1–18.

[134] MWAMAHONJE, A., Drought tolerance and application of marker-assisted selection in sorghum, Biology **10** (2021) 1–12.

[135] KUMAR, P.K.C., Development of bread wheat (*Triticum aestivum* L) variety HD3411 following marker-assisted backcross breeding for drought tolerance, Front. Genet. **14** (2023) 1–12.

[136] SANDHU, N., KUMAR, A., Bridging the rice yield gaps under Drought: QTLs, genes, and their use in breeding programs, Agronomy **7** (2017) 1–27.

[137] ANYAOHA, C.O., et al., Introgression of two drought QTLs into FUNAABOR-2 early generation backcross progenies under drought stress at reproductive stage, Rice Sci. **26** (2019) 32–41.

[138] SHAMSUDIN, N.A., Marker assisted pyramiding of drought yield QTLs into a popular Malaysian rice cultivar, MR219, BMC Genet. **17** (2016) 1–14.

[139] KUMAR, A., Breeding high-yielding drought-tolerant rice: genetic variations and conventional and molecular approaches, J. Exp. Bot. **65** (2014) 6265–6278.

[140] TAKAHASHI, F., et al., Drought stress responses and resistance in plants: From cellular responses to long-distance intercellular communication, Front. Plant Sci. **11** (2020) 1–14.

[141] SELEIMAN, M.F., et al., Drought stress impacts on plants and different approaches to alleviate its adverse effects, Plants (Basel) **10** (2021) 1–25.

[142] SAH, S.K., REDDY, K.R., LI, J., Abscisic acid and abiotic stress tolerance in crop plants, Front. Plant Sci. **7** (2016) 1–26.

[143] OGUZ, M.C., et al., Drought stress tolerance in plants: Interplay of molecular, biochemical and physiological responses in important development stages, Physiologia **2** (2022) 180–197.

[144] AHMED, R.F., Engineering drought tolerance in plants by modification of transcription and signalling factors, Biotechnol. Biotechnol. Equip. **34** (2020) 781–789.

[145] LIANG, C. "Genetically modified crops with drought tolerance: Achievements, challenges, and perspectives", Drought Stress Tolerance in Plants (HOSSAIN, M., WANI, S., BHATTACHARJEE, S., BURRITT, D., TRAN, L.S., Eds), Vol. 2, Springer, Cham, Switzerland (2016).

[146] KHAN, Z., SOHAIL, S., JAN, A., Genetic engineering approaches to understanding drought tolerance in plants, Plant Biotechnol. Rep. **14** (2020) 151–162.

[147] WANG, H., WANG, H., SHAO, H., Recent advances in utilizing transcription factors to improve plant abiotic stress tolerance by transgenic technology, Front. Plant Sci. **7** (2016) 1–13.

[148] HU, Y., CHEN, X., SHEN, X., Regulatory network established by transcription factors transmits drought stress signals in plant, Stress Biol. **2** (2022) 1–22.

[149] SHINWARI, Z.K., JAN, S.A., NAKASHIMA, K., YAMAGUCHI-SHINOZAKI, K., Genetic engineering approaches to understanding drought tolerance in plants, Plant Biotechnol. Rep. **14** (2020) 151–162.

[150] SALEHI-LISAR, S.Y., BAKHSHAYESHAN-AGDAM, H., "Drought stress in plants: Causes, consequences, and tolerance", Drought Stress Tolerance in Plants (HOSSAIN, M.A., WANI, S.H., BHATTACHARJEE, S., BURRITT, D.J., TRAN, L.S.P.), Vol. 1, Springer, Cham, Switzerland (2016) 1–16.

[151] OSAKABE, Y., WATANABE, T., SUGANO, S.S., Optimization of CRISPR/Cas9 genome editing to modify abiotic stress responses in plants, Sci. Rep. **6** (2016) 1–10.

[152] LOU, D., WANG, H., LIANG, G., YU, D., OsSAPK2 confers abscisic acid sensitivity and tolerance to drought stress in rice, Front. Plant Sci. **8** (2017) 1–15.

[153] SHI, J., GAO, H., WANG, H., ARGOS 8 variants generated by CRISPR-Cas9 improve maize grain yield under field drought stress conditions, Plant Biotechnol. J. **15** (2017) 207–216.

[154] KAUR, N., SHARMA, S., HASANUZZAMAN, M., PATI, P.K., Genome editing: A promising approach for achieving abiotic stress tolerance in plants, Int. J. Genomics **2022** (2022) 1–12.

[155] KIM, D., ALPTEKIN, B., BUDAK, H., CRISPR/Cas9 genome editing in wheat, Funct. Integr. Genomics **18** (2018) 31–41.

[156] SANTOSH KUMAR, V.V., VERMA, R.K., YADAV, S.K., CRISPR-Cas9 mediated genome editing of drought and salt tolerance (*OsDST*) gene in *Indica* mega rice cultivar MTU 1010, Physiol. Mol. Biol. Plants **26** (2020) 1099–1110.

[157] ZHANG, A., LIU, Y., WANG, F., Enhanced rice salinity tolerance via CRISPR/Cas9-targeted mutagenesis of the OsRR22 gene, Mol. Breed. **39** (2019) 1–10.

[158] DUAN, Y.-B., LI, J., QIN, R.-Y., Identification of a regulatory element responsible for salt induction of rice OsRAV2 through ex situ and in situ promoter analysis, Plant Mol. Biol. **90** (2016) 49–62.

[159] PARK, J-R., et al., Applications of CRISPR/Cas9 as new strategies for short breeding to drought gene in rice, Front. Plant Sci. **13** (2022) 1–15.

[160] NUÑEZ-MUÑOZ, L., et al., Plant drought tolerance provided through genome editing of the trehalase gene, Plant Signal. Behav. **16** (2021) e1877005 (1–12).

[161] YUE, E., CAO, H., LIU, B., OsmiR535, a potential genetic editing target for drought and salinity stress tolerance in *Oryza sativa*, Plants **9** (2020) 1–9.

[163] JOSHI, R.K., BHARAT, S.S., MISHRA, R., Engineering drought tolerance in plants through CRISPR/Cas genome editing, Biotechnol. **10** (2020) 1–14.

PART 2

DEVELOPMENT OF MUTANTS FOR DROUGHT RESILIENCE

2–1. FIELD EVALUATION OF PROMISING RICE MUTANT LINES IN DIFFERENT DROUGHT PRONE LOCATIONS OF BANGLADESH

M.M. ISLAM[1*], Md.A. HOSSAIN[2,3], M.A. HOSSAIN[3],
F. SARSU[4], M.S.R. KHANOM[1], A.C. SHARMA[3],
U. SAHA[5], S.N. BEGUM[1]

[1]Plant Breeding Division, Bangladesh Institute of Nuclear Agriculture
Mymensingh, Bangladesh

[2]Department of Biotechnology and Genetic Engineering,
Noakhali Science and Technology University,
Noakhali, Bangladesh

[3]Department of Genetics and Plant Breeding,
Bangladesh Agricultural University, Mymensingh, Bangladesh

[4]Joint FAO/IAEA Centre of Nuclear Techniques in Food and Agriculture,
International Atomic Energy Agency, Vienna

[5]Department of Biotechnology, Bangladesh Agricultural University,
Mymensingh, Bangladesh

[*]Email: mirza_islam@yahoo.com

Abstract

The study of the nature of interactions that occur between genotypes and the production environment for a certain trait is a crucial step towards the development of improved crop genotypes. The study was conducted using ten promising rice mutants (M_5) to evaluate yield and stability against drought stress at three drought prone locations in Bangladesh under rainfed condition following a randomized complete block design with three replications. Based on mean yield over locations, the mutant Binadhan-17/M_5/P-5 showed the highest yield per plant (17.2 g), while the lowest grain yield was recorded in susceptible check genotype IR-64 (5.34 g). According to the AMMI biplot and yield stability index, the Binadhan-17/M_5/P-5 mutant showed the highest yield stability followed by Binadhan-17/M_5/P-3 and NERICA-4/M_5/P-5, respectively, in the locations. Moreover, these promising mutant lines showed a regression coefficient value of around unity and less deviation from regression and also received the highest rank across the locations. Thus, these stable mutants have adapted to drought stress and can be recommended for cultivation in the drought prone areas of Bangladesh.

Key words: Drought stress, field evaluation, stability analysis, regression coefficient.

1.   INTRODUCTION

Drought stress is one of the most severe threats to global rice production, causing a significant decline in rice yield [1, 2] that has been identified as the single most critical threat to world food security [3]. The yield loss of rice due to a slight drought is from 10 to 30% and might be up to 70–90% under a severe drought [4]. The progress made in breeding for developing drought stress tolerant rice has been very slow due to the complex multitrait and polygenic control of drought tolerance, high genotype and environment (G×E) interactions (GEIs), low heritability, and difficulty in mass screening of plant traits and genes, as well as lack of variability for drought stress tolerance in the existing rice germplasms. The effect of environment on genotypes and their survivability in response to environmental factors is known as G×E interaction [5], which in turn influences the expression of functional genes of a genotype [6, 7]. Thus, the level of success depends on two factors: the stability of yield and the magnitude of GEI. It presents whether varieties are pure lines, single, double or top crosses [8]. Due to a change in the environment and rising population, it is crucial to develop climate smart agricultural crops to feed the world's population, as well as to maintain sustainability [9]. Wide adaptability of crops shows a large genotypic main effect with higher yield than lower adaptability of crops [10]. Stability analysis of crops is necessary for all agricultural researchers because it is a suitable method to find out the best one to survive across environmental conditions with better yield [11, 12]. By using this method, many researchers successfully released varieties that are widely cultivated [6, 13]. There are well established methods to analyse the GEI and stability of a genotype using models developed by Eberhart and Russell [6], Freeman and Perkin, AMMI, etc. [14]. To develop a suitable variety, we need to analyse stability among the genotypes in at least three locations to overcome this complication. If we can identify a drought tolerant stable genotype, then it can be grown successfully in different locations and finally used to release a new variety.

The response of the crops to varying environments will depend on the phenology, crop variety, and growth stage of the crop species. Significant GEI indicates that all phenotypic responses to varying agroecological conditions are not consistent. This would be due to differential performance of the genotypes from location to location and/or from year to year. GEIs have greater importance in plant breeding as they reduce the stability of genotypic values under diverse environments. Accordingly, AMMI and GGE biplot analyses were used to evaluate the agronomic performance and stability of the selected rice mutants along with their parents and checks.

2.   METHODOLOGY

2.1.   Locations and breeding materials

Three drought prone areas of Bangladesh (Magura, Rangpur and Mymensingh belonging to three different Agro-Ecological Zones (AEZs) were considered for the preliminary yield trial (between July and November 2020) of the selected rice mutants. To develop drought tolerant mutants, dry seeds of three rice genotypes: Binadhan-17 (high yielding, short duration and requires 40% less water and 30% less nitrogenous fertilizer); NERICA-4 (New Rice for Africa which was collected from Uganda and grown in upland conditions); and Galon (local land race, collected from a hill area of Bangladesh and cultivated in upland conditions) were irradiated with 200, 250, 300, 350, 400, 450, 475 and 500 Gy doses of gamma rays using the ^{60}Co source at the Bangladesh Institute of Nuclear Agriculture (BINA). At M4 generation, 27 promising rice mutants were evaluated through the assessment of morpho-physiological and biological traits related to drought stress tolerance at the seedling and reproductive stages. Finally, ten promising rice mutants (Binadhan-17/M$_5$/P-2, Binadhan-17/M$_5$/P-3, Binadhan-17/M$_5$/P-5 NERICA-4/M$_5$/P-2, NERICA-4/M$_5$/P-3, NERICA-4/M$_5$/P-5, NERICA-4/M$_5$/P-6, Galon/M$_5$/P-2, Galon/M$_5$/P-3 and Galon/M$_5$/P-6) were isolated to study their yield stability under direct field conditions along with their parents (Binadhan-17, NERICA-4 and Galon) and check varieties (IR-64 and Binadhan-19). The experiment was carried out in a two factorial randomized complete block design (RCBD) with three replications.

2.2.   Seed bed preparation and setting of the experiment

Seedlings of the tested genotypes were raised in homogeneous soil beds with urea, TSP and MP for the supply of N, P and K, respectively, at BINA, Mymensingh. For the final experiment, 38 day old seedlings were transplanted in the main field keeping one seedling per hill at a spacing of 20 cm × 15 cm. As regards the study with 15 genotypes under two treatments in three replications, homogeneous experimental plots were prepared for final transplanting. The control plot was given regular watering whenever necessary, whereas the drought treated plots were covered with a rainout shelter and moisture content was monitored and maintained using a time domain reflectometry (TDR) machine. Proper intercultural operation, fertilizer application and field management practices were carried out whenever necessary.

2.3. Screening for drought tolerance at the reproductive stage

The drought screening trials were conducted in a levelled and well drained field. The selected genotypes were transplanted in a puddled field. The trials were carried out using an RCBD design with three replications and two treatments, i.e drought stress and control (well irrigated). We used a digital soil moisture meter and recorded from five spots for a 10 m^2 field on every alternate day during the entire drought stress period. Standard agronomic management practices were followed in both stress and control plots. The recommended dose of fertilizers at 152–52–82–60 kg/ha (urea–TSP–MP–gypsum) was applied during and after field preparation if needed. Prior to beginning of drought stress, two doses of nitrogen fertilizer (basal and first split) were applied to drought trials and the third dose in adjustment with irrigation. Weeds were controlled by hand weeding. Disease and pest management was done properly using recommended agricultural practices and management. Insecticides and herbicides were applied at optimum dose. The plots used for drought treatments were drained out four weeks after transplanting for indicating drought at the reproductive stage until ten days after flowering.

In drought trials, for irrigation if severe stress appeared, the experiment was monitored (10 am) and, if the susceptible checks showed severe leaf rolling with less chance of recovery upon watering and the digital soil moisture meter showed a reading of volumetric moisture content (VMC) around 6% (equivalent to –50 kPa) soil moisture (30 cm depth), the trays/field were irrigated (40 mm). The fields were monitored regularly and scored for stress symptoms. Regular readings of water table depth, soil moisture and scoring of leaf rolling and leaf drying of drought susceptible and tolerant checks served as an index in the course of drought management.

2.4. Observation and data recording

Observations were recorded on days to first flowering, days to maturity, plant height (cm) number of total tillers/plant, number of effective tillers/plant, panicle length (cm), number of filled grains/plant, number of unfilled grains/plant, 100 seed weight (g), and grain yield/plant (g). Observations of yield and its component traits were recorded from five randomly selected plants of each treatment for each genotype.

2.5. Results

The experiment was conducted to study the effect of drought on yield and yield related traits on selected rice mutants and also determine the stable mutants based on yield performance over the locations. Different analytical methods were followed to identify the prominent effect of drought on yield and yield related traits of rice genotypes, which are presented under the following subheadings.

*2.5.1. Mean yield performance of selected rice mutant lines under drought
       stress condition in different locations/environments*

The mean yield performance of selected rice mutant lines under drought stress condition over
the locations is presented in Table 1. The results showed that higher mean grain yield was
recorded in genotype Binadhan-17/M_5/P-5 (17.20 g/plant) followed by Binadhan-17/M_5/P-3
(16.36 g/plant) and NERICA-4/M_5/P-5 (15.66 g/plant), while the lowest grain yield was
recorded for susceptible check genotype IR-64 (5.34 g/plant). At the three locations (Magura,
Rangpur and Mymensingh), Binadhan-17/M_5/P-3, Binadhan-17/M_5/P-5 and NERICA-4/M_5/P-
5 had the higher yield, which was higher than the mean yield of the parents and checks.
Therefore, the findings of the results predicted that these promising mutant lines had stable
performance across the environments and were less sensitive to the environment. Hence, these
mutant lines can be used as stable lines adapted across the environments in terms of yield.

TABLE 1. MEAN YIELD PERFORMANCE OF SELECTED RICE MUTANT LINES
UNDER DROUGHT STRESS CONDITIONS

Sl. No.	Genotypes	Yield (g/plant) in different locations			Mean yield (g/plant)
		Magura	Rangpur	Mymensingh	
01	Binadhan-17/M_5/P-2	15.07	15.77	13.60	14.82
02	Binadhan-17/M_5/P-3	16.69	16.80	15.60	16.36
03	Binadhan-17/M_5/P-5	17.21	16.89	17.52	17.20
04	Binadhan-17 (Parent)	12.84	12.73	12.50	12.69
05	NERICA-4/M_5/P-2	12.34	12.58	10.83	11.92
06	NERICA-4/M_5/P-3	12.91	12.38	14.61	13.30
07	NERICA-4/M_5/P-5	16.03	15.07	15.88	15.66
08	NERICA-4/M_5/P-6	14.61	14.59	14.85	14.68
09	NERICA-4 (Parent)	10.25	10.81	10.17	10.41
10	Galon/M_5/P-2	9.34	8.82	9.13	9.10
11	Galon/M_5/P-3	9.68	9.95	10.03	9.89
12	Galon/M_5/P-6	8.68	9.31	9.39	9.12
13	Galon (Parent)	8.89	8.85	10.47	9.40
14	IR-64 (S. check)	5.90	6.05	4.06	5.34
15	Binadhan-19 (T. check)	13.39	13.34	13.28	13.34
	Mean	12.25	12.26	12.13	12.22

2.5.2. Analysis of variance for additive main effect and multiplicative interactions

The results of analysis of variance (ANOVA) indicated that the yields differed significantly
($p \leq 0.001$) for genotypes, locations and genotypes × locations interactions (Tables 2–4). A large
sum of squares indicated that the locations were diverse. The AMMI model demonstrated the
presence of GEIs, and this has been partitioned into two highly significant ($p \leq 0.001$) interaction
principal component axes (IPCAs). The mean sum of squares indicated highly significant
differences for genotypes, locations and genotypes × locations interactions. The results also
indicated that the relative performances of the genotypes were significantly affected by the
varying environmental conditions. Pooled analysis of variance showed highly significant mean

sum of squares for genotypes and locations, indicating the presence of substantial variation among the genotypes over locations. Mean sums of squares due to genotypes found significant difference. Also, genotypes × locations (linear) suggested that the performance of different genotypes fluctuated considerably with respect to their stability for respective locations. The significant GEI suggest that the grain yield of genotypes varied across control and drought stress conditions. Significant differences for genotypes, environments and GEI indicated the effect of environments on GEI, genetic variability among entries, and the possibility of selecting stable genotypes.

TABLE 2. ANALYSIS OF VARIANCE FOR GRAIN YIELD COMBINED IN THREE LOCATIONS FOR STABILITY ANALYSIS [6]

Source of variations	df	SS	MS
Total	44	483.82	10.996
Genotypes	14	467.34***	33.382***
Locations + (genotypes × locations)	30	16.48	0.549
Locations (linear)	1	11.78	11.779
Genotypes × locations (linear)	14	3.4*	0.243*
Pooled deviation	15	1.3	0.087
Binadhan-17/M_5/P-2	1	0.01	0.01
Binadhan-17/M_5/P-3	1	0.09	0.092
Binadhan-17/M_5/P-5	1	0.07	0.066
Binadhan-17 (Parent)	1	0.12	0.118
NERICA-4/M_5/P-2	1	0.03	0.028
NERICA-4/M_5/P-3	1	0.13	0.128
NERICA-4/M_5/P-5	1	0.06	0.055
NERICA-4/M_5/P-6	1	0.01	0.007
NERICA-4 (parent)	1	0	0.001
Galon/M_5/P-2	1	0.06	0.062
Galon/M_5/P-3	1	0	0.004
Galon/M_5/P-6	1	0.01	0.007
Galon (parent)	1	0.48	0.476**
IR-64 (S. check)	1	0.24	0.241*
Binadhan-19 (T. check)	1	0	0.002
Pooled error	90	4.41	0.049

***, **, * Indicate significance at 0.1%, 1% and 5% levels of probability, respectively.

TABLE 3. ANALYSIS OF VARIANCE FOR GRAIN YIELD COMBINED IN THREE LOCATIONS FOR STABILITY PARAMETERS

Source of variations	df	SS	MS
Locations	2	35.34***	17.668***
Replications (locations)	6	0.39	0.066
Genotypes	14	1402.03***	100.145***
Genotypes × locations	28	14.09***	0.503***
Residuals	84	13.22	0.157

*** Indicates significance at 0.1% level of probability.

TABLE 4. ANALYSIS OF VARIANCE FOR GRAIN YIELD COMBINED IN THREE LOCATIONS FOR STABILITY PARAMETERS [14]

Source of variations	df	SS	MS
Total	44	1451.46	
Genotypes	14	1402.03***	100.145***
Locations	2	35.34***	17.668***
Genotypes × locations	28	14.09***	0.503***
Heterogeneity	14	10.2*	0.728*
Residual	14	3.89	0.278*
Pooled error	588		0.157

***, * Indicate significance at 0.1%, 5% levels of probability, respectively.

2.6. Stability

A yield stability study helped to identify the stable genotypes that can avoid significant fluctuation in yield over a range of environmental conditions. The stability coefficients of the tested mutant rice genotypes are presented in Table 5. The results revealed that the best rates of the parameter of b_{ij} (regression coefficient) were recorded in the rice lines Binadhan-17/M_5/P-3 (1.000), followed by Binadhan-17/M_5/P-5 (1.071) and NERICA-4/M_5/P-5 (0.960) that was equalized or close to 1 among genotypes being studied. Importantly, the deviation from regression values (Sd_{ij} = 0) of these genotypes were also close to zero. According to Eberhart and Russell [6], a stable variety is one with a regression coefficient of unity (b = 1 or near 1) and minimum deviation from the regression line (Sd_{ij} = 0 or near to zero). Therefore, based on these criteria, three mutant rice lines (Binadhan-17/M_5/P-3, Binadhan-17/M_5/P-5, and NERICA-4/M_5/P-5) have showed high genetic stability over the locations [15].

TABLE 5. STABILITY PARAMETERS OF DROUGHT STRESS TOLERANCE IN RICE GENOTYPES FOR GRAIN YIELD PER PLANT

Genotypes	b_{ij}	Sd_{ij}	Var	CV (%)	Rank	YSi
Binadhan-17/M_5/P-2	1.764	-0.042	1.230	2.49	12	7
Binadhan-17/M_5/P-3	1.000	0.040	0.439	1.35	14	17
Binadhan-17/M_5/P-5	1.071	0.014	0.484	1.38	15	18
Binadhan-17 (Parent)	0.612	0.066	0.206	1.19	8	10
NERICA-4/M_5/P-2	1.544	-0.025	0.949	2.73	7	7
NERICA-4/M_5/P-3	1.983	0.076	1.610	3.18	9	4
NERICA-4/M_5/P-5	0.960	0.003	0.390	1.29	13	16
NERICA-4/M_5/P-6	0.364	-0.046	0.055	0.53	11	10
NERICA-4 (parent)	0.737	-0.052	0.214	1.48	6	3
Galon/M_5/P-2	0.741	0.009	0.246	1.82	2	-1
Galon/M_5/P-3	0.346	-0.048	0.049	0.75	5	-2
Galon/M_5/P-6	0.836	-0.046	0.278	1.92	3	0
Galon (parent)	1.260	0.423	0.861	3.29	4	-7
IR-64 (S. Check)	1.678	0.189	1.230	6.92	1	-10
Binadhan-19 (T. Check)	0.103	0.050	0.005	0.18	10	5

b_{ij} = regression coefficient; Sd_{ij} = deviation from regression YSi = Yield stability index; Var = variance.

Based on the stability parameters (Table 5), the findings revealed that mutant line Binadhan-17/M_5/P-5 (18) showed the highest yield stability index (YSi), followed by Binadhan-17/M5/P-3 (17) and NERICA-4/M5/P-5 (16), respectively, over the locations. So, the promising mutant lines Binadhan-17/M_5/P-3, Binadhan-17/M_5/P-5 and NERICA-4/M_5/P-5 had regression coefficient values around unity and lower deviation from regression, and also had high rank as well as high yield stability index over the locations. Since this promising mutant line had stable performance across the environments and was less sensitive to the environment, it can adapt to diverse environments. Hence, these mutant lines can be used as stable lines adopted across the environments in terms of yield.

The AMMI-1 biplot of the main effect (genotype and environment effects) and IPCA1 scores were plotted against each other. The score and sign of IPCA1 reflect the magnitude of the contribution of both genotypes and environments to GEI, where scores near zero are characteristic of stability, whereas a higher score (absolute value) is considered as unstable and specifically adapted to the environment [16, 17]. The results of the AMMI-1 biplot are presented in Fig. 1 and show that Binadhan-17/M_5/P-2, Binadhan-17/M_5/P-3, Binadhan-17/M_5/P-5, Binadhan-17 (parent), NERICA-4/M_5/P-3, NERICA-4/M_5/P-5, NERICA-4/M_5/P-6 and tolerant genotype (check) had higher yield over the average grain yield while the other genotypes had below average grain yield. From these genotypes, the susceptible genotype (check) had the lowest yield genotypes. Among the tested genotypes, Binadhan-17/M_5/P-3, Binadhan-17/M_5/P-5 and NERICA-4/M_5/P-5 had low GEI effects. These mutant lines also localized near to zero, characteristic of stability. In contrast, the other genotypes had higher GEI.

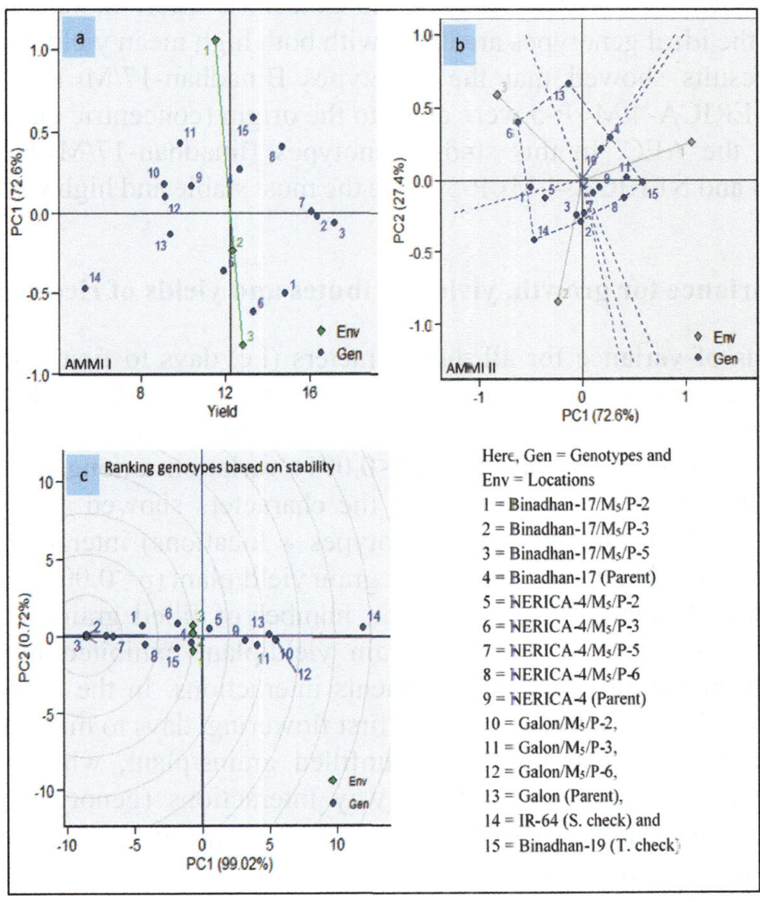

FIG. 1. AMMI-I and AMMI-II biplots for the effects of rice genotypes × locations interaction based on IPCAs (PCA1 and PCA2) stability parameters with ranking.

The AMMI-II biplot for grain yield display interaction of PC1 and PC2 of the tested genotypes in the three locations (Magura, Rangpur and Mymensingh) are presented in Fig. 1(b). In this model of environmental effects, the genotypes near the origin are not sensitive to environmental interaction while those distant from the origins are sensitive and have strong interaction with the environment. The results showed that the first principal component axis accounted for PC1 (72.60%) and the second accounted for PC2 (27.40%) of the variation. The two IPCAs together accounted for 100% of the genotype by environment interaction mean squares. This implies that the interaction of the rice genotypes was predicted by the first two components of IPCAs. As Gauch and Zobel have shown in their research, the most accurate AMMI model can be predicted from the first two IPCAs [18]. Similarly, Susanto et al. [17] reported that the first two IPCAs explained 88.8% of the total variation of genotype $\times$ location interactions on yield in rice genotypes in Indonesia under irrigated environment. In this study, genotypes Binadhan-17/M_5/P-3 (2), Binadhan-17/M_5/P-5 (3) and NERICA-4/M_5/P-5 (7) were close to the origin and hence they are less interactive to environmental differences because these genotypes were close to zero of the IPCA and located very near to polygonal region. It was found that the distances from the biplot origin are indicative of the degree of interaction between genotypes and environments.

The ranking of genotypes based on mean and stability performance are presented in Fig. 1(c). In this biplot analysis, the estimation of yield and stability of genotypes was done by using the average environment (tester) coordinate (AEC) method [19, 20]. The AEC is a line that passes through the biplot origin and can be defined by the average PC1 (mean yield) and PC2 (stability) scores for all environments [21, 22]. Closer to the concentric circle indicates a higher mean yield. For selection, the ideal genotypes are those with both high mean yield and high stability. In the biplot, the results showed that the genotypes Binadhan-17/M_5/P-5 (3), Binadhan-17/M_5/P-3 (2) and NERICA-4/M_5/P-5 were close to the origin (concentric circle) and have the shorter vector from the AEC. In this study, genotypes Binadhan-17/M_5/P-5, followed by Binadhan-17/M_5/P-3 and NERICA-4/M_5/P-5, were the most stable and high yielding genotypes based on the method.

2.7. Analysis of variance for growth, yield attributes and yields of rice genotypes

The result of analysis of variance for all the characters (i.e. days to first flowering, days to maturity, plant height, number of total tillers/plant, number of effective tillers/plant, panicle length, number of filled grains/plant, number of unfilled grains/plant, 100 seed weight, and grain yield/plant showed highly significant (p $\leq$0.001) variation among the genotypes and treatments studied. In the case of locations, all the characters showed significant variation except yield/plant (p $\leq$0.001). In two-way (genotypes $\times$ locations) interaction, all the traits studied showed highly significant variation except grain yield/plant (p $\leq$0.001). In contrast, days to first flowering, days to maturity, plant height, number of filled grains/plant, number of unfilled grains/plant, 100 seed weight, and grain yield/plant exhibited highly significant (p $\leq$0.001) variation during genotypes $\times$ treatments interactions. In the case of locations $\times$ treatments interactions, the characters are days to first flowering, days to maturity, plant height, number of filled grains/plant and number of unfilled grains/plant, which showed highly significant variation (p $\leq$0.001). During three way interactions (genotypes $\times$ locations $\times$ treatments), characters like days to first flowering, days to maturity, plant height, number of filled grains/plant, number of unfilled grains/plant, and 100 seed weight showed highly significant (p $\leq$0.001) variation.

2.8. Effect of genotypes, locations and treatments on yield and yield attributes of rice

A significant variation among the tested rice genotypes for different morphological traits was observed in well irrigated and drought stress conditions. In the case of genotypes, locations and treatments, days to first flowering, days to maturity, plant height, number of filled grains/plant, number of unfilled grains/plant, 100 seed weight, and grain yield/plant exhibited significant ($p \leq 0.001$) variation, except grain yield/plant in the case of locations. Between the growing conditions, the genotypic performance regarding yield and yield contributing traits under control (well irrigated) condition was better than for drought stress condition. Binadhan- 17/M5/P-5 (19.84 g) were found to be the best performing genotype followed by Binadhan-17/M5/P-3 (19.64 g), Binadhan-17/M5/P-2 (19.09 g), and NERICA-4/M5/P-5 (18.28 g) based on yield performance. The highest values of plant height, number of filled grains/plant, 100 seed weight, and grain yield/plant were 118.08 cm, 903.84, 2.15 g and 19.35 g, respectively, recorded under control conditions. On the other hand, the lowest values were recorded for those traits (109.89 cm, 598.24, 2.05 g and 12.77 g, respectively) under drought stress conditions. The findings of the results confirmed that drought stress significantly impacted the yield and yield attributing traits of rice.

2.9. Combined effect of genotype and treatment on growth traits, yield attributes and yields of ten rice mutant lines

The combined effect of genotype and treatment interaction showed a significant ($p \leq 0.001$, 0.01 and 0.05) difference for all the studied traits. However, panicle length showed a non-significant difference. The length of the panicle varied from 24.55 cm to 21.46 cm. The traits days to first flowering, days to maturity, number of total tillers/plant, number of effective tillers/plant were varied (95.78 days to 71.71 days), (127.33 days to 95.96 days), (10.66 to 7.51) and (9.67 to 6.91). Among the tested genotypes, Binadhan-17/M5/P-2 showed the highest number of filled grains/plant (1096.89) followed by Binadhan-17/M5/P-5 (1071.67) and Binadhan-17/M5/P-3 (1070.78) under control conditions, whereas Binadhan-17/M5/P-5 showed the highest number of filled grains/plant (860.84) followed by Binadhan-17/M5/P-3 (794.50) and Binadhan-17/M5/P-2 (713.56) under drought stress conditions. The highest number of unfilled grains/plant was recorded in Galon (parent) (878.98) and the lowest was recorded in NERICA-4/M5/P-2 (294.22) under drought stress and control conditions, respectively. Grain yield/plant varied from 23.36 g to 9.10 g. Based on yield performance, the genotype Binadhan-17/M5/P-2 showed the highest grain yield/plant (23.36 g), followed by Binadhan-17/M5/P-3 (22.92 g) and Binadhan-17/M5/P-5 (22.48 g) under control conditions, whereas Binadhan-17/M5/P-5 (17.20 g), Binadhan-17/M5/P-3 (16.36 g) and NERICA-4/M5/P-5(16.08 g), respectively under drought stress conditions. The results revealed that genotypes Binadhan-17/M5/P-3, Binadhan-17/M5/P-5 and NERICA-4/M5/P-3 showed better performance under drought stress conditions. Hence, these genotypes could be considered as drought tolerant.

2.10. Combined effect of genotypes and locations on growth traits, yield attributes and yields of ten rice mutant lines

Based on two way analysis of variance, all the traits showed significant ($p \leq 0.001$ and 0.01) differences among the tested genotypes and three locations. Significant interaction effects of genotypes and locations indicates that all the traits were varied among locations of the genotypes. The traits days to first flowering, days to maturity, plant height, number of total tillers/plant, number of effective tillers/plant varied from 101.33 days to 71.73 days, 130.17 days to 94.80 days, 147.07 cm to 94.97 cm, 10.48 to 7.85, and 9.97 to 6.97. Panicle length ranged from 25.60 cm to 21.68 cm in NERICA-4/M_5/P-5 and Galon/M_5/P-2 at Mymensingh, respectively.

The number of filled grains/plant varied from 967.17 to 543.32 and number of unfilled grains/plant ranged from 882.00 to 307.67. Among the tested genotypes, Binadhan-17/M_5/P-5 recorded the highest number of filled grains/plant (967.17) at Mymensingh followed by Binadhan-17/M_5/P-5 (966.33) at Magura, Binadhan-17/M_5/P-5 (965.27) at Rangpur, Binadhan-17/M_5/P-3 (963.50) at Magura and Binadhan-17/M_5/P-3 (947.27) at Rangpur, respectively. The highest number of unfilled grains/plant was recorded in Binadhan-17 (parent) (882.00) at Magura and the lowest was recorded in NERICA-4/M_5/P-2 (307.67) at Rangpur. Hundred seed weight ranged from 2.32 g to 1.94 g. The maximum 100 seed weight (2.32 g) was recorded in NERICA-4/M_5/P-3 at Mymensingh and the lowest 100 seed weight (1.94 g) was recorded in Galon (parent) both at Magura and Rangpur.

Grain yield/plant varied from 20.26 g to 11.78 g. Based on grain yield/plant, the genotype Binadhan-17/M_5/P-3 showed the highest grain yield/plant (20.26 g) at Magura followed by Binadhan-17/M_5/P-5 (20.17 g) at Mymensingh, Binadhan-17/M_5/P-3 (20.06 g) at Rangpur and Binadhan-17/M_5/P-5 (19.92 g) at Magura. The findings of the results showed that genotypes Binadhan-17/M_5/P-3 and Binadhan-17/M_5/P-5 showed better performance over two locations. Therefore, these genotypes could be considered as suitable genotypes at both the locations based on genotype and location interaction.

2.11. Combined effect of locations and treatments on yield and yield attributes of rice genotypes

In the case of the interaction of locations and treatments, all the studied traits exhibited significant (p $\leq$ 0.001, 0.01 and 0.05) variation. The study showed that days to first flowering varied from 89.89 days to 79.10 days and days to maturity varied from 118.01 to 108.18 days. Mean value for days to first flowering and days to maturity were 84.50 days and 113.10 days, respectively. Among the genotypes tested, early genotypes showed 'days to first flowering' (79.10 days) at Rangpur under irrigated condition, whereas late genotype showed 'for days to first flowering' (89.89 days) at Magura under drought stress conditions. At Rangpur, under irrigated conditions, the genotypes were found to mature earlier (108.18 days) ,whereas delayed maturing (118.01 days) was recorded at Magura under drought stress conditions. Drought stress had a significant effect on days to first flowering and days to maturity.

Among the genotypes, plant height (cm), number of total tillers/plant, number of effective tillers/plant and panicle length (cm) varied (122.26 cm to 106.35 cm), (10.11 to 8.18), (9.61 to 7.53), (24.10 cm to 22.01 cm), respectively. Both the highest plant height (122.26 cm) and panicle length (24.10 cm) were observed at Mymensingh under normal and drought stress conditions. The traits number of filled grains/plant, number of unfilled grains/plant ranged from 920.33 to 586.30 and 877.93 to 362.27. Mean value for number of filled grains/plant and number of unfilled grains/plant were 753.32 and 620.10, respectively. The maximum number of filled grains/plant (920.33 and 608.44) was found at Magura and Rangpur under control and drought stress conditions, respectively. In contrast, the highest number of unfilled grains/plant (527.98 and 877.93) was found at Magura under control and drought stress conditions, respectively.

Hundred seed weight and grain yield/plant varied from 2.14 g to 2.04 g and 19.62 g to 12.70 g, respectively. The highest seed weight (2.16 g) was observed at Mymensingh under irrigated conditions and 2.04 g at both Magura and Rangpur under drought stress conditions. In the case of grain yield/plant, the highest value was found (19.62 g) at Magura and (12.86 g) at Rangpur under irrigated and drought stress conditions, respectively. Among the tested genotypes, the straw yield/plant and harvest index (%) varied from 26.72 g to 20.91 g; 0.80 to 0.60, respectively. Both the maximum straw yield (26.72 g) and (22.19 g) were observed at Rangpur under irrigated and drought stress conditions. The lowest value of plant height (106.35 cm), number of total tillers/plant (8.18), number of effective tillers/plant (7.53) and panicle length (22.01 cm) were found at Magura, whereas the number of filled grains/plant (586.30) was found at Mymensingh under drought stress conditions. The minimum value of 100 seed weight (2.04 g) was observed at Magura and Rangpur, whereas the lowest yield/plant (12.70 g) was found at Mymensingh and Rangpur under drought stress condition, respectively.

2.12. Combined effect of genotypes, locations and treatments on growth parameters, yield attributes and yields of ten mutant rice lines

In the case of the interaction of genotypes, locations and treatments, traits such as days to first flowering, days to maturity, plant height, number of filled grains/plant, number of unfilled grains/plant, hundred seed weight, and yield/plant exhibited significant ($p \leq 0.001$, 0.01 and 0.05) variation. On the other hand, number of total tillers/plant, number of effective tillers/plant and panicle length exhibited non-significant variation. The study showed that days to first flowering varied from 104.67 days to 69.73 days, days to maturity varied from 133.67 to 91.93 days and plant height varied from 155.37 cm to 91.97 cm. Mean value for days to first flowering, days to maturity and plant height were 87.20 days, 112.80 days and 123.67 cm, respectively.

Among the tested genotypes, Binadhan-19 (T. check) was considered as an early genotype for days to first flowering (69.73 days) at Rangpur under irrigated condition, whereas NERICA-4/M5/P-6 considered as late genotype for days to first flowering (104.67 days) at Mymensingh under drought stress conditions. At Rangpur under irrigated conditions, the genotype Binadhan- 19 (T. check) was recorded as an early maturing genotype (91.93 days), whereas the Galon (parent) genotype recorded as a late maturing genotype (133.67 days) at Magura under drought stress conditions. In contrast, the lowest plant height (91.97 cm) was recorded in the IR- 64 (S. check) genotype at Magura under drought stress conditions, whereas the maximum plant height (155.37 cm) was recorded in Galon (parent) at Magura under control conditions. Hence, drought tress had significant effect on days to first flowering, days to maturity and plant height.

The number of filled grains/plant and unfilled grains/plant varied from 1120.33 to 203.00 and 1080.00 to 233.33, respectively. The maximum number of filled grains/plant (1120.33) was found in Binadhan-17/M5/P-2 at Magura and the minimum number of filled grains/plant (203.00) was recorded in IR-64 (S. check) at Mymensingh under irrigated and drought stress conditions, respectively. In contrast, the highest number of unfilled grains/plant (1080.00) was found in Binadhan-17 (parent) at Magura and the lowest (233.33) was in NERICA-4/M5/P-2 at Rangpur under drought stress and control conditions, respectively.

Hundred seed weight and grain yield/plant varied from 2.37 g to 1.88 g and 24.05 g to 8.67 g, respectively. The highest 100 seed weight (2.37 g) was observed in NERICA-4/M5/P-2 at Mymensingh under control conditions and the lowest 1.88 g in Galon (parent) at Magura under drought stress conditions. In the case of grain yield/plant, the highest value was found (24.05 g) in Binadhan-17/M5/P-2at Rangpur and the lowest (8.67 g) in Galon/M5/P-6 at Magura under control and drought stress conditions, respectively (Supplementary Table 5). The data show that interaction effects of genotypes, locations and treatments have a detrimental response on rice production under stress and non-stress conditions.

3.   CONCLUSIONS

Drought stress is one of the major abiotic stresses liming rice yield and productivity. Induced mutation breeding offers a significant scope to develop drought tolerant mutants as induced mutations provide new genes or recombination of new genes that cannot be found in cultivated varieties. By applying gamma irradiation, a large number of mutant populations were developed and at M_5 generation, ten promising drought tolerant mutants were isolated based on the better performance of morphological and biochemical traits under drought stress conditions. Their yield performance and stability and drought stress tolerance at the reproductive phases of growth were studied under direct field conditions in three different locations of Bangladesh. Significant genotype, treatment and location interactions were observed among the studied mutants, their parents and checks. Based on stability parameters, three mutants (Binadhan- 17/M_5/P-3, Binadhan-17/M_5/P-5 and NERICA-4/M_5/P-5) were identified as stable, high yielding mutants. Hence, these mutant lines can be used as stable lines adopted across the environments in terms of yield. Finally, these varieties can be recommended for production in the testing sites and also in similar agroecological, i.e. drought prone, areas of Bangladesh.

REFERENCES TO CHAPTER 2–1

[1]   AKTER, A., et al., Genotype × environment interaction and yield stability analysis in hybrid rice (*Oryza sativa* L.) by AMMI biplot, Bangladesh Rice J. **19** 2 (2015) 83–90.

[2]   BALAKRISHNAN, D., et al., Genotype × environment interactions of yield traits in backcross introgression lines derived from *Oryza sativa* cv. Swarna/*Oryzanivara*, Front. Plant Sci. **7** (2016) 1530.

[3]   BAYE, T.M., ABEBE, T., WILKE, R.A., Genotype-environment interactions and their translational implications, Personalized Med. 8 (2011) 59–70.

[4]   DABHOLKAR, A.R., Elements of Biometrical Genetics (Revised and Enlarged Edition), Concept Publishing Company, New Delhi (1999) 326–365.

[5]   DAVIES, N.B., KREBS, J.R., WEST, S.A., An Introduction to Behavioural Ecology, 4th edn, Wiley, New York (2012).

[6]   EBERHART, S.T., RUSSELL, W.A., Stability parameters for comparing varieties, Crop Sci. **6** (1966) 36-40.

[7]   FAROOQ, M., KOBAYASHI, N., WAHID, A., ITO, O., BASRA, S.M.A., Strategies for producing more rice with less water, Adv. Agron. **101** (2009) 351–388.

[8]   INABANGAN-ASILO, M.A., et al., Stability and G × E analysis of zinc-biofortified rice genotypes evaluated in diverse environments, Euphytica **215** 3 (2019) 1–17.

[9]   KILLI, F., HAREM, E., Genotype × environment interaction and stability analysis of cotton yield in Aegean region of Turkey, J. Environ. Biol. **37** (2006) 427–430.

[10]   KUKAL, M.S., IRMAK, S., Climate-driven crop yield and yield variability and climate change impacts on the US Great Pains agricultural production, Sci. Rep. **8** 1 (2018) 1- 18.

[11]   POLI, Y., et al., Genotype × environment interactions of Nagina22 rice mutants for yield traits under low phosphorus, water limited and normal irrigated conditions, Scientific Rep. **8** 1 (2018) 1-13.

[12]   RAHMAN, M.M., "Country report: Bangladesh", Proc. ADBI-APO Workshop on Climate Change and its Impact on Agriculture, Seoul (2011).

[13]   ROMAGOSA, I., BORRÀS-GELONCH, G., SLAFER, G., VAN EEUWIJK, F., "Genotype by environment interaction and adaptation", Sustainable Food Production (CHRISTOU P, SAVIN, R., COSTA-PIERCE, B.A., MISZTAL, I., WHITELAW, C.B.A., Eds), Springer, New York (2013).

[14]   SINGH, R.K., CHAUDHARY, B.D., Biometrical Methods in Quantitative Genetic Analysis, Kalyani, New Delhi (2004).

[15]   SAHEBI, M., et al., Improvement of drought tolerance in rice (*Oryza sativa* L.): Genetics, genomic tools, and the WRKY gene family, Biomed. Res. Int. **4** (2018) 1- 40.

[16]   SANNI, K.A., et al., Additive main effects and multiplicative interaction analysis of grain yield performance in rice genotypes across environments, Asian J. Plant Sci. **8** 1 (2009) 48–53.

[17]   SUSANTO, U., ROHAENI, W.R., JOHNSON, S.B., JAMIL, A., GGE biplot analysis for genotype x environment interaction on yield trait of high Fe content rice genotypes in Indonesian irrigated environments, AGRIVITA J. Agric. Sci. **37** 3 (2015) 265–275.

[18]   GAUCH, H.G., ZOBEL, R.W., "AMMI analysis of yield trials", Genotype by Environment Interaction (KANG, M.S., GAUCH, H.G., Eds), CRC Press, Boca Raton (1996) 85–122.

[19]   SWAMY, B.P.M., KUMAR, A., Genomics based precision breeding approaches to improve drought tolerance in rice, Biotechnol. Adv. **31** 8 (2013) 1308–1318.

[20]   YAN, W., GGE biplot-A Windows application for graphical analysis of multienvironment trial data and other types of two-way data, Agron. J. **93** 5 (2001) 1111–1118.

[21]   YAN, W., CORNELIUS, P.L., CROSSA, J., HUNT, L.A., Two types of GGE biplots for analyzing multi-environment trial data, Crop Sci. **41** 3 (2001) 656–663.

[22]   YAN, W., KANG, M.S., GGE biplot analysis: A graphical tool for breeders, geneticists and agronomists (1st edn), CRC Press, Boca Raton (2003).

2–2. MUTATION BREEDING OF RICE FOR DROUGHT TOLERANCE IN SOUTHERN BRAZIL

E. VENSKE, R. JOSEPH, V. KOPP DA LUZ, L.H.C. TEJEDA,
L. CARLOS DA MAIA, C. PEGORARO, A. COSTA DE OLIVEIRA

Plant Genomics and Breeding Center, Eliseu Maciel School of Agronomy,
Federal University of Pelotas, Brazil

Abstract

The paper presents for the first time the need and challenges of a new breeding programme aimed at addressing rice drought tolerance in southern Brazil, a region where paddy rice predominates. Rice is considerably susceptible to water deficit. Thus, any change in water management, e.g. from flooded irrigated to rainfed, requires first the development of improved and adapted cultivars. As an additional complicating factor, it has been shown that improving crops for drought tolerance is a cumbersome task, due to both the abiotic stress features and interactions, but also the plant's genetic architecture behind the tolerance. In southern Brazil the climate is temperate, presenting a shallow lowland soil profile, poor water storage, and with the drought prone period coinciding with the most sensitive phenological stages of rice, which makes this breeding target a challenge. As there has been no effort towards developing rice drought tolerance in southern Brazil, mutation breeding can be an approach which has many advantages, not only for the development of improved cultivars from already adapted germplasm, but also by generating novelty in the crop gene pool.

Key words: abiotic stress, climate change, phenotyping, target environment.

1.   INTRODUCTION

The increase in the global population, especially in developing countries, together with several abiotic and biotic factors such as ongoing climate change, unfertile arable lands and the constant evolution of crop pathogens, insects, and weeds, puts pressure on agriculture [1–4].

Rice has been found to be the crop with the highest potential to contribute to the food security of people in this complex and challenging scenario. It is the staple food for more than half of the human population, with this half comprising most of the poorest population fraction [5]. The grain is considered an important source of energy, but also provides significant amounts of protein and vitamins, amongst other beneficial compounds, especially for consumers who largely rely on a daily rice diet [6]. Brazil ranks ninth in rice production, being the most important producer outside Asia, and is also an important consumer [7].

Water scarcity and drought are well known agricultural barriers in several parts of the globe. With regard to climate change, it has been shown that the phenomena are negatively affecting the pluvial regime in diverse regions [1]. Recent trends in water scarcity threaten the cultivation of rice as we know it, making any strategy to improve water use efficiency a valuable tool in rice breeding. The yield of rice depends on traits that are expressed from as early as germination and throughout plant growth. Photosynthetic rate, transpiration rate, stomatal conductance, water use efficiency and root architecture can all have a final impact on rice yield and can be used to measure sensitive stages to drought. In southern Brazil, rice is grown as paddy rice, i.e. under a flooded irrigation system [7]. However, the need for alternatives aiming at a more optimized use of water is becoming clear. In fact, in southern Brazil studies are gradually focusing on water management, such as sprinkler irrigation [8, 9]. However, an important step in that direction is the development of adapted cultivars. In southern Brazil, virtually all available rice cultivars are strictly adapted to the paddy system [10]. However, there has been basically no research in this area.

In general, crop breeding cannot be considered an easy, fast or inexpensive task, but improving plant drought tolerance has been recognized as an even more cumbersome challenge compared with improving several other crop traits [11]. Drought tolerance cannot be considered a single trait or a generic issue [12]. Moreover, there has been a greater interplay between water deficit and several environmental factors [13]. Further, there is increasing evidence of the reduction in crop genetic variability, including rice, due to modern breeding practices. This has also been shown for Brazilian rice [14]. In this sense, the use of artificial mutagenesis has been proved to be an efficient tool for broadening the genetic base of any crop, and rice can be considered a proof of the success of this approach [15]. A remarkably positive aspect of mutation breeding is the possibility of using an adapted elite germplasm as the base of the breeding programme, and selection can be applied on the target trait, which, in the case discussed here, is drought tolerance. In fact, this strategy is the backbone of the new rice breeding programme focused on drought tolerance at the Universidade Federal de Pelotas in southern Brazil.

2. METHODOLOGY

2.1. Plant material, treatments and generation advancing of the mutant lines

Seeds of BRS Pampeira, an elite Brazilian rice cultivar, were treated with gamma ray treatment (^{60}Co) at two doses: 250 and 300 Gy. Treated seeds were sown in the field in November 2017, at Capão do Leão/RS, Brazil, to obtain M_2 seeds. A total of 2000 M_1 plants were selected and seeds were harvested in June 2018. Following this, 2000 M_2 lines, each derived from a single M_1 plant ($M_{1:2}$), were grown at Capão do Leão, RS, southern Brazil. It is important to highlight that from this set, 1000 lines were obtained with 250 Gy and the other 1000 lines were derived from the 300 Gy dose. The M_2 lines were sown in mid-November 2018 and the harvesting was done in June 2019. About 5500 individual plants were harvested and plants were selected based on visual assessment, focusing on adequate plant canopy and architecture, tillering, panicle number and size (Table 1).

TABLE 1. STAGES OF RICE DEVELOPMENT*

Stage of development	Description
V_1	Collar formed in the first leaf of the main culm
V_2	Collar formed in the second leaf of the main culm
V_3	Collar formed in the third leaf of the main culm
V_4	Collar formed in the fourth leaf of the main culm
V_5	Collar formed in the fifth leaf of the main culm
V_6	Collar formed in the sixth leaf of the main culm
V_7	Collar formed in the seventh leaf of the main culm
V_8	Collar formed in the eighth leaf of the main culm
V_9	Collar formed in the ninth leaf of the main culm. Flag leaf minus four.
V_{10}	Collar formed in the tenth leaf of the main culm. Flag leaf minus three.
V_{11}	Collar formed in the eleventh leaf of the main culm. Flag leaf minus two.
V_{12}	Collar formed in the twelfth leaf of the main culm. Flag leaf minus one.
V_{13}	Collar formed in the thirteenth leaf of the main culm (flag leaf).
R_0	Panicle initiation
R_1	Panicle differentiation
R_2	Collar formation in the flag leaf
R_3	Panicle exsertion
R_4	Anthesis
R_5	Elongation of one or more grains in the husk
R_6	Expansion of one or more grains in depth
R_7	At least one grain with the husk of the typical colour of the cultivar
R_8	Maturity of at least one grain
R_9	Complete maturity of the grains (harvesting point)

* According to technical recommendations [7].

3.   INITIAL EXPERIMENTS FOR SELECTION OF DROUGHT TOLERANCE

Two pilot experiments were carried out in this season aiming to refine procedures for selecting drought tolerant mutant lines. The first aimed at selection at seedling stage and the second, and more important, at the reproductive period. In general, both experiments followed the same design, only differing for the plant stages in which drought stress was applied. Thus, four cultivars were chosen for this study, two tolerant (developed for rainfed conditions and released as improved for this trait) and two sensitive to drought (developed for paddy, i.e. flooded irrigation system). The cultivars were grown under a rainout shelter to ensure drought stress (avoid rainfall) at the predefined stages. The experimental design was entirely randomized with four replications. For both experiments, the drought stress was imposed by keeping the plants for 10 days at tension of 50 kPa at 10 cm deep, through the aid of tensiometers. In the seedling assay, the stress started when the seedlings were well stabilized, i.e. 25 days after sowing. For the reproductive experiment, the stress started at the booting stage, on average, as expressive variation regarding plant cycle was verified among genotypes. For both experiments, trait evaluations were performed on plant growth and development, including photosynthesis and root related traits.

The results have suggested that the water deficit condition applied was not effective for tolerance assessment, since the water tension applied (50 kPa at 10 cm) was not high enough, and considering that it was not possible to differentiate the cultivars regarding performance under drought (Tables 2 and 3). Thus, the preliminary results suggested that the drought stress period and water tension needed to be increased to allow better selection. The phenotypic evaluations of the mutant lines were continued in further experiments in which the water tension was increased to 100 kPa, at 15 cm, and the drought stress was kept from stage R2 to 10 days after R4 (Table 4).

TABLE 2. SUMMARY OF THE ANALYSIS OF VARIANCE FOR RELATIVE PERFORMANCE TRAITS (CONTROL/WATER DEFICIT) ASSAYED IN FOUR CONTRASTING RICE VARIETIES DURING THE VEGETATIVE STAGE

SV[1]	DF	Mean square					
		RL	SL	NL	NT	SDW	RDW
Genotype	3	867.20*	475.95*	507.07	229.61*	813.82	779.521
Block	2	627.8	26.89	207.6	28.27	565.34	842.6
CV %	-	18.47	14.55	20.92	15.5	72.85	68.96

[1]SV = Source of variation; DF= degrees of freedom; CV% = coefficient of variation; RL = root length; SL = shoot length; NL = number of leaves; NT = number of tillers; SDW = shoot dry weight; RDW = root dry weight; *significant ($p \leq 0.05$) according to the F-test.

TABLE 3. MEANS OF RELATIVE PERFORMANCE TRAITS OF FOUR CONTRASTING GENOTYPES UNDER CONTROL AND WATER DEFICIT DURING THE VEGETATIVE STAGE

Genotypes	RL[1]	SL	NL	NT	SDW	RDW
BRS Pampeira	62.0* AB	57.6 AB	45.7 A	46.09 A	44.4 A	55.64 A
IRGA 424	83.7 A	53.8 AB	61.0 A	29.49 B	45.2 A	51.96 A
BRSGO Serra Dourada	46.0 B	45.7 B	38.1 A	26.19 B	12.6 A	21.41 A
BRS Esmeralda	78.1 AB	75.5 A	66.0 A	35.55 AB	46.8 A	52.88 A
General mean	67.5	58.1	52.7	34.33	37.2	45.47

[1]RL = root length; SL = shoot length; NL = number of leaves; NT = number of tillers; SDM = shoot dry weight; RDW = root dry weight; * = means followed by the same letter do not differ statistically according to the Tukey test ($p \leq 0.05$).

TABLE 4. SUMMARY OF THE ANALYSIS OF VARIANCE FOR RELATIVE PERFORMANCE TRAITS (CONTROL/WATER DEFICIT) ASSAYED IN FOUR CONTRASTING RICE ACCESSIONS DURING THE REPRODUCTIVE STAGE

SV	DF	Mean square								
		Chlo	NBI	Flav	PH	PanL	PanW	NPan	NG	TGW
Genotype	3	109.07	226.19	24.48	105.26	8446	2235.46	478.87	1195.74	30.83
Block	2	437.12	161.9	249.73*	163.17	6891	1500.66	1134.71	207.88	12.59
CV %	-	11.67	10.36	7.42	6.47	5.35	25.75	25.02	18.92	3.26

[1]SV = Source of variation; DF= degrees of freedom; CV % = coefficient of variation; Chlo = chlorophyll; NBI = nitrogen balance index; Flav = flavonoids; PH = plant height; PanL = panicle length; PanW = panicle weight; NPan = number of panicles; NG = number of grains; TGW = thousand grain weight; * = significant ($p \leqslant 0.05$) according to the F-test.

4. GENERATION ADVANCEMENT AND SELECTION OF MUTANT LINES

From the individual M_2 plants harvested, a total of 4947 M3 lines (families), each derived from a single plant ($M_{2:3}$), were sown at Capão do Leão, RS, southern Brazil in December 2019. Each line was 0.5 m long. The sowing has been concluded. Given the large number of mutant lines, the experimental design applied was the Federer augmented blocks, with intercalary controls (checks). The intercalary check was the parent cultivar, which was the origin of the mutant lines (BRS Pampeira) at a 2% rate (i.e. one control every 50 mutant lines) (Table 2.5). In general, the experiment was managed according to the southern Brazilian recommendations for rice crop, except regarding irrigation, which was managed differently (imposing drought stress, etc.).

At M_3, a selection pressure for drought tolerance was applied to the mutant lines. For screening and selection, the drought stress (~100 kPa tension at 15 cm) was applied at the reproductive period (from R2–booting to 10 days after R4 stage–anthesis). Tensiometers were placed every 10 m^2. There were no expressive rainfalls during this period.

The following drought responsive traits were used as criteria for selection:

(1) Early flowering;
(2) Low spikelet sterility;
(3) Good visual phenotype (stay-green, plant architecture, less leaf rolling, good tillering and number and size of panicles, no lodging, good blast resistance and cold tolerance at the flowering stage);
(4) High grain yield in the test location versus the parental cultivar.

In fact, even though these traits were taken into consideration for selection, in practical terms, as grain sterility was extremely high, probably due to the drought stress applied but also due to the occurrence of cold (low temperatures) at the reproductive stage, the major criterion for selection was harvesting mutant lines which produced any seed. Thus, about 20–25% of the lines were discarded due to the negative selection procedure applied.

TABLE 5. EVALUATION OF AGRONOMIC TRAITS UNDER DROUGHT CONDITIONS IN THE M_3 MUTANT LINES OF BRS PAMPEIRA OBTAINED FROM A 250 Gy GAMMA RADIATION DOSE

Traits	Pop[1]	CD	CI	Min	Max	SD	S	K	CV (%)
VS	3.93	5.77	1.00	1.00	9.00	2.34	0.21	-1.03	59.60
				54.0					
NT	96.55	97.24	167.28	0	167.28	21.04	0.39	0.10	21.79
	117.6			97.0					
DF	6	120.00	97.00	0	143.00	7.78	1.90	4.46	6.61
				46.1					
PH	68.32	66.57	71.06	0	82.90	6.81	-0.75	1.18	9.98
				14.2					
PanL	22.85	22.38	25.83	6	26.70	1.98	-1.86	8.14	7.38
				43.0					
NPan	91.43	96.78	142.33	0	142.33	16.74	-0.04	0.31	18.31
PanW	0.74	0.64	1.27	0.27	1.61	0.25	1.40	2.54	33.83
				25.7					
NSSPan	81.69	88.44	72.16	0	129.50	21.27	-0.63	-0.02	26.04
NFSPan	13.64	12.63	36.10	1.34	43.90	11.57	0.93	-0.15	84.80
	1166.	2469.1	12845.3		12845.0	2405.9			
Yield	60	0	0	0.00	0	8	2.19	4.69	206.23

[1]Pop = population mean; DC = control cultivar (BRS Pampeira) under drought mean; CI = control cultivar (BRS Pampeira) irrigated mean; Min = minimum value; Max= maximum value; SD = standard deviation; S = distribution asymmetry coefficient; K = curtosis; CV (%) = coefficient of variation; VS = visual score; NT = number of tillers; DF= days to flowering; PH = plant height (cm); PanL = panicle length (cm); NPan = number of panicles per 0.5 m row; PanW = panicle weight (g); NSSPan = number of sterile spikelets per panicle; NFSPan = number of fertile spikelets per panicle; Yield = grain yield (kg ha^{-1}).

TABLE 6. DESCRIPTIVE STATISTICS OF AGRONOMIC TRAITS ASSAYED UNDER DROUGHT CONDITIONS AT THE REPRODUCTIVE STAGE IN A SET OF M_3 MUTANT LINES DERIVED FROM THE BRAZILIAN CULTIVAR BRS PAMPEIRA OBTAINED FROM THE 300 Gy GAMMA RADIATION DOSE

Traits	Pop[1]	CD	CI	Min	Max	SD	S	K	CV (%)
VS	5.14	6.80	1.00	1.00	9.00	1.64	0.08	0.07	32.04
	107.5								
NT	102.25	0	167.28	67.00	167.28	17.33	0.56	0.91	16.24
	123.0			112.0					
DF	124.9	0	97.00	0	143.00	11.52	0.79	-0.89	9.22
PH	58.88	59.83	71.06	33.60	79.90	8.70	-0.41	0.16	14.78
PanL	21.62	21.50	25.83	13.80	25.83	2.00	-1.65	3.66	9.29
NPan	93.51	94.40	142.33	43.00	142.33	16.99	-0.12	0.36	18.17
PanW	0.58	0.61	1.27	0.15	4.46	0.23	1.57	2.76	39.94
NSSPa									
n	82.29	83.87	72.16	32.30	141.20	18.18	-0.12	0.55	22.10
NFSPa									
n	10.34	9.70	36.10	1.00	44.20	10.65	1.61	1.74	103.02
	2051.	12845.3				2188.			
Yield	704.60	10	0	0.00	12845.00	60	3.72	14.35	310.60

[1]Pop = population mean; CD = control cultivar (BRS Pampeira) under drought mean; CI = control cultivar (BRS Pampeira) irrigated mean; Min = minimum value; Max = maximum value; SD = standard deviation; S = distribution asymmetry coefficient; K = curtosis; CV (%) = coefficient of variation; VS = visual score; NT = number of tillers; DF= days to flowering; PH = plant height (cm); PanL= panicle length (cm); NPan=Number of panicles per 0.5 m row; PanW = panicle weight (g); NSSPan = number of sterile spikelets per panicle; NFSPan = number of fertile spikelets per panicle; Yield = grain yield (kg ha^{-1}).

5.   PHENOTYPING OF MUTANT LINES UNDER DROUGHT STRESS

In the M_3 generation (2019–2020), half of the lines derived from the 250 Gy and the other half from the 300 Gy population were considered for the phenotyping under rainout shelter conditions (Table 6). The experimental design was Federer augmented block, with intercalary checks, with the mutant's parental cultivar being the control used. The experiment was monitored on a daily basis, with growth accounted for (data were taken from an official station located near the experiment site) in order to precisely apply the drought stress at the defined phenological stages. Additionally, three replications of the parental cultivar were kept in a normal (flooded) irrigation regime to be used as an additional type of control. The mutant lines of this experiment were phenotyped for a large number of traits, from agronomic to photosynthesis related.

Drought stress was applied from the R2 stage up to 10 days after R4 with ~100 kPa water tension at 15 cm. In order to achieve this tension, the area was drained three weeks before the R2 stage (estimated through the degree-days calculation method). Irrigation, when needed, was performed using buckets when the soil water tension reached more than 100 kPa and/or the susceptible check showed severe leaf rolling at 10 a.m. (following the project roadmap recommendations) through visual assessment. The irrigation was stopped when the tension dropped back to 100 kPa, and the leaf rolling stopped.

During the stress period, chlorophyll content was assayed using portable SPAD equipment. Three measurements were taken at: (1) beginning of the stress (first day of stress); (2) mid-point of stress (5 days); and (3) 10 days after stress (recovery). The mutant lines were evaluated agronomically for days to flowering and days to maturation; plant height; number of tillers per row; number of panicles per row; panicle weight; number of grains per panicle; grain sterility; weight of a thousand grains and grain yield. Most of the mutant lines did not show drought tolerance, displaying high sterility, and hence only the top 10% of the lines (most fertile lines) were continued for the next cycle under the rainout shelter. Thus, the 300 line set was restabilized taking 270 lines from the large set.

6.   GENERATION ADVANCEMENT AND SELECTION OF M_4 MUTANT LINES

Around 300 M_4 mutant lines were grown in the 2020–2021 crop season. Approximately half of the panel was composed of 250 Gy treated lines and the other half were from the 300 Gy treated set. All agromanagement practices were followed similar to the same procedures of the previous generation. Harvest was concluded by the end of May 2021. For the stress evaluation, 75.17% of lines showed better visual grade than the parental variety under stress. Regarding tiller number, around 30% of the lines showed a higher number of tillers than the parental variety. For DFL, around 37.5% of lines were earlier than the parental variety. Another trait highly influenced by mutation induction was plant height, with the parental line showing 67.5 cm on average and 30-40% of the lines displaying between 36 and 62.5 cm. Panicle length also displayed variation in the mutant lines, but around 60% of the lines showed longer panicles than the parental variety. The panicle weight of mutated lines was superior to the parental variety in 51.34% and 30.61% of the lines from the 250 Gy and 300 Gy populations, respectively. When the number of fertile spikelets was measured, 44.12% of lines (250 Gy) and 40.98% (300 Gy) were superior to the parental variety. Regarding yield, 8.12% (250 Gy) and 22.66% (300 Gy) of lines displayed higher yields than the parental variety.

7.   RAINOUT SHELTER AND PHENOTYPING OF MUTANT LINES

Around 300 mutant lines were assayed under a rainout shelter for drought tolerance assessment at the reproductive stage. This experiment was performed exactly in the same way as the previous growing season and generation, except that other compositions of the mutant lines were set, i.e. only the top 10% of the mutant lines from the previous experiment were kept in the set, and the set was re-composed with lines from the large group. All phenotypic evaluations similar to those performed in the previous season are being done and the analysis is under way (Tables 7–10).

TABLE 7. CHLOROPHYL INDICES (SPAD) OF 250 Gy–M_3 MUTANT LINES UNDER DROUGHT CONDITIONS AT THE REPRODUCTIVE STAGE

Traits	Pop[1]	CD	CI	Min	Max	SD	S	K	CV (%)
Beginning of the stress period									
Chl a	30.58	31.22	31.01	23.34	39.91	2.19	0.37	3.75	7.17
Chl b	10.62	10.90	10.01	7.45	13.46	0.88	-0.21	1.86	8.37
Chl a/b	2.89	2.86	3.10	2.12	3.67	0.25	0.00	0.54	8.91
Chl total	41.21	42.16	41.14	31.88	51.31	2.62	-0.01	2.78	6.37
Mid-point of the stress period									
Chl a	30.03	30.58	32.21	23.66	35.53	2.14	-0.06	0.73	7.15
Chl b	9.87	9.73	11.79	6.28	12.65	1.04	-0.22	0.69	10.58
Chl a/b	3.06	3.14	2.73	2.41	4.11	0.28	0.80	1.52	9.38
Chl total	39.90	40.41	44.01	30.85	47.33	2.84	-0.31	0.73	7.14
Ten days after the stress period									
Chl a	24.39	24.39	24.45	18.32	34.28	2.57	0.00	0.84	10.56
Chl b	7.03	6.61	7.44	4.68	9.40	0.99	-0.10	-0.41	14.08
Chl a/b	3.50	3.69	3.29	2.80	4.79	0.33	1.26	3.54	9.48
Chl total	31.42	31.06	31.91	23.10	41.44	3.39	-0.20	-0.13	10.79

[1]Pop = population mean; CD = control cultivar (BRS Pampeira) under drought mean; CI = control cultivar (BRS Pampeira) irrigated mean; Min = minimum value; Max= maximum value; SD = standard deviation; S = distribution asymmetry coefficient; K = curtosis; CV (%) = coefficient of variation; Chl a = chlorophyll a (SPAD index); Chl b = chlorophyll b (SPAD index); Chl a/b = ratio chlorophyll a/b (SPAD index); Chl total = chlorophyll a+b (SPAD index).

TABLE 8. CHLOROPHYL INDICES (SPAD) OF 300 Gy–M$_3$ MUTANT LINES UNDER DROUGHT CONDITIONS AT THE REPRODUCTIVE STAGE

Traits	Pop	CD	CI	Min	Max	SD	S	K	CV (%)
Beginning of the stress period									
Chl a	29.5	28.83	31.01	14.66	36.13	3.01	-1.60	5.64	10.20
Chl b	9.53	9.23	10.01	4.65	14.00	1.17	-0.79	4.24	11.82
Chl a/b	3.11	3.12	3.10	1.86	4.34	0.28	0.46	5.78	9.11
Chl total	39.03	38.08	41.14	19.31	46.42	3.85	-1.77	6.29	9.84
Mid-point of the stress period									
Chl a	29.37	28.34	32.21	12.99	37.45	2.85	-1.76	7.90	9.71
Chl b	9.66	9.20	11.79	4.32	12.10	1.22	-0.90	1.96	12.72
Chl a/b	3.06	3.08	2.73	2.03	6.76	0.40	5.16	44.9	13.35
Chl total	39.05	37.59	44.01	19.37	47.73	3.79	-1.52	5.13	9.70
Ten days after the stress period									
Chl a	24.01	24.46	24.45	13.04	33.43	3.36	-0.76	1.16	13.96
Chl b	6.80	7.17	7.44	3.83	11.09	1.20	0.18	0.74	16.58
Chl a/b	3.55	3.41	3.29	2.33	5.07	0.36	0.06	2.49	10.26
Chl total	30.82	31.67	31.91	16.87	41.79	4.30	-0.64	0.62	13.96

Pop = population mean; CD = control cultivar (BRS Pampeira) under drought mean; CI = control cultivar (BRS Pampeira) irrigated mean; Min = minimum value; Max= maximum value; SD = standard deviation; S = distribution assymetry coefficient; K = curtosis; CV (%) = coefficient of variation; Chl a = chlorophyll a (SPAD index); Chl b = chlorophyll b (SPAD index); Chl a/b = ratio chlorophyll a/b (SPAD index); Chl total = chlorophyll a+b (SPAD index).

TABLE 9. PEARSON LINEAR CORRELATION BETWEEN TRAITS UNDER DROUGHT STRESS FOR THE 250 Gy MUTANT LINE POPULATION

	VS	NT	DF	PH	PanL	NPan	PanW	NSSPan	NFSPan	Yield	ChlA1	ChlB1	ChlA2	ChlB2	ChlA3	ChlB3
VS	-	0.20*	0.51*	0.31*	0.41*	-0.20*	-0.12	-0.06	0.14	-0.06	0.10	0.24*	-0.11	-0.05	-0.09	-0.13
NT		-	-0.08	0.17*	0.05*	0.29*	0.14	-0.10	0.01	0.19*	-0.03	0.15	0.06	0.05	-0.01	0.01
DF			-	0.59*	0.63*	-0.21*	0.24*	-0.02	0.09	0.19*	0.34*	-0.11	-0.11	0.17*	0.00	-0.01
PH				-	0.53*	0.30*	0.40*	-0.15	-0.02	0.21	0.26*	0.12	0.15	0.22*	0.15*	0.20*
PanL					-	0.28*	0.45*	-0.01	0.02	0.28*	-0.14	0.13	0.12	0.05	0.11	0.13
NPan						-	0.35*	-0.02	-0.07	0.19*	0.05*	-0.04	0.17*	0.16*	0.03	0.12
PanW							-	0.41*	-0.04	0.55*	0.15*	-0.03	-0.14	-0.13	0.08*	0.00
NSSPan								-	0.06	0.45*	0.21*	0.08*	0.12	0.17*	0.07*	0.01*
NFSPan									-	0.02	0.00	-0.06	0.06	-0.06	-0.13	-0.10
Yield										-	0.23*	-0.09	-0.13	-0.09	-0.11	0.00
ChlA1											-	0.33*	0.12	0.10	0.08	0.08
ChlB1												-	0.08	-0.01	0.10	0.06
ChlA2													-	0.53*	0.25*	0.28*
ChlB2														-	0.24*	0.23*
ChlA3															-	0.76*
ChlB3																-

VS = visual score; NT = number of tillers; DF= days to flowering; PH = plant height; PanL = panicle length; NPan = Number of panicles per area; PanW = panicle weight; NSSPan = number of sterile spikelets per panicle; NFSPan = number of fertile spikelets per panicle; Yield = grain yield; ChlA1 = chlorophyl A (SPAD index) at the beginning of the drought stress; ChlB1 = chlorophyl B (SPAD index) at the beginning of the drought stress; ChlA2 = chlorophyl A (SPAD index) at the mid-point of the drought stress; ChlB2= chlorophyl B (SPAD index) at the mid-point of the drought stress; ChlA3 = chlorophyl A (SPAD index) ten days after the drought stress; ChlB3= chlorophyl B (SPAD index) ten days after the drought stress. *Significant (p≤0.05) according to the t test.

TABLE 10. PEARSON LINEAR CORRELATION BETWEEN TRAITS UNDER DROUGHT STRESS FOR THE 300 Gy MUTANT LINE POPULATION

	300 Gy															
	VS	NT	DF	PH	PanL	NPan	PanW	NSSPan	NFSPan	Yield	ChlA1	ChlB1	ChlA2	ChlB2	ChlA3	ChlB3
VS	-	0.21*	0.13	-0.15	0.20*	-0.13	0.23*	-0.12	-0.12	-0.11	0.22*	-0.10	0.06	-0.11	0.15*	-0.05
NT		-	0.02	-0.07	0.12	0.40*	0.01	0.08	0.04	0.13	0.03	-0.02	0.06	0.25*	-0.05	-0.03
DF			-	-0.65*	0.55*	-0.39*	0.31*	-0.01	-0.36*	-0.28*	0.21*	-0.10	0.17*	0.18*	0.17*	-0.12
PH				-	0.70*	0.51*	0.37*	0.01	0.40*	0.33	0.29*	0.20*	0.27*	0.27*	0.27	0.19
PanL					-	0.61*	0.29*	0.19*	0.26*	0.21*	0.22*	0.18*	0.30	0.37	0.10	0.13
NPan						-	0.27*	0.03	0.22*	0.13*	0.06	0.06	0.13	0.20*	-0.03	0.04
PanW							-	0.03	0.53	0.48	0.16*	0.16*	0.18*	0.13	0.05*	0.07*
NSSPan								-	-0.29*	-0.24*	0.21*	0.22*	0.09	0.19*	0.13	0.01
NFSPan									-	0.90*	0.01	0.01	0.14	-0.01	0.01	0.07
Yield										-	0.04	0.05	0.01	-0.02	-0.02	0.05
ChlA1											-	0.66*	0.11	0.08	0.06	0.01
ChlB1												-	0.14	0.18	0.05	0.03
ChlA2													-	0.67*	0.02	-0.01
ChlB2														-	0.13	0.08
ChlA3															-	0.77*
ChlB3																-

VS = visual score; NT = number of tillers; DF= days to flowering; PH = plant height; PanL= panicle length; NPan=Number of panicles per area; PanW = panicle weight; NSSPan = number of sterile spikelets per panicle; NFSPan = number of fertile spikelets per panicle; Yield = grain yield; ChlA1 = chlorophyl A (SPAD index) at the beginning of the drought stress; ChlB1= chlorophyl B (SPAD index) at the beginning of the drought stress; ChlA2 = chlorophyl A (SPAD index) at the mid point of the drought stress; ChlB2= chlorophyl B (SPAD index) at the mid point of the drought stress; ChlA3 = chlorophyl A (SPAD index) ten days after the drought stress; ChlB3= chlorophyl B (SPAD index) ten days after the drought stress. *Significant (p≤0.05) according to the t test.

8.   EARLY FLOWERING PROMISING MUTANT LINE EVALUATION TRIALS

A total of five mutant lines, showing highlighted promising performance at the M3 generation, were selected for advanced trials, started in the 2020–2021 growing season. These lines showed early flowering and high performance regarding yield components. The trial was designed in replicated blocks, with the parental cultivar (BRS Pampeira) as control (check cultivar). The results on flowering and height of some promising mutants are shown in Table 2.11.

TABLE 11. FLOWERING DATE AND PLANT HEIGHT OF SELECTED PROMISING RICE MUTANT GENOTYPES OF BRS PAMPEIRA

Genotype	Flowering(days)	Height (cm)
BRS Pampeira (parental)	108 a	83.1 a
250G/1 (636)	87 c	77.1 bc
250G/1 (519)	88 c	75.3 bc
250G/1 (498)	84 cd	79.7 ab
250G/1 (425)	94 b	82.8 a
300G/2 (267)	82 d	73.3 c

9.   CONCLUSIONS

This chapter presented the needs and challenges of a breeding programme aimed at enhancing rice drought tolerance in southern Brazil. Brazil is an important rice producer worldwide and the southern states of Rio Grande do Sul and Santa Catarina play a key role in this regard. In this region, paddy rice predominates largely in a system in which the area is flood irrigated during most of the growing season. Paddy rice is usually high yielding and the water layer used for irrigation plays several other agronomic roles. However, flooding the area has also disadvantages and, more importantly, climate changes have brought uncertainty regarding water availability for agriculture. These issues represent reasons for the beginning of research efforts towards rice production with optimized use of water, namely the rainfed system, which is, however, prone to drought. It is important to mention that any change in a crop management system also requires breeding efforts for the development of adapted cultivars.

As drought interacts with other factors, such as climate, soil conditions, biotic factors and also with the target germplasm, all these aspects, related to the region under analysis were presented and discussed in this chapter. In summary, it is clear that any method of drought assessment already employed in published studies has first to be tested thoroughly in this region for drought tolerance before it is routinely applied in a rice breeding programme. This chapter does not suggest a complete and rapid change in the rice management system in southern Brazil, from paddy to rainfed. It aims, on the other hand, to support the development of alternative systems, which will ultimately save water.

REFERENCES TO CHAPTER 2–2

[1]   GUO, H., WANG, R., GARFIN, G. M., ZHANG, A., LIN, D., Rice drought risk assessment under climate change: Based on physical vulnerability a quantitative assessment method, Sci. Total Environ. **751** (2021) 141–481.

[2]   RAY, D. K., RAMANKUTTY, N., MUELLER, N.D., WEST, P.C., FOLEY, J.A., Recent patterns of crop yield growth and stagnation, Nature Commun. **3** 1 (2012) 1– 7.

[3]   RAY, D.K., MUELLER, N.D., WEST, P.C., FOLEY, J.A., Yield trends are insufficient to double global crop production by 2050, PloS One **8** 6 (2013) e66428.

[4]   WAGENA, M.B., EASTON, Z.M., Agricultural conservation practices can help mitigate the impact of climate change, Sci. Total Environ. **635** (2018) 132–143.

[5]   PANDEY, S., et al., Rice in the Global Economy, Strategic Research and Policy Issues for Food Security, International Rice Research Institute, Manila (2010).

[6]   SEN, S., CHAKRABORTY, R., KALITA, P., Rice — Not just a staple food: A comprehensive review on its phytochemicals and therapeutic potential, Trends Food Sci. Technol. **97** (2020) 265–285.

[7]   SOSBAI, R., Arroz irrigado: recomendações técnicas da pesquisa para o Sul do Brasil. XXXI Reunião Técnica da Cultura do Arroz Irrigado (2018).

[8]   PINTO, M.A.B., PARFITT, J.M.B., TIMM, L.C., FARIA, L.C., SCIVITTARO, W.B., Produtividade de arroz irrigado por aspersão em terras baixas em função da disponibilidade de água e de atributos do solo, Pesquisa Agropecuária Brasileira **51** 9 (2016) 1584–1593.

[9]   PINTO, M.A.B., et al., Sprinkler irrigation in lowland rice: Crop yield and its components as a function of water availability in different phenological phases, Field Crops Res. **248** (2020) 107–714.

[10]   STRECK, E.A., et al., Genetic progress in 45 years of irrigated rice breeding in Southern Brazil, Crop Sci. **58** 3 (2018) 1094–1105.

[11]   TUBEROSA, R. Phenotyping for drought tolerance of crops in the genomics era, Front. Physiol. **3** (2012) 347.

[12]   PASSIOURA, J.B., Phenotyping for drought tolerance in grain crops: When is it useful to breeders? Funct. Plant Biol. **39** 11 (2012) 851–85.

[13]   CAIRNS, J.E., IMPA, S.M., O'TOOLE, J.C., JAGADISH, S.V.K., PRICE, A.H., Influence of the soil physical environment on rice (*Oryza sativa* L.) response to drought stress and its implications for drought research, Field Crops Res. **121** 3 (2011) 303–310.

[14]   BUSANELLO, C., et al., Is the genetic variability of elite rice in southern Brazil really disappearing? Crop Breeding Appl. Biotechnol. **20** 2 (2020).

[15]   VIANA, V.E., PEGORARO, C., BUSANELLO, C., DE OLIVEIRA, A.C., Mutagenesis in rice: The basis for breeding a new super plant. Front. Plant Sci. **10** (2019).

2-3. SCREENING FOR DROUGHT TOLERANCE IN RICE MUTANTS

J. BAGRI[1], R.N. SAHOO[1],
S.L. SINGLA-PAREEK[2] A. PAREEK[1,3]

[1]Jawaharlal Nehru University, New Delhi, India
[2]International Centre for Genetic Engineering and Biotechnology, New Delhi, India
[3]National Agri-Food Biotechnology Institute, Mohali, India

Abstract

Crop yield stability requires an attenuation of the reduction of yield losses caused by environmental stresses such as drought. Improving tolerance in crop plants to drought via genetic engineering techniques appears to be challenging. The availability of adequate donor drought tolerant lines is a prerequisite for the classical breeding technique for this stress tolerance trait. However, developing this trait through mutation breeding still remains an attractive approach due to the process being simple and the product acceptable to people. A population of gamma induced mutant lines of rice in the background of an elite rice cultivar was analysed for variation in complex drought adaptive traits. A set of M_2 and M_3 lines were screened for variation in chlorophyll content, shoot length, root length, fresh weight, and dry weight at the seedling stage for two consecutive generations. Drought tolerant M_3 shortlisted lines, the potential mutant lines, were selected for stabilization and screening of various drought adaptive traits in the M_3 and M_4 generations for tolerance towards drought at the pre-flowering or post-flowering stage. Mutants differing in morphophysiological, biochemical and molecular parameters were analysed further under a managed drought environment and open field conditions to assess the relevance of the traits compared with the wild type.

Key words: Rice, mutagenesis, drought stress, reproductive stage.

1. INTRODUCTION

Drought is a major abiotic stress, limiting yield in many crop plants. About 34 million ha of shallow, rainfed, lowland rice and 8 million ha of upland rice are subject to intermittent or regular drought stress in Asia. This is approximately one-third of the total Asian rice growing region, for example, in India, Bangladesh, Pakistan, and China [1–3]. Drought is a common occurrence in India, especially during the monsoon season (pre-monsoon and post-monsoon), leading to a major reduction in rice yield. India and bordering parts of Nepal are some of the world's largest drought prone locations among the many rainfed, rice growing ecologies. Eastern Indian states account for around 16.2 million ha of India's total 20.4 million ha of rainfed rice, with 6.3 million ha of upland and 7.3 million ha of lowland areas being drought prone [4, 5]. Drought in India in 2002, 2003, 2009 and 2010 decreased rice production significantly, especially in the eastern Indian states of Jharkhand, Bihar, Uttar Pradesh, Chhattisgarh and Odisha. Similarly, the severe drought that ravaged portions of eastern Uttar Pradesh and Bihar in 2015 wreaked havoc on the rice crop [4, 5]. Moreover, 40% of the rice producing region of the Indian subcontinent is rainfed, of which 70% is present in different states of India — north, western, and central India. This region's total rainfed area comprises 77% of lowlands and 23% of uplands, of which 52% of lowlands and all uplands are affected by drought [6]. In 2017, India had an area of about 43 mha under rice cultivation [7], nearly 60% of which grows in eastern India, according to the International Rice Research Institute.

The rainfed, rice growing area of eastern India alone accounts for 12.9 mha [8]. These regions are often marked by a scarcity of rain or a long duration of rain. As a result, drought stress can appear at any stage of crop growth, such as seedlings, active tillering and reproductive stages during any period [9]. According to data reports for 2018–2019, India has around 43 million hectares of rice growing area and produces about 115.60 million tonnes of milled rice.

Water scarcity is projected to be a key concern for long term rice production in the near future due to the continued negative consequences of climate change [10]. Speculation suggests that the frequency and intensity of droughts would rise, posing a severe danger to rice production and, by extension, to world food security [11]. Droughts have a far-reaching impact on the output and productivity of crops planted later in the season. Farmers growing rice in drought prone environments are fully aware of the risks and are therefore hesitant to use expensive agricultural inputs, resulting in a further drop in yield potential [12]. Furthermore, the majority of farmers in drought prone areas come from socioeconomically disadvantaged backgrounds. As a result, drought has a disproportionate impact on poor farmers, forcing them to limit their consumption, withdraw their children from school, force them to migrate for work and liquidate assets to fulfil immediate requirements [3, 13, 14].

One of the major goals for rice breeders includes breeding for drought tolerance, which requires selection of drought tolerant breeding lines from large segregation populations [15]. Drought screening in the field is arduous as soil moisture is dynamic; the soil moisture level horizontally and vertically changes with time. Drought stress is the most complex among abiotic stresses because of the magnitude at which it varies at different growth stages in the rice life cycle [16]. Comparatively, rice is most susceptible to drought stress at various developmental stages, such as seedling, vegetative growth and grain filling. It has been reported that the two stages, i.e. seedling and grain filling, are the most sensitive [17].

Climate change adaptability, genetic change/mutation, and crop improvement have all been exploited by humans in the agricultural process of domestication and crop development. In plant breeding programmes, simple selection of desired offspring based on desirable phenotype was most commonly used, and thus relied on the occurrence of random mutation. Gregor Johann Mendel's genetic rules, on the other hand, have exerted a key influence on inbreeding since the beginning of the twentieth century, which effectively led to a shift from a theoretical to an empirical or research-based approach. Over 3400 new varieties created by mutation induction were documented by the FAO/IAEA Mutant Variety Database, and these have significantly boosted food security in many countries [18, 19]. The use of radiation techniques to create mutants is now becoming increasingly relevant [20. 21]. Conducting field screening assessments is one of the important strategies, particularly in the context of crop breeding programmes. To enhance grain yields and impart resistance or tolerance to potential existing climate threats, research institutes around the world are trying to identify mutants with suitable traits incorporated into existing genotypes/cultivars [22, 23].

2.   METHODOLOGY

2.1.   Plant growth conditions and stress treatment for the seedling stage screening of mutants

Uniform and disease free seeds of four standard genotypes (Table 1), along with mutant genotypes, were used for the study. Initially, seeds were surface sterilized and germinated for two–three days in the dark at 28/23°C (day/night) and 70% relative humidity [24, 25]. Germinated seeds (four seeds per genotype/mutant) were then sown into 8 cm diameter and

12 cm deep pots containing equal amounts of sieved soil mixed with NPK (nitrogen, phosphorous, potassium) in 10:6:6 ratios. After sowing, germinated seeds were kept in a plant growth chamber with a day and night temperature of 28/23°C and a relative humidity of 70±5%. After 10 days, the seedlings were subjected to drought by withdrawal of water until soil water content (SWC) reached 40–50%, while control plants were well-watered and grown in a controlled environment as mentioned (Fig. 1). A gravimetric method was used to assess the SWC.

TABLE 1. STANDARD GENOTYPES USED FOR THE CLASSIFICATION OF THE POPULATION (SENSITIVE AND DROUGHT)

S. N.	Standard genotypes	Drought tolerant/susceptible	Lowland/ upland	Germplasm group	Origin	Reference
1	Elite rice cultivar	Drought susceptible	Lowland	Indica	Philippines (IRRI)	[26]
2	Samba Mashuri	Drought susceptible	Lowland	Indica	India (Andhra Pradesh)	[27]
3	Nagina_22	Drought tolerant	Upland	Indica	India (Uttar Pradesh)	[26, 28]
4	Vandana	Drought tolerant	Upland	Aus/japonica	India (Odisha)	[26]

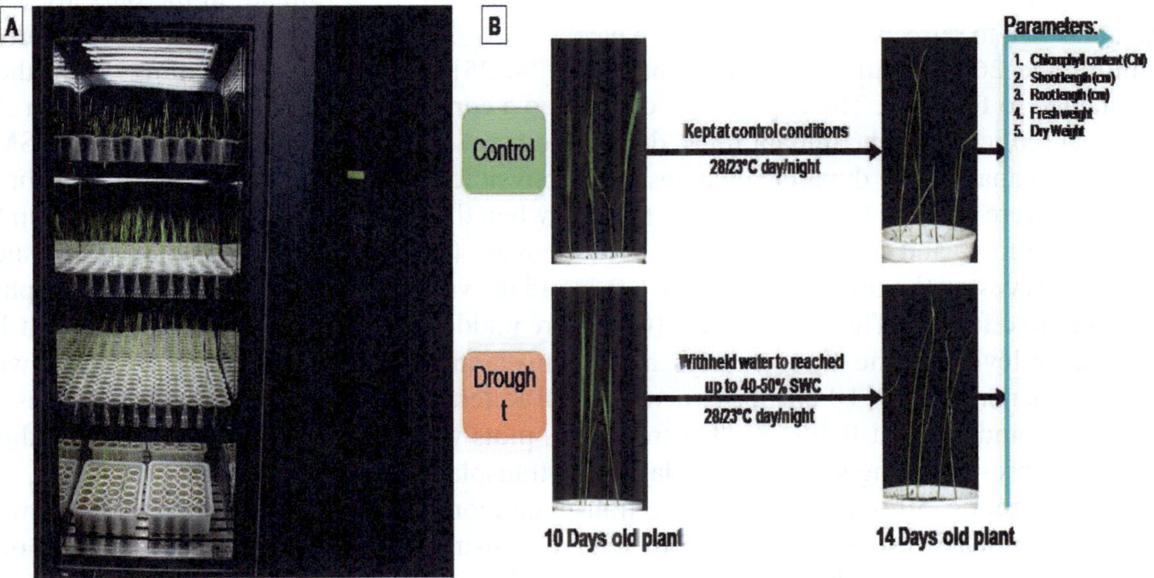

FIG. 1. (A) Gamma induced different rice mutant lines showing germination under control conditions (growth chamber set-up: 28°/23°C day/night and 70±5% humidity). For drought stress (DS), gamma induced different mutant lines were germinated in a plant growth chamber and further exposed to stress. (B) Experimental conditions for DS. Watering was withheld for the plants so they could reach an SWC of between 40 and 50%.

2.2. Assessment of morpho-physiological traits

Seedling morpho-physiological parameters (shoot length (SL), root length (RL), total fresh weight (FW), and total dry weight (DW)) were evaluated under control and DS conditions, with three replicates for each genotype/mutant. Seedling weight was immediately measured to record FW and kept in the oven at 70°C for 48 hours prior to determining total dry weight.

2.3. SPAD measurement

SPAD values of the fully expanded leaf of 10-day old seedlings were measured. A chlorophyll meter was used to measure the chlorophyll content.

2.4. Assessment of drought tolerance index

After assessing morpho-physiological traits, the drought tolerance index (DTI) was calculated based on the measured morpho-physiological parameters to classify the mutant population as tolerant, moderate, or susceptible compared with the reference genotype. Fernandez [29] gave the following formula for calculating the stress tolerance index:

$$DTI = (\text{value under control/value under stress}) *100$$

3. FIELD SCREENING OF INDUCED MUTANTS FOR DROUGHT TOLERANCE

The study included a field technique that was successfully employed in lowlands at two different locations to evaluate rice genotypes/mutants over two generations (M_3, M_4). Rice seeds (*Oryza sativa* L.) were surface sterilized and allowed to germinate in the dark for 2–3 days and sown to raise a nursery bed for two weeks at 28°/23°C (day/night) with 70%±5 relative humidity and 260–350 µE m^2 s^{-1} light intensity [24, 25]. Two-week old seedlings were then transplanted to the field. The experiment comprised a control field and DS field, each with six biological replicates of the mutant lines, drought tolerant check (N22), susceptible check (SM) and WT. To analyse the dynamic changes in the physiology of the rice plants under DS, three treatments were chosen: (i) control (CO); (ii) DS, when the soil moisture content reached up to severe conditions; and (iii) 10 days of drought recovery (DR). Before the onset of DS, tissues were also harvested (beginning stress (BS)), to avoid the variation arise due to any other edaphic or topographic factors. The control plots (CO) were puddled and continuously flooded with 1–3 cm water level for the first 15 days after transplanting. Thereafter, the water level was gradually increased to 5–10 cm for two weeks. NPK fertilizer was applied in 10:6:6 ratios to both control and drought fields. For DS treatment, plots were induced drought by withholding water at the pre-flowering stage, i.e. 60 days after transplantation and post-flowering stage, i.e. 70 days (targeting 50% flowering) after transplantation (Fig. 2). Soil water potential was monitored regularly using a soil tensiometer (five instruments in each plot or at least one tensiometer/10 m^2 as per the protocol given in the layout) in order to achieve the soil water potential at −35 ±5 kPa level at 30 cm depth. The slow progression of the drought environment with time in both the experimental fields is shown in Fig. 3. The entire screening study was carried out during the 2019 and 2020 paddy seasons at two different field locations following standard agronomic practices.

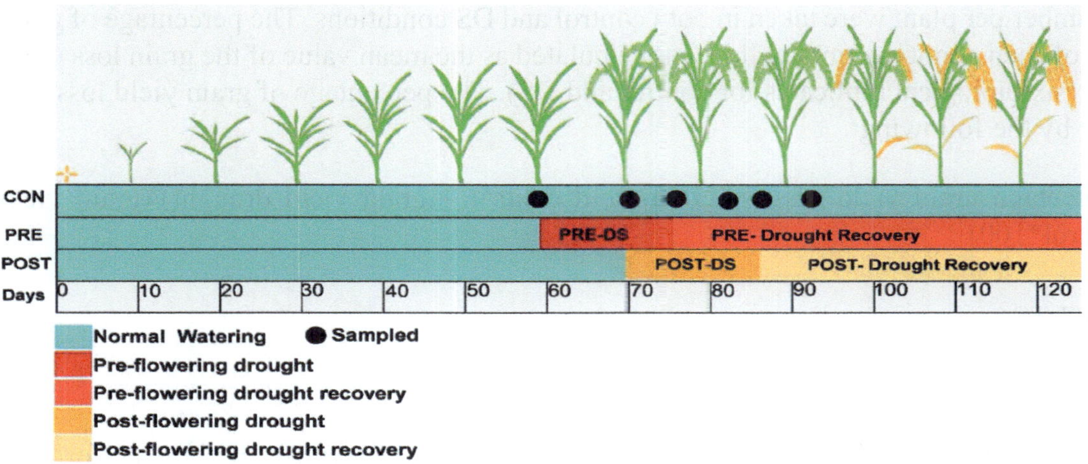

FIG. 2. Schematic overview of the experimental design for control (CON), pre-flowering (PRE) and post-flowering (POST) drought. Black dots represent whether plants were sampled for the specified treatment/day, and the colour of the boxes reflects the irrigation status for the plants, i.e. watered; pre-flowering drought; pre-flowering drought recovery; post-flowering drought; and post-flowering drought recovery. Samples from pre-flowering and post-flowering; four treatments were chosen for detailed investigation, namely (1) beginning stress (BS); (2) control (CON); (3) drought stress (DS); and (4) 10 days of drought recovery (DR).

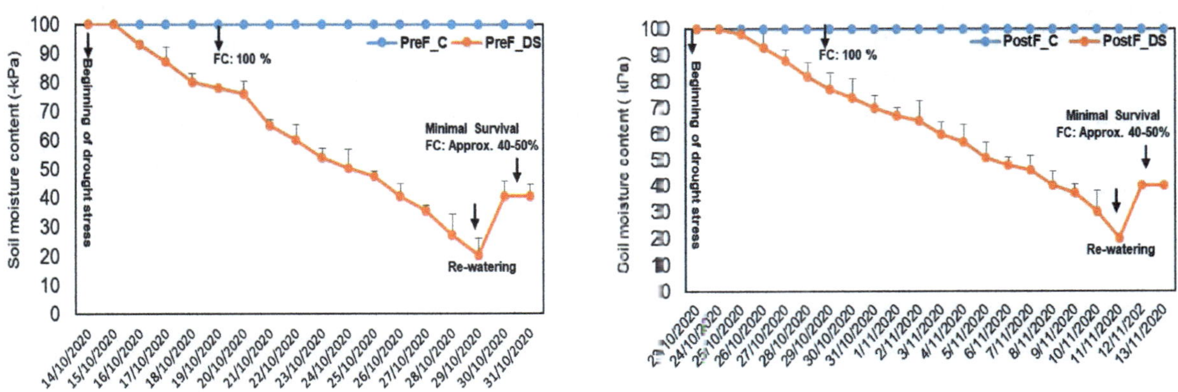

FIG. 3. Graphs depicting the progressive onset of DS. The soil moisture content (-kPa) is shown as a function of time. For both sets of fields experiments: (a) the left panel shows the graphs for pre-flowering, and (b) the right panel shows the graphs for the post-flowering of the experiments. The control (CON) field showed 100% field capacity while under drought stress (DR), moisture contents were allowed to drop until ~-30 kPa, after that the fields were irrigated as a part of the recovery process (DR).

Tolerance to DS in rice is a complex trait that depends on various environmental and physiological factors. Therefore, a comprehensive understanding of the plant's responses to DS may provide a means to identify and confer tolerance in agronomic, molecular and genetic aspects. Quantitative trait data, such as plant height (cm), the number of panicles, tiller number, grain yield, and 1000 seed weight (g) were recorded.

Under DS, grain yield is a parameter that has been widely used to select drought tolerant genotypes of rice [30, 31]. Within genotypes/mutant lines, we applied the same criteria for selecting pre-flowering and post-flowering drought tolerance. The total grain weight and 1000 seed weight was used to screen drought tolerance of genotype/mutant lines. Plant height and panicle number per plant were taken in both control and DS conditions. The percentage of grain yield loss of each genotype/mutant line was calculated as the mean value of the grain loss of all replicates (six biological replicates for control and DS). The percentage of grain yield loss was calculated by the following:

Percentage grain yield loss = (grain yield control − grain yield drought)/(grain yield control*100)

4. RESULTS

4.1. Seedling–stage screening of mutants

A total of 4000 M_2 and M_3 rice mutant lines were screened to identify the best performing mutant lines under DS. The M_2 and M_3 mutant populations were screened at the seedling stage through quick assessment for physiological traits such as chlorophyll content (SPAD), shoot length, root length, total fresh weight, and total dry weight. The results showed that there was significant variation among the mutants and genotypes. The decrease in chlorophyll levels of seedlings is considered a symptom of oxidative stress resulting from pigment photo-oxidation and chlorophyll degradation [32]. The higher chlorophyll content is generally associated with a higher tolerance to stress [33]. The SPAD based estimation of chlorophyll in fully expanded second leaves of seedlings was carried out after 24 h of stress, followed by the regular measurements carried out every 24 h until soil moisture content (SMC) reached up to 40–50%. In this experiment, a continuous decrease in the SPAD value was recorded in drought tolerant genotypes (N22 and Vandana), mutant lines (M_1 and M_2), and the susceptible genotypes (SM and WT) with the imposition of DS. However, the tolerant genotypes and mutant lines showed a relatively smaller decrease (15–20%) in SPAD value under DS in comparison with the susceptible genotypes (Fig. 4(B)).

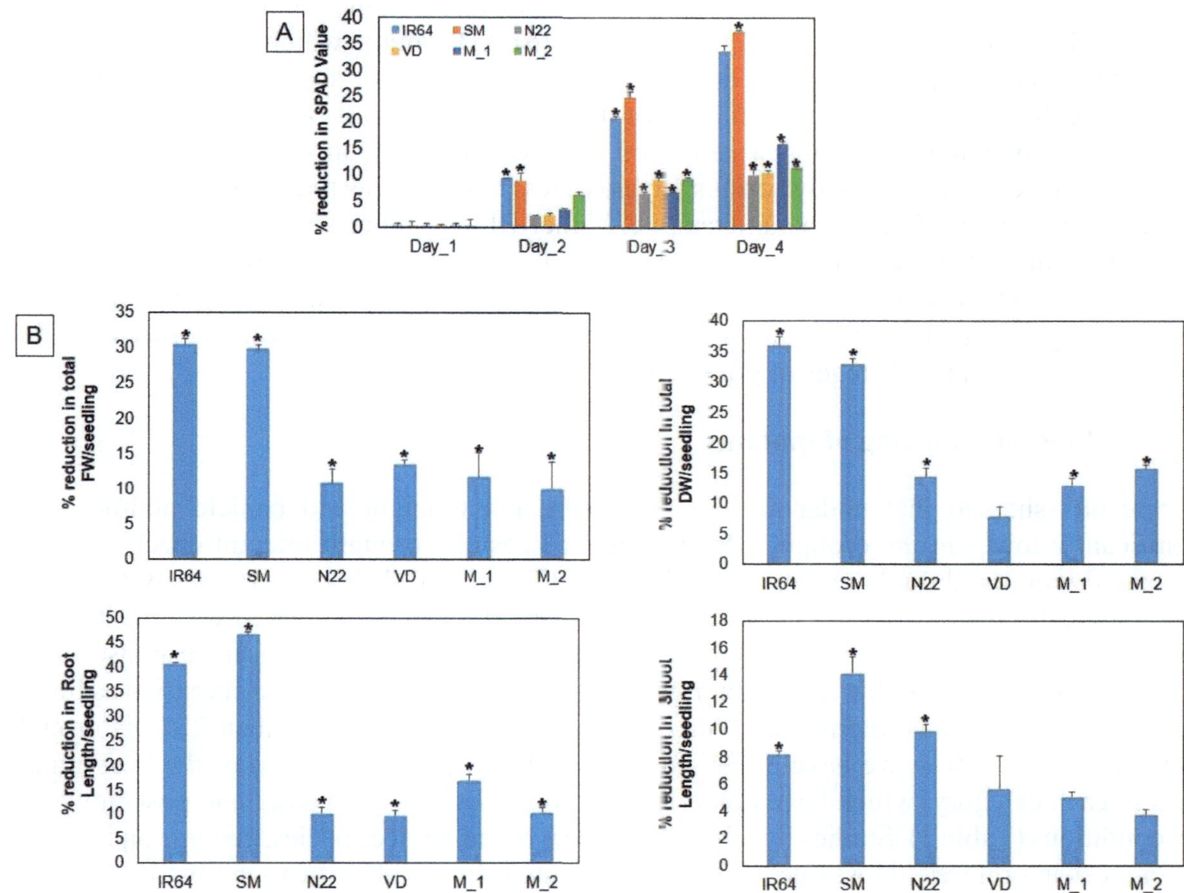

FIG. 4. *Morpho-physiological parameters of all genotypes. (A) The chlorophyll content (SPAD value) in drought tolerant genotypes (N22 and Vandana), drought tolerant mutant lines (M_1 and M_2) and the susceptible genotype (Samba Masuhri and IR64) seedlings under DS. (B) Shoot length (SL), root length (RL), total fresh weight (FW) and total dry weight (DW) under control and drought condition data are the means (±SE) of three biological replicates. The asterisk represents significant differences between genotype/mutant and the WT (IR64).*

The data on SL, RL, FW and DW showed that the tolerant (Nagina 22 and Vandna) genotypes and mutant lines showed comparatively less decrease in SL (5–10%), RL (10–15%), FW (10– 15%), and DW (10–20%) than the susceptible ones R64 and Samba Masuhri under DS conditions (Fig. 4(B)). The preliminary phenotyping led to the identification of the genotype/mutant lines with a robust phenotype under stress. Based on these parameters, DTI was calculated and mutant lines were classified (Table 2).

TABLE 2. CLASSIFICATION BASED ON THE DROUGHT TOLERANCE INDEX OF RICE MUTANTS AT THE SEEDLING STAGE (~2000 M_2 AND ~200 M_3) IN COMPARISON TO THE STANDARD GENOTYPES
(Note: Mutant lines mentioned here are examples)

More susceptible than reference WT	Similar to reference WT	More tolerant than reference WT
#M_4 #M_5	#M_11, #M_17	#M_1, #M_2

Drought tolerance is a multigenic trait and is reflected by various morpho-physiological parameters. In this study, we investigated a combination of traits, including physiological traits (SL, RL, FW and DW) and chlorophyll content (SPAD value), which contributed toward drought tolerance in a gamma induced mutant population set of rice seedlings. Phenotypic data under DS exhibited large variation in most of the physiological traits and chlorophyll content. Rice mutant populations having a range of sensitive and tolerance levels under drought conditions were characterized by various physiological traits such as reductions in SL, RL, FW, DW and Chl content. Growth parameters, i.e. higher RL and higher FW, are useful to enable survival during DS by enabling improved access to soil moisture. On the other hand, high chlorophyll content increases the chance of maintaining the maximum photosynthesis rate during drought. Together, these traits enable plants to always have better water homeostasis and photosynthesis during all stages of growth under stress conditions.

4.2. Field level screening of mutants

Our results showed that under DS, grain weight, plant height and panicle number were significantly lower in the drought tolerant mutant lines and drought tolerant check (N22) as compared with the drought susceptible (SM) and WT in both M_3/M_4 generations. Overall, the physiological traits in response to DS were drastically affected as compared with plants under non-stress conditions. There was a significant variation among the traits in response to DS (Fig. 5). Mutant lines with a reduction in yield loss of more than 60% were classified as highly susceptible, 25–60% as intermediate (moderately susceptible), and less than 25% as drought tolerant. Based on this, we have classified the mutants and presented two of the best mutant lines in each category, which have outperformed under both pre-flowering and post-flowering DS conditions (Table 3). Studies by other researchers based on rice species, ecotype and cultivar differences in grain yield loss due to DS also suggested similar classification [34, 35].

TABLE 3. CLASSIFICATION OF DROUGHT TOLERANCE MUTANTS BASED ON OBSERVATIONS MADE UNDER FIELD CONDITIONS
(Note: Only two of the best mutant lines are presented which have outperformed under both pre-flowering and post-flowering DS conditions)

More susceptible than reference WT	Similar to reference WT	More tolerant than reference WT
#M_6, #M_7,	#M_11, #M_17,	#M_1, #M_2,

Apart from gain yield loss, significant differences were observed under control and stress conditions in the above quantitative traits, i.e. plant height (cm), the number of panicles, tiller number, grain yield, and 1000 seed weight (g) among the tested lines (Fig. 5). In general, plant height and tiller number were decreased under DS among the genotypes/mutant lines. On average, significant differences existed within the tolerant or susceptible genotypes and mutant lines. In most cases, the panicle number among selected genotypes/mutant lines was either the same or less than under DS conditions. However, several mutant lines showed significantly higher panicle numbers than stress conditions under control. In the case of leaf morphology, earlier studies have reported an increase in the leaf area, reduced reflectance and improved light transmission, enhanced photosynthetic ability, delayed leaf senescence and enhanced nitrogen storage during grain filling [34, 35]. Several gamma mutant rice lines showed significant phenotypic variome in the leaf length, leaf width, and enhanced photosynthesis ability in many mutant rice lines under both control and stress conditions. It has been shown that these quantitative traits play a vital role under abiotic stress, particularly in the case of DS. Taken together, these results suggest the occurrence of genetic variability for drought tolerance among the gamma induced mutant rice lines, which could be useful in the rice breeding programme.

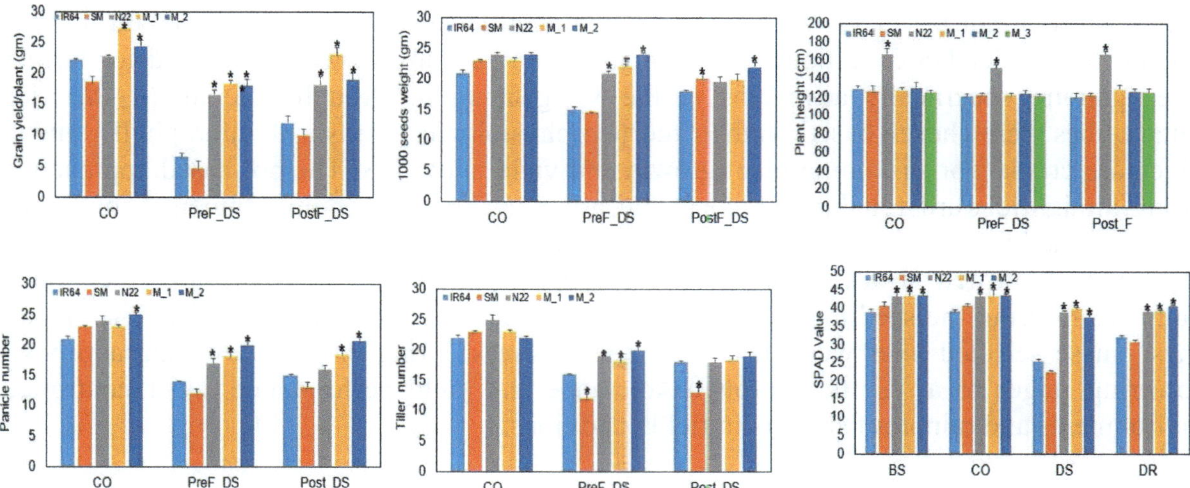

FIG. 5. Assessment of morpho-agronomic traits in gamma induced rice (M₄) mutant lines at pre-flowering and post-flowering under control and drought stress conditions. The following traits such as grain yield (g), 1000 seed weight (g), plant height (cm), panicle number, tiller number and chlorophyll content (SPAD value) were measured under control, pre-flowering and post-flowering stress. The bar graph represents control (CO) pre-flowering DS (PreF-DS) and post-flowering DS (Post-DS). Data are means (±SE) of three independent biological replicates. The asterisk (P<0.05) represents significant differences between mutant/genotypes and the WT (IR64).*

Several efforts have been made to improve crop plants for drought tolerance by conventional breeding, but the success rate has been limited due to the multigenic and lack of genetic information. However, advances in induced mutation and molecular technologies have paved the way for precise plant characterization and have assisted in the development of drought tolerant crops [36]. There is an urgent need to develop a simple, rapid, reliable and cost effective high throughput screening protocol for screening large scale germplasm/mutant populations under drought stress. Several morpho-physiological traits have been taken into account and are strongly associated with tolerance to drought, for example, deep root system, water use efficiency, leaf water potential, characteristics of stomata, and osmotic adjustment [37–39]. From this perspective, the preliminary screening was performed at the seedling stage in the growth chamber and further subjected a selected population to field experiments at the reproductive stage. After preliminary analysis, selected populations were grown in the monsoon season to avoid the variation caused by the environmental factors that were present during the dry season (off-season). To impose DS, we used a rainproof shelter to protect plants from rainfall and also withdrew water. Conversely, experimenting with the rainproof shelters results in approximation of near-natural drought scenarios in rice and describes phenotyping protocol processes that use the rainproof shelter to screen the small number of rice genotypes in field conditions during the paddy growing season under DS. The the genotypes have been classified into three different categories: sensitive, moderate, and drought tolerance [40, 41]. The second protocol presented here was optimized to screen for drought tolerance in rice at the pre-flowering and post-flowering stages with an open rainproof shelter to screen the mutant population. Pre-flowering and post-flowering are the most drought sensitive stages that can decrease photosynthesis and increase yield loss. Under stress, grain yield stability is a parameter that has been successfully used to select drought tolerance in rice. Furthermore, detailed characterization of these genotypes or mutant lines should be based on their reproductive, physio-morphological, agronomical and biochemical traits.

5. CONCLUSIONS

In this study, a pot based system was developed as a dependable screening method for rice mutant lines under DS conditions. These mutants can be transplanted to the field for further investigations, specifically advancing to the M_3 generation. Drought tolerant M_2–M_3 rice mutant lines were chosen based on their morpho-physiological traits and chlorophyll content, which are crucial for sustaining plant growth and yield under DS. The pot based approach is strong, straightforward, reliable and repeatable, allowing control over environmental conditions for reproducible results. Consequently, these screened mutants hold promise for the development of new drought tolerant rice cultivars. After analysing the aforementioned parameters during the seedling stage, we have identified the most promising M_3–M_4 mutant lines. These selected lines were then subjected to stress during the pre flowering and post-flowering stages to assess their drought tolerance and yield performance under controlled environments and rainproof shelters in field conditions.

REFERENCES TO CHAPTER 2–3

[1] PAREEK, A, SOPORY, S.K., BOHNERT, H., GOVINDJEE, J., Abiotic Stress Adaptation in Plants: Physiological, Molecular and Genomic Foundation, Springer, Dordrecht (2010).

[2] DIXIT, S., et al., Multiple major QTL lead to stable yield performance of rice cultivars across varying drought intensities, BMC Genetics **15** (2014) 1–13.

[3] DAR, M.H., et al., Drought tolerant rice for ensuring food security in Eastern India, Sustainability **12** 6 (2020) 2214.

[4] PANDEY, G.K., GRANT, J.J., CHEONG, Y.H., KIM, B.G., LUAN, S.,. Calcineurin-B-like protein CBL9 interacts with target kinase CIPK3 in the regulation of ABA response in seed germination, Mol. Plant **1** 2 (2008) 238–248.

[5] SINGH, N., et al., Genetic diversity trend in Indian rice varieties: An analysis using SSR markers, BMC Genetics **17** (2016) 1–13.

[6] PATHAK, H., et al., Rice research for enhancing productivity, profitability and climate resilience (2018).

[7] KEELERY, S., Area of cultivation for rice in India 2013–2018. Statista (2020); https://www.statista.com/statistics/765691/india-area-of-cultivation-for-rice/

[8] LAL, B., et al., Crop and varietal diversification of rainfed rice based cropping systems for higher productivity and profitability in Eastern India,. PLoS One **12** 4 (2017) p.e0175709.

[9] SERRAJ, R., et al., Drought resistance improvement in rice: An integrated genetic and resource management strategy, Plant Prod. Sci. **14** 1 (2011) 1–14.

[10] WASSMANN, R., et al., Regional vulnerability of climate change impacts on Asian rice production and scope for adaptation, Adv. Agron. **102** (2009) 91–133.

[11] MACKILL, D.J., et al., Stress tolerant rice varieties for adaptation to a changing climate, Crop Environ. Bioinformatics **7** (2010) 250–259.

[12] COURTOIS, B., et al., Mapping QTLs associated with drought avoidance in upland rice, Mol. Breeding **6** (2000) 55–66.

[13] SERRAJ, R., BENNETT, J., HARDY, B., Drought Frontiers in Rice: Crop Improvement for Increased Rainfed Production, International Rice Research Institute, Manila (2008).

[14] DAR, M.H., DE JANVRY, A., EMERICK, K., RAITZER, D., SADOULET, E., Flood tolerant rice reduces yield variability and raises expected yield, differentially benefitting socially disadvantaged groups, Sci. Rep. **3** 1 (2013) 3315.

[15] WANG, ZH, JIA, Y., Development and characterization of rice mutants for functional genomic studies and breeding. Mutagenesis: Exploring novel genes and pathways, (2014) 307–332.

[16] KRASENSKY, J., JONAK, C., Drought, salt, and temperature stress-induced metabolic rearrangements and regulatory networks, J. Exp. Bot. **63** 4 (2012) 1593–1608.

[17] HATFIELD, J.L., PRUEGER, J.H., Weather Climate Extremes **10** (2015) 4–10.

[18]   FAO/IAEA Mutant Variety Database, IAEA, Vienna (2022); https://nucleus.iaea.org/sites/mvd/SitePages/Home.aspx.

[19]   JANKOWICZ-CIESLAK, J, TILL, B.J., "Forward and reverse genetics in crop breeding", Advances in Plant Breeding Strategies: Breeding, Biotechnology and Molecular Tools (AL-KHAYRIJM, JAIN, S.M., JOHNSON, D.V., Eds), Vol. 1, Springer, Berlin (2015) 215–240.

[20]   AHLOOWALIA, B.S., MALUSZYNSKI, M., NICHTERLEIN, K., Global impact of mutation-derived varieties, Euphytica **135** 2 (2004) 187–204.

[21]   SARSU, F., FORSTER, B.,. Report of the FAO/IAEA coordinated research project on climate proofing crops: Genetic improvement for adaptation to high temperatures in drought-prone areas and beyond, Austral. J. Crop Sci. **15** 9 (2021) 5–10.

[22]   EL-DEGWY, I.S., Mutation induced genetic variability in rice (*Oryza sativa* L.), Int. J. Agric. Crop Sci. **5** 23 (2013) 2789–2794.

[23]   KUROWSKA, M., et al., TILLING — A shortcut in functional genomics, J. Appl. Genetics **52** (2011) 371–390.

[24]   KUMARI, S., SINGH, P., SINGLA-PAREEK, S.L., PAREEK, A., Heterologous expression of a salinity and developmentally regulated rice cyclophilin gene (OsCyp2) in *E. coli* and *S. cerevisiae* confers tolerance towards multiple abiotic stresses, Mol. Biotechnol. **42** (2009) 195–204.

[25]   LAKRA, N., KAUR, C., ANWAR, K., SINGLA-PAREEK, S.L., PAREEK, A., Proteomics of contrasting rice genotypes: Identification of potential targets for raising crops for saline environment, Plant Cell Environ. **41** 5 (2018) 947–969.

[26]   JAGADISH, S.V.K., CRAUFURD, P.Q., WHEELER, T.R., Phenotyping parents of mapping populations of rice for heat tolerance during anthesis, Crop Sci. **48** 3 (2008) 1140–1146.

[27]   LACHAGARI, V.R., et al., Uncovering genome wide novel allelic variants for eating and cooking quality in a popular Indian rice cultivar, Samba Mahsuri, Curr. Plant Biol. **18** (2019) 100111.

[28]   PRASAD, P.V.V., BOOTE, K.J., ALLEN Jr., L.H., SHEEHY, J.E., THOMAS, J.M.G., Species, ecotype and cultivar differences in spikelet fertility and harvest index of rice in response to high temperature stress, Field Crops Res. **95** 2–3 (2006) 398– 411.

[29]   FERNANDEZ, G.C., "Effective selection criteria for assessing plant stress tolerance", Adaptation of Vegetables and other Food Crops in Temperature and Water Stress (Proc. Int. Symp. Shanhua, 1992), Asian Vegetable Research and Development Center, Taipei (1992) 257–270.

[30]   VENUPRASAD, R., LAFITTE, H.R., ATLIN, G.N., Response to direct selection for grain yield under drought stress in rice, Crop Sci. **47** 1 (2007) 285–293.

[31]   VENUPRASAD, R., et al., Response to two cycles of divergent selection for grain yield under drought stress in four rice breeding populations, Field Crops Res. **107**(3) (2008) 232–-44

[32]   ESPINOZA-CORRAL, R., SCHWENKERT, S., SCHNEIDER, A., Characterization of the preferred cation cofactors of chloroplast protein kinases in *Arabidopsis thaliana*, FEBS Open Bio. **13** 3 (2023) 511–518.

[33]   WU, J.L., Chemical- and irradiation- induced mutants of indica rice IR64 for forward and reverse genetics, Plant Mol. Biol. **59** 1 (2005) 85.

[34]   MELANDRI, G., et al., Biomarkers for grain yield stability in rice under drought stress, J. Exp. Bot. **71** 2 (2020) 669–83

[35]   KADAM, N.N., STRUIK, P.C., REBOLLEDO, M.C., YIN, X, JAGADISH, S.K., Genome-wide association reveals novel genomic loci controlling rice grain yield and its component traits under water-deficit stress during the reproductive stage, J. Exper. Botany **69** 16 (2018) 4017–4032.

[36]   BARNABAS, B., JAGER, K., FEHER, A., The effect of drought and heat stress on reproductive processes in cereals, Plant Cell Environ. **31** (2008) 11–38.

[37]   HU, Y., Silencing of OsGRXS17 in rice improves drought stress tolerance by modulating ROS accumulation and stomatal closure, Sci. Rep. **7** 1 (2017) 15950.

[38]   TURNER, T.R., et al., Comparative metatranscriptomics reveals kingdom level changes in the rhizosphere microbiome of plants, ISME J. **7** 12 (2018) 2248–2258.

[39]   SARSU, F., Pre-Field Screening Protocols for Heat-Tolerant Mutants in Rice, Springer Nature, Cham (2018).

[40]   VENUPRASAD, R., et al., Identification and characterization of large-effect quantitative trait loci for grain yield under lowland drought stress in rice using bulk-segregant analysis, Theor. Appl. Genetics **120** 1 (2009) 177–190.

[41]   VIKRAM, P., qDTY 1.1, a major QTL for rice grain yield under reproductive-stage drought stress with a consistent effect in multiple elite genetic backgrounds, BMC Genetics **12** 1 (2011) 1–5.

2-4. MUTATION BREEDING AND PRE-FLOWERING STAGE FIELD SCREENING FOR DROUGHT TOLERANCE IN UPLAND RICE

A. KUSUMA DEWI, W.M. INDRIATAMA
Research Center for Radiation Process Technology,
National Research and Innovation Agency, Jakarta, Indonesia

REFLINUR
Research Center for Genetic Engineering,
National Research and Innovation Agency, Jakarta, Indonesia

Abstract

A simple protocol is presented for field screening drought tolerance in upland rice at the pre-flowering stage. The method is based on a plastic house (rainout shelter) field test in which drought is imposed during the critical period for rice in the reproductive phase, which is 16 days before flowering and 10 days after flowering. The experiment was an augmented block design with control plants (parent and check varieties) in each block. The plots were irrigated only when susceptibility check showed leaf rolling with a score of 9 (as per the International Rice Research Institute) and the groundwater potential reached more than −35 kPa at 30 cm soil depth. Information on the responses of standard genotypes (parent, tolerant and susceptible) is given to which test plants are compared. After this phase, the plants are watered again. Enough water should be applied to saturate the root zone.

Key words: Drought, upland rice, pre-flowering stage.

1. INTRODUCTION

For more than half of the world's population, especially in Asia, rice (*Oryza sativa* L.) is the most widely consumed staple food. According to the Food and Agriculture Organization of the United Nations (FAO), it is estimated that over 50% of world rice production is contributed by India, China and Indonesia, and rice is one of the most rapidly growing food crops in terms of production and consumption. In recent years, Asia has been at the top in terms of production and consumption of rice. In the future, the production of rice must be doubled to feed the rapidly growing human population. However, this projected increase in rice production is being restrained by several biotic and abiotic stresses [1]. The FAO states that the average production of rice is estimated to be 5.0×10^8 t, and due to the rise in population, the requirement is expected to increase up to 2.0×10^9 t by 2030. The world population is expected to rise to 10 billion by 2050; 763 million tonnes of rice were required in 2020, while 852 million tonnes will be required by 2035 [2].

In recent years, agricultural areas worldwide have been affected by drought. Droughts are slow onset events resulting from prolonged periods of deficient rainfall lasting from a few weeks to several years and, thus, are difficult to characterize and manage. Drought frequency, severity and magnitude have increased in the South-East Asia region, often triggered by El Niño events, particularly over the past two decades. As a slow onset disaster, drought can have devastating impacts on agriculture, environment, socioeconomics and livelihoods, as well as degrading land and increasing the prospects of violent conflict [3]. In the ASEAN region, drought has gained attention in recent years due to its increasingly frequent occurrence, magnitude and severity.

The ASEAN region is susceptible to droughts and there will be many more dry years ahead, with more parts of the region exposed to extreme drought conditions [4].

As projected by global climate models, recurrent drought events under future climate change may result in increased losses when droughts overlap with sensitive stages of crop growth [5]. Drought has remained one of the most prominent and persistent environmental issues, severely affecting plant growth, development and yield. The Global Drought Information System reveals that drought is becoming progressively severe and intense across the globe [6]. From the agricultural aspect, drought is a time span with low average precipitation/poor rain or higher evaporation rates, causing a reduction in crop growth and yield [7]. The intensity/severity of drought is very complex and is dependent on different reasons, such as frequency of rainfall, evaporation and soil moisture [8, 9]. More than one third of the world's total cultivated area is affected by drought stress. In Asia alone, about 3.4×10^7 hm^2 of rainfed lowland and 8.0×10^6 hm^2 of upland rice is exposed to drought stress [10].

In South Asia, the worst droughts in the future are likely to be more intense and widespread. Droughts over South Asia are mainly driven by the failure of the summer monsoon (June–September), which is a lifeline for the millions of people in the region and provides about 80% of the total annual rainfall [11]. In Indonesia, droughts have severe consequences for rice cultivation, as these occur annually and increase during El Niño phenomena. Indonesia, with a tropical climate, is highly sensitive to the climate anomaly El-Nino Southern Oscillation (ENSO), which is a source of drought. In addition to El Nino, the drought in Indonesia is also influenced by a positive Indian Ocean Dipole (IOD), which is a regional climate phenomenon in the Indian Ocean [12].

At present, rice (*Oryza sativa* L.) is one of the prominent cereal crops serving more than three billion people. Modelling simulations by FAO estimate that agricultural production will need to double by 2050 to sustain the growing population. Among abiotic stress, drought is the most imperative and major limitation for rice production in rainfed ecosystems [13, 14]. Drought and the limited availability of water serve as a serious limitation for rice production in rainfed ecosystems. About 34 million ha of rainfed lowland and 8 million ha of upland rice in Asia are being affected every year from drought stress of varying intensities [10]. Rice is highly susceptible to drought due to shallow rooting behaviour [10]. The production of rice is highly water intensive and is, therefore, grown under flooded conditions [15]. Rice cultivation consumes 24–30% of the world's available fresh water [16]. Therefore, a water deficit in the form of drought can result in huge yield losses of rice. Almost 18 million tonnes of rice yield are lost globally per year due to drought [17]. Breeding rice varieties with tolerance to drought stress offers an economically viable and sustainable option to improve rice productivity [18].

Drought is induced by the absence of water due to irregular rainfalls or insufficient irrigation, but it can be caused by other factors such as soil salinity and physical properties and high air or soil temperature. Likewise, drought is insufficiency of water availability, including precipitation and soil moisture storage capacity, in quantity and supply through the life cycle of a crop to restrict the maximum genetic grain yield possibility. Drought stress on plants occurs when the available water lags continuous plant loss of water by transpiration [14]. When water stress occurs, plants react by slowing down or stopping their growth. This is a normal plant reaction to lack of water and it acts as a survival technique [19]. Drought resistance is the capability of a plant to produce its maximum economic yield under water limited conditions [7]. It is a complex trait that depends on the action and interaction of different morphological, biochemical and physiological responses [20]. Drought resistance can be classified into four types based on plant responses to drought stress: drought avoidance; drought tolerance; drought recovery; and drought escape [21].

Further, drought mitigation, through the development of drought resistant varieties with higher yields suitable for water limiting environments, will be the key factor to improve stable plant crop production. Efforts to develop drought tolerant and high yielding varieties require a good knowledge of the physiological mechanisms, yield under drought related components and the genetic control of traits contributing to drought resistance. Another important factor to consider is improving secondary traits such as root architecture, leaf water potential, panicle water potential, osmotic adjustment and relative water content [22].

2. FIELD SCREENING FOR DROUGHT TOLERANCE IN UPLAND RICE AT THE REPRODUCTIVE STAGE (PRE- AND POST-FLOWERING PHASE)

Crop growth and production are strongly affected by abiotic and biotic stresses [21]. When crops are subjected to drought stress, numerous changes will occur at the physiological, metabolic and molecular levels in comparison with crops grown under non-stressed conditions [23, 24]. Of the rice cultivated around the world, approximately 27 million ha of rice are grown in upland rather than paddy fields, and is subject to drought stress. Drought is an abiotic stress that can drastically decrease grain yields [25–27], especially in rainfed ecosystems. Further, climate change is predicted to increase the frequency and severity of drought, which will likely result in increasingly serious constraints on rice production worldwide [28]. To counter this stress, it is desirable to breed new rice cultivars with improved drought tolerance trait. For breeding purposes, especially for upland rice breeding programmes, it is desirable to develop simple and accurate protocols to evaluate rice drought tolerance in the field.

Rice (*Oryza sativa* L.) growth and development are sensitive to water limited conditions due to the lower ability of taking up resources compared with other crops. Drought is one of the most severe climate related risks for rice production, affecting more than 23 million ha of rainfed areas in South East Asia [29], and the occurrence of drought at any developmental stage of rice can cause significant yield loss in those areas [30].Yield losses caused by drought stress are estimated to be 34–53% under moderate drought stress and up to 65–88% under severe drought stress when compared with irrigated lowland rice [31].

Drought can occur at any stage of the rice crop in any year in rainfed areas [32] and the response of rice plants to drought varies [33], depending on the severity and stage of drought. Generally, modern rice varieties are highly sensitive to drought stress at seedling, vegetative and reproductive stages, and even mild drought stress can result in a significant yield reduction in rice [34]. At the seedling stage, drought affects crop establishment and seedling survival rates. At the vegetative stage, drought reduces leaf formation and tillering, which subsequently reduces the development of panicles per plant, thus causing a yield loss. However, drought at the vegetative stage was earlier predicted to have a relatively small effect on grain yield in rice [35]. Drought in the reproductive stage is more severe and many changes involve morphological [36], physiological and biochemical [23], as well as various agronomic changes [37, 38]. At the reproductive stage, drought causes a reduction in the number of grains per panicle, increases grain sterility and reduces grain weight [39]. Drought scoring is used as a primary criterion for screening rice genotypes for drought tolerance at later growth stages or under field conditions [40].

3. METHODOLOGY

Two upland rice varieties, Towuti and Situguntung, were irradiated with gamma rays at doses of 200 Gy and 300 Gy, respectively. The purpose was to develop drought tolerant varieties in upland and rainfed rice ecosystems. Situgintung is a mutant upland rice variety and has low yield potential, and the level of tolerance to drought stress is unknown. Towuti is an upland rice variety with medium yield potential; the level of drought tolerance is also unknown.

A total of 1482 rice mutant lines, derived from 114 genotypes of 200 Gy gamma irradiation of the Towuti variety, and 478 rice genotypes as parent, tolerant and susceptible check plants, respectively, were screened for evaluation of drought tolerance at the M_3 generation. A total of 1298 rice mutant lines derived from 72 genotypes of 200 Gy gamma irradiation of the Situgintung variety and 600 rice genotypes as parent, tolerant and susceptible check plants, respectively, were screened for evaluation of drought tolerance at the M_3 generation. Each of the two varieties of drought tolerant and susceptible checks and the parent plant were used in this drought screening, as described below. The seeds used for this screening were M_3 seeds from the selection based on agronomic characters in the M_2 population in the field:

— Parent check: Towuti variety, Situgintung variety, Indonesian upland variety;
— Positive check: International drought tolerant genotype: Salumpikit;
— Positive check: National drought tolerant variety: Limboto;
— Negative check: International drought susceptible genotype: IR 20;
— Negative check: National drought susceptible genotype: IR 64.

3.1. Field experiments

Rice mutant lines were screened in the greenhouse under controlled environmental conditions. The green house is covered by 14% UV plastic with a thickness of 0.2 mm. The experiment was an augmented block design with control plants (parent and check varieties) in each block. Standard rice management practices were applied to the land with respect to soil, fertilizer and watering until the application of drought stress. There were 13 plants of each genotype in each row with row to row and plant to plant spacing of 20 cm × 20 cm. Nitrogen, phosphorus, and potassium (NPK) are applied according to upland rice cultivation standards. Phosphorus and potassium were applied as basal, and nitrogen was applied in three splits, the first as basal, the second at maximum tillering and the third at panicle initiation. A soil tensiometer at 30 cm depth was placed on each block used to measure groundwater potential.

3.2. Drought stress imposition

Drought treatment was applied to the critical period of rice in the reproductive phase, which was 16 days before flowering to 10 days after flowering (Figs 1 and 2). Up to 45 days after sowing (DAS), the trials were irrigated by sprinkler every day during establishment and early vegetative growth. Stress was initiated after this period (54–57 days after sowing with an estimated average harvest time of all plants screened of 110 days) by withholding irrigation. The plots were irrigated only when susceptibility check showed leaf rolling with a score of 9 [41, 42] and the groundwater potential reached more than –35 kPa at 30 cm soil depth. After this phase, the plants were watered again. Enough water should be applied to saturate the root zone. This was likely to require 40 mm of water. The irrigated control received the same cultural practices as the stress trials, except that irrigation was continued every day up to 10 days before harvest.

The experiment was accomplished using two types of stress imposition: Towuti mutant lines with the cycle of stress and irrigation repeated until harvest; and the Situgintung mutant lines, which had only one stress cycle, and for which irrigation was carried out until harvest.

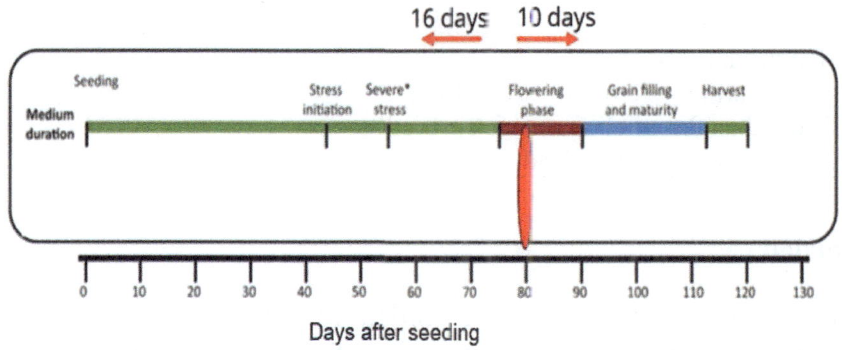

FIG. 1. Outline of the drought treatment schedule for direct seeded upland cultivar.

FIG. 2. Drought screening stages. (A) Field preparation; (B) seedling stage, 14 DAS; (C) vegetative stage, 33 DAS; (D) control plants; (E) drought treatment; (F) installation of soil tensiometer at 30 cm depth.

TABLE 1. SCALE OF LEAF ROLLING BASED ON STANDARD EVALUATION SYSTEM FOR RICE BY THE IRRI [40]

Scale	Symptoms	Categorize
0	Leaves healthy	Very tolerant
1	Leaves start to fold (shallow)	Tolerant
3	Leaves folding (deep V shape)	Moderately tolerant
5	Leaves fully cupped (U shape)	Moderately susceptible
7	Leaf margins touching (0 shape)	Susceptible
9	Leaves tightly rolled V shape)	Very susceptible

TABLE 2. SCALE OF LEAF DRYING BASED ON STANDARD
EVALUATION SYSTEM FOR RICE FROM THE INTERNATIONAL
RICE RESEARCH INSTITUTE [41, 42]

Scale	Symptoms
0	No symptoms
1	Slight tip drying
3	Tip drying extended up to ¼
5	One-fourth to 1/2 of all leaves dried
7	More than 2/3 of all leaves fully dried
9	All plants apparently dead, most leaves fully dried

3.3. Data recorded

The morpho-physiological traits data were recorded. Morphological traits: Plant height (cm);
tillering number; panicle length (cm). The number of filled and empty grains was measured
from five random plants of each plot and then the mean was calculated. Physiological traits:
Leaf rolling index (scores); leaf drying index (scores); chlorophyll content (SPAD); and soil
moisture. Data on daily rainfall, relative humidity, and maximum and minimum temperature
were also recorded (Tables 1 and 2).

The rice mutation breeding scheme for drought tolerance is shown in Fig. 3.

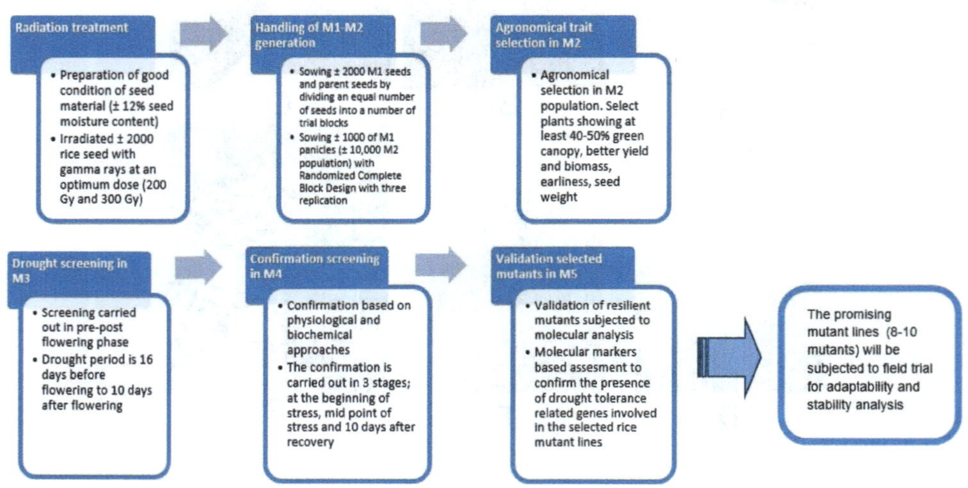

FIG. 3. Schematic of mutation breeding for drought tolerance.

4. RESULTS

4.1. Assessment of soil moisture content, air temperature and humidity

The average temperatures and humidity during drought stress treatment under plastic houses in
2020 are represented in Fig. 4. Data show that the temperature in the plastic house is quite high
during the drought stress in the pre- and post-flowering phase, especially from 10:00 to 13.00.
This condition is also supported by the average temperature around the experimental area
(control treatment), which is 28–29°C. The stresses obtained by plants are not only from the
drought stress treatment given, but also from heat stress from the plastic housing itself. Based
on the estimated average flowering time of all plants, i.e. mutant lines, parent and check
varieties, the drought was imposed in the third week of August 2020 and the first watering was

after 23 days of exposure to dryness and the susceptible check had shown a score of 9 (second week of September 2020).

The weather conditions during the trial season in the experiment area showed that the maximum temperature during the experimental period (June–December 2020) was more than 30°C, with an average temperature of 28°–29°C and low rainfall.

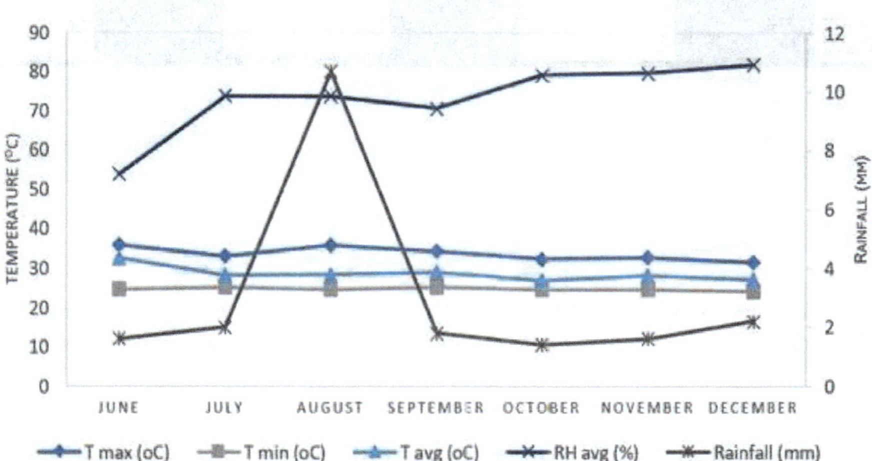

FIG. 4. Weather condition at control treatment and around the experimental area, June–December 2020. Tmax: maximum temperature; Tmin: minimum temperature; Tavg: average temperature; RH: relative humidity.

During the experiment, the soil water potential was measured using a neutron probe at a depth of 25 cm and 50 cm from the ground surface (Fig. 5). Measurement of groundwater content was carried out twice, i.e. 13th and 23th day during the stress treatment. Soil water potential also was confirmed through tensiometer reading. On the 23th day after stopping irrigation, the soil water potential at the 30 cm depth reached −35 kPa.

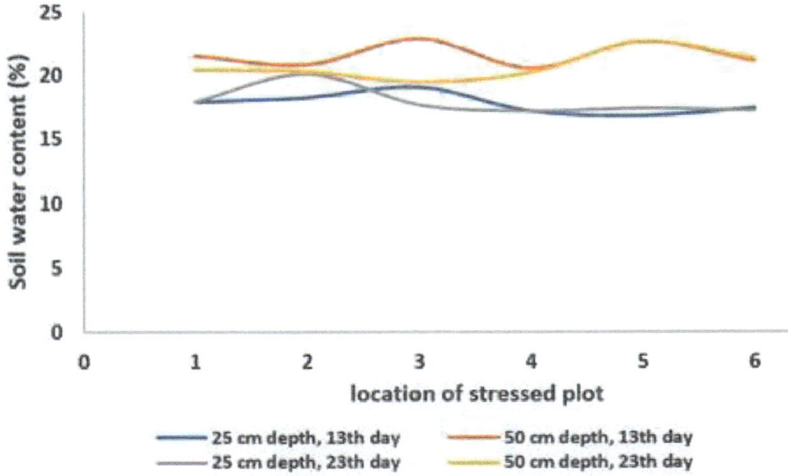

FIG. 5. Data of soil water content measurement with a neutron probe.

4.2. Assessment of drought tolerance at the reproductive stage

Plant performance during drought stress is shown in Fig. 6.

FIG. 6. From left: Plant performance at the seventh day of drought imposition; plant performance at the 24th day of drought imposition; plant performance at first watering; plant performance one day after watering.

According to our observations, at the 7th day of drought imposition, in general, the appearance of the plants was still good, but had begun to show a response to stress, especially in the check negative plants that are susceptible to drought (IR-20 and IR-64), where leaf rolling was found. In this experiment, the leaf rolling and drying was observed and scored. On the 24th day of drought imposition, positive check plants gave a response to rolling and drying of leaves at scores of 1 and 3, while negative check plants showed scores of 7 and 9. For mutant plants, scores varied from 1 to 9, while the parent check plants showed scores in the range of 5 and 7.

The following is the chlorophyll data of the Towuti mutant lines (Fig. 7) with drought stress during pre- and post-flowering, with the cycle of stress and irrigation repeated until harvest. The chlorophyll content of the mutant lines showed a higher value than the parent and both negative check varieties.

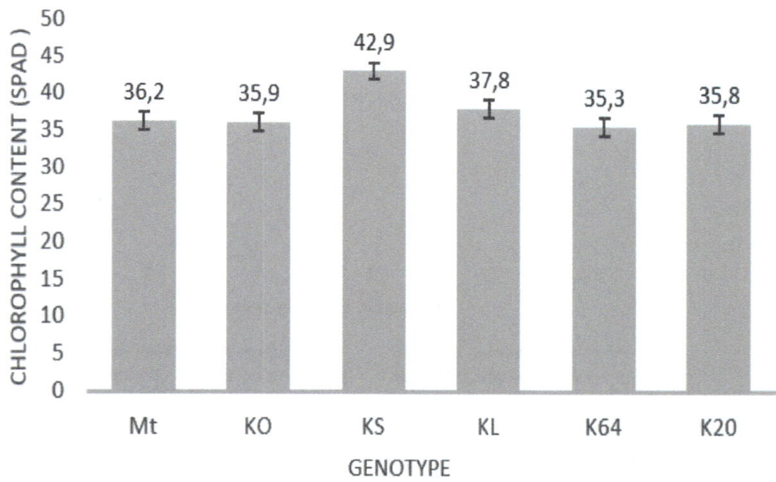

FIG. 7. Chlorophyll data of Towuti mutant lines.

Based on correlation, path and PCA analysis, yield component traits that can be used to select mutant lines with the best performance during drought stress include: panicle length; panicle density; and grain weight. Yield and yield components data of selected Situgintung mutant lines, parent and check varieties under drought treatment during pre-post flowering phase are represented in Fig. 8. Based on the three character traits of these yield components, 20 Situgintung mutant lines were selected (Fig. 9), for validation of their drought tolerance in the M_4 generation.

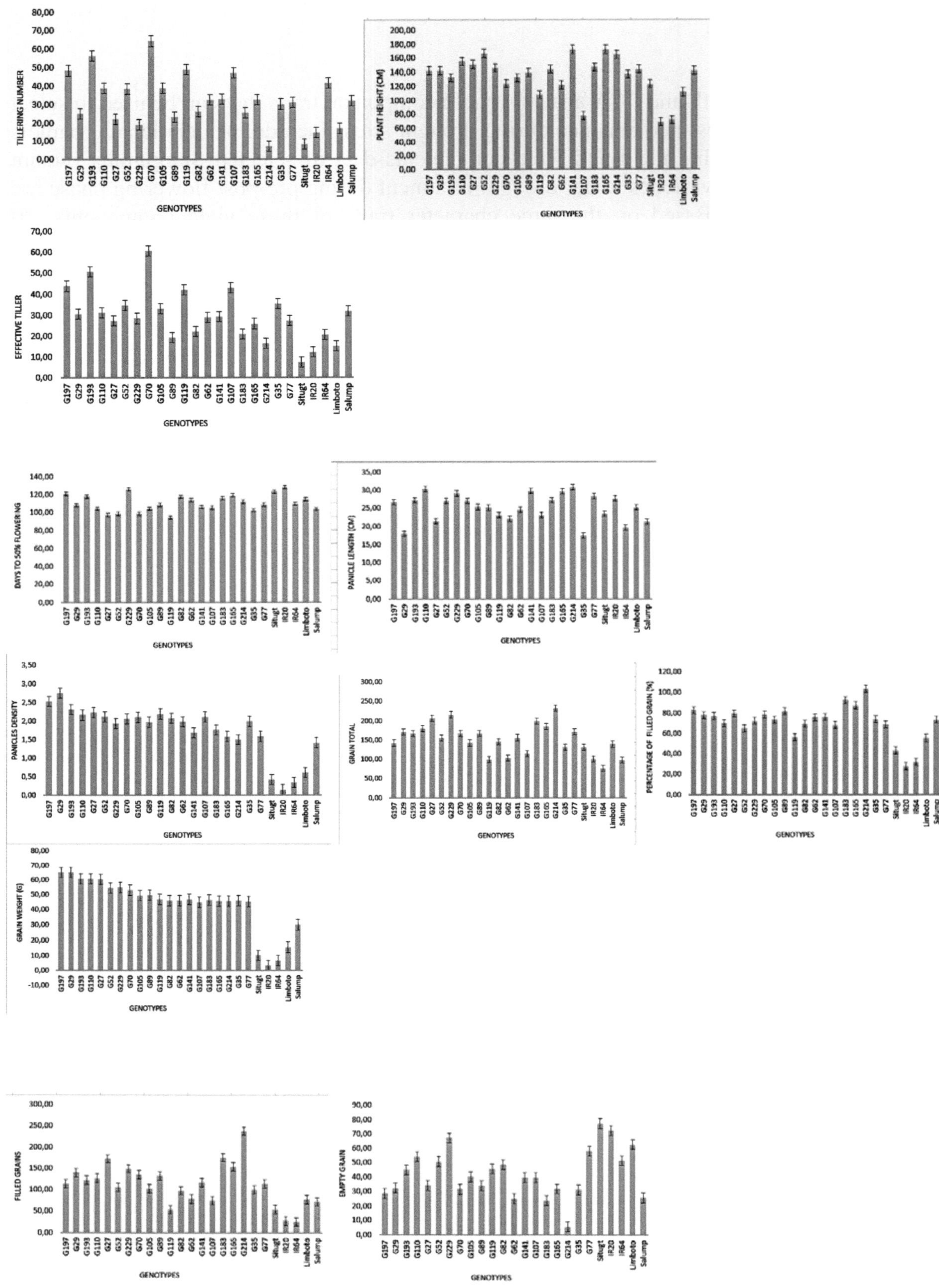

FIG. 8. *Yield and yield component data of the Situgintung mutant lines, parent and check varieties after drought screening in the M₃ generation.*

A: parent check, B: mutant plants, C: negative check
Every three rows of mutant plants there is one row of check plants

Plant performance under drought screening in plastic house.
A: Check plant, B: mutant plants

A: Check plant, B: mutant plants. Circle marks are mutant plants that are able
to flower during drought stress

FIG. 9. Mutant performance under drought stress.

5. CONCLUSIONS

The treatment of drought stress in the pre- and post-flowering phase with the cycle of stress and irrigation was repeated until harvest showed more severe plants and lower grain yields than one drought stress cycle and irrigation was carried out until harvest. In total, 24 Towuti mutant lines and 20 Situgintung mutant lines were selected for validation drought screening in the M_4 generation.

REFERENCES TO CHAPTER 2–4

[1] SAMAL, R., et al., Morphological and molecular dissection of wild rices from eastern India suggests distinct speciation between *O. rufipogon* and *O. nivara* populations, Sci. Rep. **8** 1 (2018) 2773.

[2] GANIE, S.A., GOLAM, J.S., Dynamics of cell wall structure and related genomic resources for drought tolerance in rice, Plant Cell Rep. **40** (2021) 437–459 https://doi.org/10.1007/s00299-020-02649-2.

[3] ZHANG, H., LI, Y.Y., ZHU, J.K., Developing naturally stress resistant crops for a sustainable agriculture, Nat. Plants **4** 12 (2018) 989–996.

[4] UNITED NATIONS, Ready for the Dry Years: Building resilience to drought in South-East Asia. A Joint Study Between ASEAN and ESCAP, 1st edn, UN, New York (2020); https://www.unwater.org/building-resilience-to-drought-insouth-east-asia/

[5] ASSOCIATION OF SOUTHEAST ASIAN NATIONS, ASEAN Regional Plan of Action for Adaptation to Drought 2021–2025, ASEAN, Jakarta (2021); www.asean.org

[6] KUMAR, A., et al., Breeding high-yielding drought-tolerant rice. Genetic variations and conventional and molecular approaches, J. Exp. Bot. **65** (2014) 6265–6278.

[7] NAUMANN, G., et al., Global changes in drought conditions under different levels of warming, Geophys. Res. Lett. **45** (2018) 3285–3296.

[8] ROLLINS, J.A., et al., Leaf proteome alterations in the context of physiological and morphological responses to drought and heat stress in barley (*Hordeum vulgare* L.), J. Exp. Bot. **64** 11 (2013) 3201–3212.

[9] OLADOSU, Y., Drought resistance in rice from conventional to molecular breeding: A review, Int. J. Mol. Sci. **20** 14 (2019) 3519.

[10] PANDA, D., SWATI, S.M., PRAFULLA, K.M., Drought tolerance in rice: Focus on recent mechanisms and approaches, Rice Sci. **28** 2 (2021) 119–132.

[11] SINGH, R., et al., From QTL to variety-harnessing the benefits of QTLs for drought, flood and salt tolerance in mega rice varieties of India through a multi-institutional network, Plant Sci. **242** (2016) 278–287.

[12] AADHAR, S., VIMAL, M., On the occurrence of the worst drought in South Asia in the observed and future climate, Environ. Res. Lett. **16** (2021) 024–050; https://doi.org/10.1088/1748-9326/abd6a6

[13] MURSIDI, A., DEFFI, A.P.S., Management of drought disaster in Indonesia, J. Terapan Manajemen dan Bisnis **3** 2 (2017) 165–171.

[14] PANDEY, V., SHUKLA, A., Acclimation and tolerance strategies of rice under drought stress, Rice Sci. **22** 4 (2015) 147–161.

[15] XU, G.W., LU, D.K., WANG, H.Z., LI, Y., Morphological and physiological traits of rice roots and their relationships to yield and nitrogen utilization as influenced by irrigation regime and nitrogen rate, Agric. Water Manage. **203** (2018) 385–394.

[16] XU, J., HENRY, A., SREENIVASULU, N., Rice yield formation under high day and night temperatures — A prerequisite to ensure future food security, Plant Cell Environ. **43** (2020) 1595–1608.

[17] BAHUGUNA, R.N., TAMILSELVAN, A., MUTHURAJAN, R., SOLIS, C.A., JAGADISH, S.V.K., Mild preflowering drought priming improves stress defences,

assimilation and sink strength in rice under severe terminal drought, Funct. Plant Biol. **45** (2020) 827–839.

[18]   DHAKREY, R., et al., Physiological and proteomic analysis of the rice mutant *cpm2* suggests a negative regulatory role of jasmonic acid in drought tolerance, Front. Plant Sci. **8** (2017) 1903.

[19]   LOBELL, D.B., et al., Greater sensitivity to drought accompanies maize yield increase in the U.S-Midwest, Science **344** (2014) 516–519.

[20]   BHANDARI, U., et al.. Morpho-physiological and biochemical response of rice (Oryza sativa L.) to drought stress: A review. Heliyon, 9(3), **(2023).** e13744. https://doi.org/10.1016/j.heliyon.2023.e13744.

[21]   KUMAR, A., BASU, S., RAMEGOWDA, V., PEREIRA, A., "Mechanisms of drought tolerance in rice", Achieving Sustainable Cultivation of Rice (SASAKI, T., Ed.), Vol. 11, Burleigh Dodds Science Publishing, London (2016); http://dx.doi.org/10.19103/AS.2106.0003.08

[22]   LUO, L., et al., Water-saving and drought-resistance rice: From the concept to practice and theory, Mol. Breeding **39** (2019) 145.

[23]   SHANKER, A.K., et al., Drought stress responses in crops, Funct. Integr. Genomics **14** (2014) 11–22.

[24]   GILL, S.S., TUTEJA, N., Reactive oxygen species and antioxidant machinery in abiotic stress tolerance in crop plants, Plant Physiol. Biochem. **48** (2010) 909–930.

[25]   FANG, Y.J., XIONG, L.Z., General mechanisms of drought response and their application in drought resistance improvement in plants, Cell. Mol. Life Sci. **72** (2015) 673–689.

[26]   AHADIYAT, Y.R., HIDAYAT, P., SUSANTO, U., Drought tolerance, phosphorus efficiency and yield characters of upland rice lines, Emir. J. Food Agric. **26** (2014) 25– 34.

[27]   MAISURA, M.A., CHOZIN, I., LUBIS, A., JUNAEDI, H., EHARA, Some physiological character responses of rice under drought conditions in a paddy system, J. Int. Soc. Southeast Asian Agric. Sci. **20** (2014) 104–114.

[28]   SWAPNA, S., SHYLARAJ, K.S., Screening for osmotic stress responses in rice varieties under drought condition, Rice Sci. **24** (2017) 253–263.

[29]   WASSMANN, R., et al., Regional vulnerability of climate change impacts on Asian rice production and scope for adaptation, Adv. Agron. **102** (2009) 91–133.

[30]   INTERNATIONAL RICE RESEARCH INSTITUTE, Climate-Smart Rice, IRRI, Manila (2019); https://www.irri.org/climate-smart-rice (accessed 26 September 2019).

[31]   KUMAR, S., et al., Anatomical, agro-morphological and physiological changes in rice under cumulative and stage specific drought conditions prevailing in eastern region of India, Field Crop Res. **245** (2020) 107658. doi: 10.1016/j.fcr.2019.107658.

[32]   VERULKAR, S.B., et al., Breeding resilient and productive genotypes adapted to drought-prone rainfed ecosystem of India, Field Crops Res. **117** (2010) 197–208.

[33]   SWAIN, P., ANITA, R., SINGH, S.P., ARVIND, K., Breeding drought tolerant rice for shallow rainfed ecosystem of eastern India, **209** (2017) 168–178.

[34]   JI, K., et al., Drought-responsive mechanisms in rice genotypes with contrasting drought tolerance during reproductive stage, J. Plant Physiol. **169** 4 (2012) 336–344.

[35]   TORRES, R.O., HENRY, A., Yield stability of selected rice breeding lines and donors across conditions of mild to moderately severe drought stress, Field Crops Res. **220** (2016).

[36]   HASSAN, M.A., et al., Drought stress in rice: Morpho-physiological and molecular responses and marker-assisted breeding, Front. Plant Sci. **14** (2023) 1215371; doi: 10.3389/fpls.2023.1215371.

[37]   SAUD, S, et al., Comprehensive impacts of climate change on rice production and adaptive strategies in China, Front. Microbiol. **13** (2022) 926059; doi: 10.3389/fmicb.2022.926059.

[38]   BHATTARAI, U., SUBUDHI, P.K., Genetic analysis of yield and agronomic traits under reproductive-stage drought stress in rice using a high-resolution linkage map, Gene **669** (2018) 69–76; doi: 10.1016/j.gene.2018.05.086.

[39]   ZHANG, J., et al., Effect of drought on agronomic traits of rice and wheat: A meta-analysis, Int. J. Environ. Res. Public Health **15** 5 (2018); doi: 10.3390/ijerph15050839.

[40]   AKRAM, H.M., ALI, A., SATTAR, A., REHMAN, H., BIBI, A., Impact of water deficit stress on various physiological and agronomic traits of three Basmati rice (*Oryza sativa* L.) cultivars, J. Anim. Plant Sci. **23** 5 (2013) 1415–1423.

[41]   INTERNATIONAL RICE RESEARCH INSTITUTE, Standard Evaluation System for Rice. 5th edn, IRRI, Los Baños (2014).

[42]   INTERNATIONAL RICE RESEARCH INSTITUTE, Standard Evaluation System for Rice, 5th edn, IRRI, Los Baños (2002).

2–5. FIELD SCREENING PROTOCOL OF REPRODUCTIVE DROUGHT TOLERANT UPLAND RICE MUTANTS

N. TEME, A. YEBEDIÉ, K. THÉRA, Md. L. TEKETÉ,
F. KONATÉ, A. MAÏGA, A. SISSOKO, S. KAMISSOKO,
S. SANOGO, M. KOURESSY

Institut d'Economie Rurale,
Bamako, Mali

Abstract

Identification of drought resilient and productive upland rice mutants is essential for sustainable agriculture when a validated drought tolerant screening protocol is available at the breeder site. Reproductive stage drought is a major constraint that compromises upland rice yield. Mutants released through this validated screening protocol, can sustainably alleviate producer worries at the end of the cropping season. To achieve this goal, NERICA8 derived M_4 mutants, obtained after gamma ray irradiation, were subjected to control and water stress conditions on-station in 2020. The objective was to develop an efficient reproductive stage, drought resistant screening protocol to identify mutants combining drought tolerance and yield potential. A completely randomized block design with three repetitions was used for each environment under control and water stress conditions in the 2020 cropping season at Sotuba research station in Mali. Ten morphological and five physiological traits and eight drought tolerance indices were analysed. Linear correlations were established to weight relationships between yields and its contributing parameters. Multivariate analysis was performed for similarity among mutants and drought tolerant checks and among mutants themselves. Yield gaps were higher among mutants under control than under water stress conditions. Mutants showed differences for most of their traits in a single environment and across environments. Irrigation was significant for most traits and showed less interaction with mutants in paddy yield. Higher, average and lower yielding mutants compared to the parental line and NERICA4 were identified. Drought susceptible and tolerant mutants were also identified. Drought tolerance indices identified less productive and stable mutants than higher yielding ones under control condition. There were mutants closer to NERICA4 and Moroberekan stipulating that these mutants are drought resistance due to their closer similarity with these two reproductive drought resistant checks. The similarity between mutants depended mainly on yield potential and sensitivity to yield loss levels. Data from morphological, physiological and drought tolerance indices across environments are solid assets to validate this protocol developed under Sotuba conditions for future utilization by breeders.

Key words: mutation, irrigation, drought, resilience, index, identification.

1.   INTRODUCTION

One of the major abiotic stresses influencing crop productivity is water deficit [1]. It affects approximately 27 million ha of upland rice worldwide [2]. Drought is a daily concern in southern Mali for farmers where upland rice is mainly grown due to favourable climatic conditions. Breeding strategies were developed by breeders to mitigate reproductive stage drought stress through desirable new rice lines. Recently, New Rice for Africa (NERICA) has attracted major attention from producers supported by policy makers in regions where climatic conditions are favourable for upland rice in Mali. NERICA 8, one of the popular upland rice types was irradiated to identify mutants resistant to reproductive drought.

Drought manifestation at a given genotype varies from avoidance, escape and tolerance. Avoidance is the ability of plants to maintain normal physiological processes under moisture stress conditions by maintaining high water potential. Drought stress is exhibited by higher temperature, lower water potential and reduced soil moisture (Fig. 1). Plants may reduce water loss by rapid stomatal closure, leaf rolling, increased wax accumulation on the leaf and stem surface in rice, sorghum and maize and enhancing water uptake by profuse root system. An escape mechanism is a plant's response to drought stress by adjusting its crop cycle before the onset of severe moisture stress.

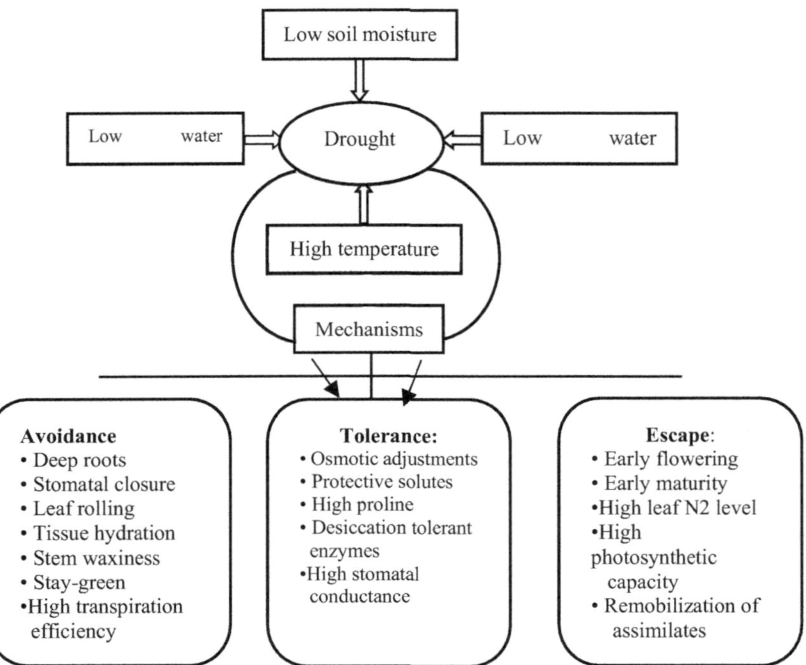

FIG. 1. Drought is manifested by high temperature, low water potential and reduced availability of soil moisture during post flowering phase.

Depending on availability and moisture levels in the soil regime, reproductive growth is completed with small sized panicle. This is also accompanied by remobilization of leaf nitrogen to maintain adequate carbohydrate. Tolerance to drought is the plant's ability to sustain physiological activities under moisture stress conditions. It can withstand stress conditions through osmotic adjustment and production of proteins, such as proline. Plants invariably reduce stomatal conductance to save water and thus may lead to a reduced rate of CO_2 fixation and growth. Improving osmotic adjustments, increasing cell wall elasticity to maintain tissue turgidity and altering metabolic pathways are the physiological mechanisms adapted by plants under severe drought stress conditions [3].

To this end, upland rice mutants were used to develop drought tolerant screening protocols resulting from contrasting water regimes for reproductive drought spells. The specific objectives were to: (1) determine mutant's response to grain yield under contrasting water environments; and (2) identify the most suitable drought tolerant mutants.

2.   DROUGHT INDUCED EFFECTS

Drought induced effects have several implications for various plant combined functions on morphological, biochemical and molecular characters [4]. To achieve a full understanding of drought-response mechanism in rice and to produce rice with improved drought tolerance, the two authors reviewed several rice studies on drought and outlined how various traits which affect rice productivity respond to water deficit on morphological, physiological, biochemical, molecular, yield and its associated traits, as well as acclimation and tolerant mechanism of upland rice to drought stress. Drought induced various responses in rice which ultimately affected yield: water use efficiency (WUE); relative water content (RWC); abscisic acid (ABA) [4] detail each of the causes of drought stress.

Morphological changes included reduction in germination, plant height, elongation and expansion growth, plant biomass, number of tillers, leaf number and size, and increased leaf rolling. At the molecular level, there are altered expressions of genes which encode defence related proteins, protein kinases, transcription factors. The biochemical changes include accumulation of osmoprotectants like proline and sugars, polyamines, and antioxidants. At the physiological level, reduced chlorophyll content, photosystem II activity, photosynthesis, transpiration, stomatal conductance, WUE, RWC, membrane stability, carbon isotope discrimination, ABA content were recorded. The yield attributes include, impaired assimilate translocation, increased spikelet sterility, reduced rate of grain filling, grain size, weight and yield.

3.   REPRODUCTIVE STAGE DROUGHT EFFECTS

Upland rice mutants used in this study completed their life cycle between July and October of 2020. During this time, minimum temperatures ranged from 23.31° to 21.24°C and the maximum ranged from 31.14° to 33.63°C. Sarsu et al. reported that flowering is one of the most susceptible stages in the life cycle of rice, and rice spikelets at anthesis exposed to more than 35°C for 4–5 days induce sterility, with no seed produced [5]. Temperatures above 35°C at flowering cause failure of anther dehiscence, and thus less pollen, resulting in incomplete fertilization in rice. The use of high temperatures (32–36°C) with low (60%) and high (85%) relative humidity recorded high spikelet sterility. Flowering (meiosis, anthesis and fertilization) is considered to be one of the most sensitive stages for temperature stress in rice. The threshold temperature for success in flowering in rice is 33°C. Therefore, the fertility of spikelets at high temperature can be used as a screening tool for high temperature tolerance [5].

4.   DROUGHT TOLERANCE INDICES

Studies on rice resistance to drought have frequently used indirect selection indicators such as morphological and physiological responses [6–10]. Nevertheless, such indirect selection methods are more labour and resource intensive, as many characters need to be evaluated and the selected tolerant genotypes may not necessarily be high yielding. For this reason, selection for drought tolerance based on indices developed from grain yield is considered a more rapid and effective approach to selecting genotypes that combine drought tolerance with general high yield potential.

To date, more than 20 drought tolerance indices (DTIs) have been developed: stress susceptibility index (SSI) [11]; relative drought index (RDI); mean productivity (MP) [12]; drought response index (DRI) [13]; stress tolerance index (STI) and geometric mean productivity (GMP) [14]; relative efficiency index (REI) and mean relative performance (MRP) [15]; modified STI1 (MISTIk1) and modified STI2 (MSTIK2) [16]. Recently drought tolerance indices were approved comprising abiotic tolerance index (ATI), stress susceptibility percentage index (SSPI), and stress/non-stress production index (SNPI) [17]; harmonic mean yield (HARM) [18]; and relative decrease in yield (RDY) [19]. Drought tolerant indices used in the current study are listed in Table 1.

TABLE 1. ESTIMATION OF DTIs DURING THE 2020–2021 CROPPING SEASON AT SOTUBA FOR M$_4$ UPLAND RICE MUTANT REPRODUCTIVE DROUGHT SCREENING

Formula used	Reference	Significance/direction
Stress tolerance: TOL = Yp – Ys	[12]	Lower index means stable in two different environments, water stress and no water stress conditions
Mean productivity: MP = (Yp + Ys)/2	[12]	Higher index means tolerance to drought between water stress and no water stress conditions
Stress Susceptibility Index: SSI = [(1 – (Ys/Yp)]/ SI	[20]	Lower index across stress levels indicate lower differences in yield, or more resistant in drought
Stress Index: SI = 1 – (Ys/Yp)	[20]	
Stress Tolerance Index: STI = (Yp × Ys)/(Yp)^2	[20]	Higher index means tolerance to drought
Geometric mean productivity: GMP = (Yp*Ys)$^{0.5}$	[14]	Higher index means tolerance to drought across water stress and no water stress
Yield Index: YI = Ys/Ys	[21]	Higher index (>1) means tolerance to drought; lower index (<1) susceptible
Yield stability index: YSI = Ys/Yp	[22]	Higher index means tolerance to drought under water stress and no water stress
Per cent action: PYR= (Yp — Ys)/Yp*100	[23]	Lower index is preferred between water and water stress conditions

5. SCREENING PROTOCOL FOR REPRODUCTIVE DROUGHT TOLERANCE FOR UPLAND RICE MUTANTS

5.1. Background on parental line used in the development of mutants

Upland rice is an important crop produced in areas with between 800 and 1200 mm rainfall in Sikasso, south of Kayes, Segou and Koulikoro regions in Mali. Weather stations from these regions indicated one year of drought out of two from 1950 to 2015. The driest years were 1972, 2008 and 2013 [24]. Following this series of droughts, the Malian Government set up a rice initiative to boost and satisfy domestic production mainly targeting rainfed New Rice for Africa (NERICA). This is a promising African upland rice in West Africa which was developed crossing African rice species resistant to disease and drought and high yielding Asian rice species. Kang and Futakuchi [25] claimed that drought stress, as expected, negatively affected all evaluated agronomic traits of upland adapted NERICA lines; in terms of yield loss. All NERICA lines except UN3 showed lower yield (23–78%) under moderate drought stress condition compared with that under the wet control condition. Among the NERICA lines (UN1-UN8), UN1 was found to have the highest yield performance under drought conditions. This means that NERICA8 (UN8) is susceptible to reproductive stage drought. Other studies stipulated that modern rice varieties are highly sensitive to drought stress at seedling, vegetative and reproductive stages and even mild drought stress can result in a significant yield reduction in rice [26, 27]. The reproductive stage is recognized as the most critical stage at which drought stress can cause a high yield reduction [28]. Based on these findings, upland rice NERICA8 was irradiated to identify mutants for Malian drought prone zone cultivation.

Mutants were derived from NERICA8. NERICA is an interspecific hybridization between *Oryza sativa* L. and *O. glaberrima* Steud, which produced several NERICA lines, including NERICA8. This rice, an upland ecosystem, was introduced in 1996 by WARDA (AfricaRice) and is now widely grown in Mali. Two lots of seeds from NERICA8 were irradiated in 2017 at the IAEA Seibersdorf Laboratories, Austria, using gamma ray doses of 250 Gγ and 300 Gγ. Irradiated seeds (M_1) were grown during the 2017–2018 offseason to generate M_2 mutants. Coarse morphological phenotyping was done during the 2018–2019 cropping season on flowering, plant height, tillering, hundred seed weight and panicle seed setting. Selected M_3 mutants were advanced in M_4 generation during the 2019–2020 offseason. One hundred homogeneous M_3 mutants in height with enough M_4 seeds were selected and phenotyped during the normal 2020–2021 cropping season. From M_2 until M_4 generation, single seed descent was applied as selection technique (Table 2).

TABLE 2. LIST OF M_4 MUTANTS EVALUATED FOR REPRODUCTIVE DROUGHT TOLERANCE INDEX STUDIES DURING THE NORMAL CROPPING OF 2020–2021

Mutant No.	Pedigree	Mutant No.	Pedigree
2	NERICA 8_250 Gγ-M4-9-1-1	51	NERICA 8_250 Gγ-M4-33-3-3
3	NERICA 8_250 Gγ-M4-9-1-2	52	NERICA 8_250 Gγ-M4-34-1-1
4	NERICA 8_250 Gγ-M4-9-1-3	53	NERICA 8_250 Gγ-M4-34-1-2
5	NERICA 8_250 Gγ-M4-9-2-1	55	NERICA 8_250 Gγ-M4-34-1-4
7	NERICA 8_250 Gγ-M4-9-2-3	56	NERICA 8_250 Gγ-M4-37-17-1
8	NERICA 8_250 Gγ-M4-11-1-2	57	NERICA 8_250 Gγ-M4-37-17-2
9	NERICA 8_250 Gγ-M4-11-2-1	58	NERICA 8_250 Gγ-M4-37-17-3
10	NERICA 8_250 Gγ-M4-11-2-2	59	NERICA 8_250 Gγ-M4-37-17-4
11	NERICA 8_250 Gγ-M4-11-2-3	61	NERICA 8_250 Gγ-M4-37-18-2
12	NERICA 8_250 Gγ-M4-11-2-4	62	NERICA 8_250 Gγ-M4-37-19-1
13	NERICA 8_250 Gγ-M4-12-1-1	63	NERICA 8_250 Gγ-M4-37-19-2
14	NERICA 8_250 Gγ-M4-12-1-2	64	NERICA 8_250 Gγ-M4-37-19-3
15	NERICA 8_250 Gγ-M4-12-1-3	65	NERICA 8_250 Gγ-M4-37-19-4
16	NERICA 8_250 Gγ-M4-13-1-1	66	NERICA 8_250 Gγ-M4-39-10-1
17	NERICA 8_250 Gγ-M4-13-1-2	67	NERICA 8_250 Gγ-M4-39-10-2
18	NERICA 8_250 Gγ-M4-13-1-3	68	NERICA 8_250 Gγ-M4-85-1-1
19	NERICA 8_250 Gγ-M4-13-1-4	70	NERICA 8_250 Gγ-M4-85-1-3
20	NERICA 8_250 Gγ-M4-15-2-1	71	NERICA 8_250 Gγ-M4-85-2-1
22	NERICA 8_250 Gγ-M4-15-2-4	72	NERICA 8_250 Gγ-M4-85-2-2
23	NERICA 8_250 Gγ-M4-21-2-1	73	NERICA 8_250 Gγ-M4-85-12-1
24	NERICA 8_250 Gγ-M4-21-2-2	74	NERICA 8_250 Gγ-M4-85-12-2
25	NERICA 8_250 Gγ-M4-23-1-1	75	NERICA 8_250 Gγ-M4-85-12-3
26	NERICA 8_250 Gγ-M4-23-1-2	76	NERICA 8_250 Gγ-M4-85-13-1
27	NERICA 8_250 Gγ-M4-23-1-3	77	NERICA 8_250 Gγ-M4-85-13-2
28	NERICA 8_250 Gγ-M4-23-2-1	78	NERICA 8_250 Gγ-M4-85-13-3
29	NERICA 8_250 Gγ-M4-23-2-2	79	NERICA 8_300 Gγ-M4-3-1-1
31	NERICA 8_250 Gγ-M4-23-2-4	80	NERICA 8_300 Gγ-M4-7-1-2
32	NERICA 8_250 Gγ-M4-24-11-1	81	NERICA 8_300 Gγ-M4-8-7-1
33	NERICA 8_250 Gγ-M4-24-11-2	82	NERICA 8_300 Gγ-M4-8-7-2
34	NERICA 8_250 Gγ-M4-24-11-3	83	NERICA 8_300 Gγ-M4-8-7-3
35	NERICA 8_250 Gγ-M4-24-11-4	84	NERICA 8_300 Gγ-M4-8-7-4
36	NERICA 8_250 Gγ-M4-24-20-1	85	NERICA 8_300 Gγ-M4-9-1-1
38	NERICA 8_250 Gγ-M4-25-1-1	86	NERICA 8_300 Gγ-M4-13-9-1
39	NERICA 8_250 Gγ-M4-25-1-2	87	NERICA 8_300 Gγ-M4-14-1-1
40	NERICA 8_250 Gγ-M4-25-1-3	88	NERICA 8_300 Gγ-M4-14-1-2

41	NERICA 8_250 Gγ-M4-25-1-4	90	NERICA 8_300 Gγ-M4-15-10-1
42	NERICA 8_250 Gγ-M4-30-3-1	91	NERICA 8_300 Gγ-M4-15-10-2
43	NERICA 8_250 Gγ-M4-30-3-2	93[1]	MOROBEREKAN(drought tolerant check)
44	NERICA 8_250 Gγ-M4-30-3-3	94[2]	NERICA 4 (drought tolerant check)
45	NERICA 8_250 Gγ-M4-30-3-4	96	NERICA 8_300 Gγ-M4-4-1-1
46	NERICA 8_250 Gγ-M4-31-1-1	97	NERICA 8_250 Gγ-M4-11-1-1
47	NERICA 8_250 Gγ-M4-32-2-1	98[3]	NERICA8-M0-(parental line)
48	NERICA 8_250 Gγ-M4-32-2-2	99	NERCICA8-250 Gγ-M4-9-1-4
49	NERICA 8_250 Gγ-M4-33-3-1	100	NERCICA8-250 Gγ-M4-9-1-5

[1,2]: Moroberekan and NERICA 4 reproductive drought tolerant checks, respectively, provided by AfricaRice based in Bouaké, Cote d'Ivoire.

[3]: NERICA8: parental line of M_4 mutants.

6. CHARACTERISTICS OF THE EXPERIMENTAL SITE

Sotuba is one of the main Malian Institute of Rural Economy (IER) Agricultural Research Stations. It is located at 12°39' N and 07°56' W; 320 m altitude. Experiments were carried out on sandy silt soil (96.84%) with low clay content (3.85%), water pH (5.75), organic matter (0.37%), nitrogen (0.12%), assimilable phosphorus (10.77 ppm) and exchangeable potassium (0.13 meq/g) [29]. The soil sample data were obtained from 0 to 40 cm depth and were analysed by the Soil Water Plant Laboratory of the IER in Sotuba, Mali [29].

7. EXPERIMENTAL DESIGN AND WATER REGIMES

A randomized block design was used for each control and water stress experiment. Two water regimes, control (Co), no water stress and water stress (WS), conditions were used to screen in the 2020–2021 cropping season field conditions for M_4 mutants for drought tolerance. The number of M4 mutants was 85. There were 72 and 13 mutants at doses 250 Gγ and 300 Gγ, respectively. Three checks were used of which two were reproductive drought susceptible (Moroberekan and NERICA4) and one (NERICA8-0 Gγ) was the mutant's parental line. Drought susceptible and tolerant checks were advised by AfricaRice from Bouaké, Côte d'Ivoire in 2019.

8. CULTURAL PRACTICES

The two experiments were planted on 15 July 2020 after seed bed preparation. Poultry manure (63 kg/ha) was applied in each experiment after planting. Cereal complex (^{15}N–^{15}P–^{15}K) fertilizer (100 kg/ha) was applied as basal. Urea (100 kg/ha), split into 50 kg/ha each, was applied. The first split was applied as top dressing and the second was applied before flag leaf appearance. Weeding was manual with a hoe and the followings were dealt with by hand pulling. Chemical treatment (Oxymil 50g/kg GR) was applied to control termites and nematodes. Experiments were harvested from 16 to 24 November 2020 and thrashing took place between 16 and 24 December 2020 for both conditions.

A drip irrigation system was installed which was filled by drilling through a 20 m^3 water tower. Homemade water proof plastic rainout shelter was designed to protect the water stress experiment. This experiment was protected at 19:00 and opened at 06:00 in the morning, then covered during rainy periods and opened when the rain stopped. The drip irrigation system provided water with a rain gauge to record the amount applied per irrigation (2.4 m^2) or 10 mm and then recorded. Control and water stress (WS) experiments were under both rainfed and drip irrigation conditions concomitantly, while WS had its water withheld for 34 days starting on 22/09/2020 and replenished on 26/10/2020 with 80 mm for mutant recovery. Water stress imposition started when the flag leaves of most of the mutants appeared.

9.   DATA COLLECTION

9.1.   Morphological and physiological data

The list of morphological and physiological traits collected is given in Table 3. Leaf firing, rolling, paddy (PY g/m²), biomass (BY g/m²) yields and harvest index (HI) were on a whole plot basis. Each harvested mutant was sun dried for 30 days prior to thrashing, which was done manually and seed weighted individually for each mutant. Samples figures were pooled per plot while keeping labelled seed and biomass separate.

10.   DATA ANALYSIS

Simple and combined analysis of variance (ANOVA) was performed for each trait. Homogeneity of variance per trait across water stress and normality were checked before combining trait data. Least significant difference (LSD$_{0.05}$) was used for mean separations. Significant level was set at P =0.05. The MINITAB-18 software for statistics with general linear model and multivariate analyses was used.

Pearson linear correlation coefficients were done to appreciate the relationship between grain yields and morphological, physiological and drought stress indices. Selection pressure was set at 20% to identify drought index values linked to higher paddy yield mutants. Then the identified best 20 mutants per index were checked for their appearance in the 20% best PY from water stress and no water stress conditions. Once the index identified a mutant in both water stress and control conditions, it was retained as a valuable index.

TABLE 3. KEY AND MEASUREMENTS OF MORPHOLOGICAL AND PHYSIOLOGICAL PARAMETERS FOR FIELD LEVEL DROUGHT TOLERANCE SCREENING IN THE 2020–2021 CROPPING SEASON

No.	Morphological traits	Measurements procedures	Abbreviations
1	50% flowering	Number of days from planting to 50% flowering of the labelled plant	NDPF
2	No. of secondary branches	No. of total number of secondary branches bearing seeds at physiological maturity	NSB
3	Plant height (cm)	Plant height (cm) was measured from the ground to the tip of each panicle on the selected plant	PH cm
4	Panicle length	Panicle length was measured from the bottom of the panicle to its tip on the labelled plant	PLcm
5	Total leaves produced	No. of total leaves produced per labelled plant was counted every week until flag leaf appearance.	NTLP
6	Plant height (cm)	Panicle length was deduced from plant height to have stem length	SLcm
7	Paddy yield (g/m²)	Paddy yield (PY g/m²) of each plot was calculated	Py·g/m²
8	Biomass yield (g/m²)	Total biomass yield was estimated from dried PY + stover yield and extrapolated to g/m²	B g/m²
9	Harvest index (%)	Obtained from total paddy yield/total biomass (paddy yield plus stover yield)	HI
10	Hundred seed weight	Bulk seeds of labelled plant/family were cleaned; hundred seeds were hand counted and weighted with 2 decimal scale	HSW/r
	Physiological data		
11	Leaf rolling scale	(1) No leaf rolling, (2) two leaves rolling, (3) intermediate (more than 3 leaves rolling), 7 (half leaves rolling) and (9) all leaves rolling, when grains have become fully ripened	LR
12	Leaf firing scale	(1) Late and slow senescence, (2) two or more leaves retain their green colour at maturity, 3 (less than average) retain green colour, (5) ¾ leaves retain, and (9) early and fast senescence , leaves are dead when the grains have become fully ripened	LF
13	Stalk lodging scale	Stalk lodging resistance was measured at physiological maturity by estimating the number of stalks fallen on the ground (= 0–20%) no lodging, (3 = 20–40%) moderately strong, (5 = 41–60%) intermediate, moderately lodged), (7 = 61–80%) weak (most plants nearly flat), and (9 = 81–100%) very weak (all plants flat)	STL
14	Spikelet abortion scale	Spikelet abortion was recorded as: (1) highly fertile (>90%); (3) fertile (75-90%); (5) partly sterile (50-74%); (7) highly sterile (<50% to trace); (9) completely sterile (0%),	SA
15	Green leaves at physiological maturity	Number of green leaves after water replenishment was counted at physiological maturity	NGLPMFP

11.   RESULTS

Under control condition and water stress condition mutants responded differently on flowering time (NDPF), number of total fertile tillers produced (NTFTP), panicle length (PL·cm), paddy (PY g/m²) and biomass (BY g/m²) yields, harvest index (HI) and hundred seed weight (HSW)

on both control and waters stress conditions. Mutants discriminated themselves under control condition, but did not under water stress conditions, in plant height. Water deficits were the limiting factor for mutants to express their full plant height. Mutants produced the same number of leaves (NTLP) and number of secondary branches bearing seeds (NSB) under control and water stress conditions (Table 4).

TABLE 4. ANALYSES OF SELECTED TRAITS OF M_4 MUTANTS DURING THE NORMAL CROPPING SEASON OF 2020–2021

Variables	Source	DF	Control condition			Water stress condition		
			σ^2	F value	P value	σ^2	F value	P value
NTLP	Rep	2	1.822	1.550	0.215	2.253	2.970	0.054
	Fam	87	1.502	1.280	0.087	0.942	1.240	0.115
	Error	174	1.174			0.759		
NDPF	Rep	2	159.280	11.670	0.000	56.050	4.420	0.013
	Fam	87	35.470	2.600	0.000	23.540	1.860	0.000
	Error	174	13.650			12.670		
NTFTP	Rep	2	9312,300	10,300	0.000	10216,000	8,300	0,000
	Fam	87	3390,200	3750	0.000	1949,000	1,580	0,006
	Error	174	903,800			1231,000		
PH·cm	Rep	2	72.870	0.890	0.411	213.270	2.890	0.058
	Fam	87	86.200	1.060	0.373	139.550	1.890	0.000
	Error	174	81.500			73.690		
PL·cm	Rep	2	5.027	1.440	0.240	0.345	0.060	0.939
	Fam	87	6.378	1.830	0.000	8.581	1.560	0.007
	Error	173	3.495			5.502		
NSB	Rep	2	0.5568	0.190	0.831	41.527	10.890	0.000
	Fam	87	3.7767	1.260	0.101	3.617	0.950	0.604
	Error	174	2.9974			3.814		
PY·g/m²	Rep	2	17356.000	1.210	0.302	69737.000	12.440	0.000
	Fam	87	26896.000	1.870	0.000	10225.000	1.820	0.000
	Error	174	14391.000			5607.000		
BY·g/m²	Rep	2	19512.000	0.46	0.629	136182.000	3.550	0.031
	Fam	87	83242.000	1.98	0.000	67920.000	1.770	0.001
	Error	174	42043.000			38308.000		
HI%/g	Rep	2	172.440	4.81	0.009	290.580	5.960	0.003
	Fam	87	169.360	4.72	0.000	108.160	2.220	0.000
	Error	174	35.860			48.750		
HSW/g	Rep	2	0.049	2.510	0.084	0.1091	4.640	0.011
	Fam	87	0.062	3.160	0.000	0.091	3.860	0.000
	Error	174	0.019			0.024		

NTLP: number of leaves produced; NDPF: number of days from planting to 50% flowering; NTFTP: number of total fertile tillers produced (m²); PH: plant height; PLcm: panicle length (cm); NSB: number of secondary branches bearing seeds; PY (g/m²); mean paddy yield: BY(g/m²) , mean biomass yield; HI %: harvest index; HSW , hundred seed weight.

11.1. Performance of mutants across environments

Mutant screening for drought tolerance under water stress regimes assumes significance. Water stress impacted all mutant traits except the number of total leaves produced (NTLP) and number of secondary branches bearing seeds (NSB). Response of the mutants varied on flowering time (NDPF), harvest index (HI) and hundred seed weight (HSW) (Table 5).

TABLE 5. COMBINED ANALYSIS OF SELECTED TRAITS AND GENOTYPE X ENVIRONMENT INTERACTIONS OF M_4 MUTANTS

Variables	Source	DF	σ^2	F value	P value	σ	$\bar{x}$	CV%	$LSD_{0.05}$
NTLP	Irrigation	1	15.171	7.450	0.053				
	Mutants	87	1.714	1.770	0.000			6.612	0.820
	Irrigation*Fam	87	0.730	0.760	0.942	0.855	12.924	7.607	
	Error	348	0.967			0.983			
NDPF	Irrigation	1	1914.830	17.790	0.014				
	Mutants	87	39.450	3.000	0.000			5.670	4.245
	Irrigation*Fam	87	19.560	1.490	0.007	4.423	78.000	4.651	3.454
	Error	348	13.160			3.628			
NTFTP/m²	Irrigation	1	135,800	0,010	0,914				
	Mutants	87	3739,000	2,340	0,000		129,835	30,771	53,858
	Irrigation*Fam	87	1599,700	1,500	0,006	39,996		25,137	43,996
	Error	348	1067,500			32,673			
PH (cm)	Irrigation	1	3908.210	27.320	0.006				
	Mutants	87	134.750	1.740	0.000			8.496	9.157
	Irrigation*Fam	87	91.000	1.170	0.162	9.539	112.281	7.846	
	Error	348	77.600			8.809			
PL (cm)	Irrigation	1	33.000	12.290	0.025				
	Mutants	87	10.212	2.280	0.000			8.011	2.112
	Irrigation*Fam	87	4.839	1.080	0.315	2.200	27.458	7.715	
	Error	348	4.488			2.118			
NSB	Irrigation	1	0.121	0.010	0.943				
	Mutants	87	3.709	1.090	0.294			12.911	
	Irrigation*Fam	87	3.684	1.080	0.308	1.919	14.867	12.413	
	Error	348	3.406			1.845			
PY/g m²	Irrigation	1	2716674.000	64.960	0.001				
	Mutants	87	24942.000	2.450	0.000			32.442	106.022

Variables	Source	DF	σ^2	F value	P value	σ	$\bar{x}$	CV%	LSD$_{0.05}$
	Irrigation*Fam	87	12200.000	1.200	0.130	110.454	340.467	29.619	
	Error	348	10169.000			100.841			
BY/g m²	Irrigation	1	716923.000	9.660	0.036				
	Mutants	87	106478.000	2.600	0.000			43.396	201.636
	Irrigation*Fam	87	44127.000	1.080	0.315	210.064	751.937	27.936	
	Error	348	40919.000			202.284			
HI% (g)	Irrigation	1	25385.700	109.650	0.000				
	Mutants	87	211.000	4.990	0.000			18.090	7.828
	Irrigation*Fam	87	66.500	1.570	0.002	8.155	45.080	14.427	6.193
	Error	348	42.300			6.504			
HSW (g)	Irrigation	1	1.510	19.040	0.012				
	Mutants	87	0.121	5.590	0.000			6.542	0.173
	Irrigation*Fam	87	0.032	1.500	0.006	0.180	2.750	5.347	0.140
	Error	348	0.022			0.147			

NTLP: Number of total leaves produced; NDPF: number of days from planting to flowering; NTFTP: number of total fertile tillers produced (m²); PH·cm: plant height: PL·cm: panicle length; NSB: number of secondary branches bearing seeds; PY g/m²: mean paddy yield; BY g/m²: mean biomass yield; HI%: harvest index; HSW/g: hundred seed weight.

11.2. Physiological traits

Scarcity of water negatively affects physiological characteristics, such as decrease in net photosynthetic rate, transpiration rate, stomatal conductance, water use efficiency, internal CO_2 concentration, photosystem II (PSII) activity, relative water content and membrane stability index [30]. Positive correlations have been observed between flag leaf traits and yield under drought [31]. Flag leaf is very important in grain filling under drought for maintaining the synthesis and transport of photo assimilates. Drought stress in rice impairs assimilate translocation, increased spikelet sterility, reduced rate of grain filling, grain size, weight and yield morphological changes. Leaf rolling is one of the acclimation responses of rice and is used as a criterion for scoring drought tolerance. Leaf rolling reduces light interception, transpiration and leaf dehydration. It may help in maintaining internal plant water status [4]. All the physiological parameters analysed in this study were collected during water stress experiments and the mean data were drawn from three replications, including visual scores obtained at the end of the water stress imposition. Maintenance of green leaves at maturity is a physiological manifestation of their chlorophyll functioning. Selected mutants showed variability in maintaining green leaves, while some of the mutants maintained greener leaves compared with the parental lines. NERICA4, the known drought tolerant line and Moroberekan, another drought check, also displayed more green leaves at physiological maturity. Spikelets abortion is an important physiological manifestation in rice under drought conditions. Some mutants and drought tolerant check, NERICA4 and the parental lines had almost the same higher spikelets abortion with a note of 6 (50–75% sterility) or above. Some mutants (Nos 70, 85) did not abort their spikelets compared with NERICA4 and NERICA8. Mutants (Nos 5, 42, 70) also showed less leaf rolling due to drought stress compared with NERICA4 and NERICA8. Leaf firing was also varied among mutants, which had less leaf firing than the tolerant check (NERICA4) and NERICA8, and Moroberekan which showed no leaf firing (Table 6).

TABLE 6. MEAN PHYSIOLOGICAL TRAITS OF SELECTED M_4 UPLAND RICE UNDER WATER STRESS (WS) CONDITIONS COLLECTED DURING THE 2020–2021 CROPPING SEASON AT SOTUBA, IN MALI

Mutants	NGLPMFP-WS	SA-WS	LR-WS	LF-WS	STL-WS
2	0.667	5.000	4.330	2.833	1.000
5	1.333	5.667	2.000	2.167	1.000
7	1.000	5.000	5.330	2.333	1.000
20	2.000	6.333	5.000	2.000	1.000
36	1.000	7.000	4.330	1.667	1.000
42	1.333	6.333	2.670	1.500	1.000
58	0.333	7.000	7.000	2.667	1.667
62	2.000	6.333	3.670	2.000	1.000
65	0.667	6.333	7.000	2.500	1.000
66	2.000	6.333	2.670	2.000	1.000
70	1.000	4.333	2.000	2.167	1.000
73	0.000	5.000	6.330	2.667	1.333
74	0.667	5.000	5.670	2.667	1.333
76	0.000	5.667	6.670	4.000	2.000
77	0.333	5.667	7.670	2.500	1.667
78	1.333	6.000	6.000	2.000	1.000
80	1.667	5.667	5.000	2.167	1.000
81	1.333	6.333	6.670	1.833	1.000
84	1.000	5.000	7.000	2.833	1.667
85	0.667	3.000	5.000	2.833	1.000
87	1.667	7.000	5.670	2.667	1.333
90	0.667	7.000	4.000	2.833	1.000
93[1]	3.333	-	7.670	1.000	1.000
94[2]	1.500	7.000	4.670	2.333	1.000
98[3]	0.667	6.333	7.670	3.333	2.667
99	0.667	5.000	6.330	2.667	1.333
Mean	1.343	5.862	4.820	2.225	1.231

WS: water stress condition; SA: spikelet abortion; LR: leaf rolling; LF: leaf firing first reading; STL: stalk lodging; NGLPM: number of green leaves recorded at physiological maturity.
[1] Moroberekan and [2]NERICA4 are reproductive drought tolerant checks.
[3] NERICA8 (M_0 Gγ) original mutant parental line. Observations were made at the end of stress imposition but prior to recovery.

12.   DROUGHT TOLERANCE INDEX (DTI)

The ability of crop cultivars to perform reasonably well in drought stressed environments is paramount for stability of production. The relative yield performance of genotypes in drought stressed and non-stressed environments can be used as an indicator to identify drought resistant varieties in breeding for drought prone environments. Several drought indices have been suggested on the basis of a mathematical relationship between yield under drought conditions and non-stressed conditions. These indices are based on either drought resistance or drought susceptibility of genotypes [32].

Mutants used for the estimation of DTIs were mentioned earlier. Paddy yield in mutants under control and water stress environment was used to estimate DTIs using the formula given in Table 1. Selection pressure at 20% was applied to identify the index involved in screening criteria for isolating higher paddy yielding mutants. The best 20 families per index were identified for their appearance in the 20% best PY mutants from water stress and no water stress conditions. Once the index has identified a mutant in both water stress conditions, it was retained as a valuable index combining yield and DTIs.

Half of the DTIs, tolerance indices (TOLs), yield stability indices (YSIs), stress susceptibility indices (SSIs) and percentage yield reductions (PYRs) exhibited no variation in paddy among mutants under water stress. The remaining half of the mutant indices, mean productivities (MP), stress tolerance indices (STIs), yield indices (YIs) and geometric mean productivities (GMPs) discriminated mutants based on their indices (Table 7), suggesting that such indices are significant for screening and breeding for drought tolerance. Researchers have reported that indices that discriminated against the mutants are useful since 8 out of 15 indices were the most effective for selection of drought tolerant, high yielding upland rice genotypes.

13. MUTANT VALIDATION

Mutants differed on MP, STI, YI and on GMP, while they did not show variation in TOL, YSI, SSI and on percentage yield reduction (%YR) (Table 8). Mutants showed higher mid-parent productivity performance. Average paddy yields between control and water stress were, respectively, 411.137 g/m² and 268.737 g/m², or a yield loss of 34.637% of the control. Mutants showed better performance compared with tolerant checks (NERICA4, Moroberekan) and their parental line (NERICA8).

TABLE 7. DTI ANALYSIS OF VARIANCES OF M_4 UPLAND RICE

DTI	Source	DF	σ^2	F value	P value	σ	$\bar{X}$	CV%	LSD 0.05
TOL	Rep	2	83686.000	4.460	0.013				
	Mutants	87	24135.000	1.280	0.083		142.400	96.244	
	Error	174	18783.000			137.052			
MP	Rep	2	22625.000	4.270	0.016				
	Mutants	87	12527.000	2.360	0.000		339.937	21.423	98.324
	Error	174	5303.000			72.825			
STI	Rep	2	0.516	6.190	0.003				
	Mutants	87	0.165	1.980	0.000		0.669	43.168	0.390
	Error	174	0.083			0.289			

YI	Rep	2	0.966	12.440	0.000				
	Mutants	87	0.142	1.820	0.000		1.000	27.864	0.376
	Error	174	0.078			0.279			
YSI	Rep	2	0.960	6.690	0.002				
	Mutants	87	0.191	1.330	0.059		0.739	51.253	
	Error	174	0.144			0.379			
SSI	Rep	2	8.005	6.690	0.002				
	Mutants	87	1.588	1.330	0.059		0.753	145.318	
	Error	174	1.197			1.094			
GMP	Rep	2	29649.000	6.140	0.003				
	Mutants	87	11766.000	2.440	0.000		325.200	21.373	93.847
	Error	174	4831.000			69.506			
PYR	Rep	2	9603.000	6.690	0.002				
	Mutants	87	1905.000	1.330	0.059		32.008	145.317	
	Error	174	1436.000			37.890			

TOL: stress tolerance index; MP: mean productivity; STI: stress tolerance index; YI: yield index; YSI: yield stability index; SSI: stress susceptibility index; GMP: geometric mean productivity; PYR: yield reduction percentage; LSD_{005}: one tail (upper tail).

Genotypes (mutants) with low TOL are more stable in two different conditions and suitable for drought tolerance screening, while mutants with higher TOL values are not fit for drought conditions [33]. The mutants performing better across water regimes with low TOL indices were mutants 58, 76, 80, 81, 90, 93 (check) and 97. These mutants are grown in drought prone environments with minimum yield loss compared to normal ones (Table 8). A higher STI value indicates better water stress tolerance. Mutants 77, 74, 93, 87, 70, 62, 78, 90 sowed more tolerance with higher indices. In terms of yield potential under both treatment conditions, three mutants (74, 77, 87) which had a higher STI ranking were linked to their higher yield. Paddy yields of better STI varied from 430.481 g/m² (mutant 93 = check) to 632.652 g/m² (mutant 70) under control and from 274.272 g/m² (mutant 42) to 436.484 g/m² (mutant 93) under water stress condition (Table 8).

Upland rice with stable yield under control and water stress environments is a very valuable asset for farmers in enhancing yield. Mutants responding to both environments with higher YIs were identified. Yields between control and water stress were very close, meaning they provided reasonable paddy yield under both environments. These were mutants 93, 78, 81, 90, 62, 77, 73, 87, 80, 76, 99, 20, 74, 58, and 96. Mutants with the best YI were not better yielding under water stress and control conditions. Six mutants with best YI were 93, 78, 81, 90, 62 and their yields varied from 511.657 g/m² (mutant 81) to 506.167 (mutant 78) under control and from 436.484 g/m² (mutant 93 = check) under water stress to 430.481 g/m² (mutant 93) under control condition. These mutants were better yielding than their parental line under both water conditions, thus securing yield in a water shortage during a reproductive drought period (Table 8). Mutants with higher YSI values are stable and desired under water stress and control conditions. Mutants with lower SSI values are considered tolerant to drought. These mutants were 93, 81, 80, 58, 78. They were not all among the highest yielding under water stress conditions. However, reproductive drought resistant check (93) and mutants 80, 81 were among the highest yielding under water stress conditions at 20% selection pressure considering their lower SSI values. Stress susceptibility index tended to identify mutants performing better under water stress, thus this parameter could not distinguish well performing mutants under control (Table 8).

Mutants with greater GMP are considered as being tolerant to reproductive drought. Mutants with better GMP were identified as 74, 77, 62, 87, 93, 70, 78, 90, 99, 73; their GMP values ranged from 443.819 g/m² for mutant 74 to 402.039 g/m² for the first mutant (73). Five mutants (74, 77, 99, 78, 62) produced well under controls, while two mutants (77 and 78) performed well under both water conditions (Table 8).

Losing yield under water shortage conditions compared with the control environment during the upland rice reproductive stage is desirable to secure yield for drought environment farmers. General mean loss of all 88 mutants, including checks was 34.636% between control and water stress environments. Mutants with lower yield loss were 93, 81, 80, 58, 90, 78, 73, 62, 97 and 76. They were not among the highest yielding mutants under control, but most were among better performing mutants under water stress condition, suggesting the utility of this index for securing yield, especially under a drought environment (Fig. 2) (Table 8).

TABLE 8. DROUGHT TOLERANCE INDICES OF SELECTED M₄ UPLAND RICE UNDER CONTROL AND WATER STRESS CONDITIONS

Mutants	PY/m² Col	PY g/m² WS	MP	TOL	STI	YI	YSI	SSI	GMP	PYR
2	576.963	231.985	404.474	344.978	0.781	0.863	0.413	1.694	361.660	59.792
5	514.825	231.625	373.225	283.200	0.680	0.862	0.501	1.442	337.699	55.009
7	530.327	226.020	378.174	304.307	0.686	0.841	0.461	1.555	340.003	57.381
20	453.948	329.156	391.552	124.793	0.877	1.225	0.766	0.676	383.450	27.490
36	495.683	280.637	388.160	215.046	0.824	1.044	0.566	1.252	372.527	43.384
42	585.990	274.272	430.131	311.718	0.939	1.021	0.485	1.486	396.990	53.195
58	390.519	315.733	353.126	74.785	0.748	1.175	0.816	0.530	350.525	19.150
62	495.846	362.326	429.086	133.520	1.088	1.348	0.731	0.777	423.506	26.928
65	518.507	299.059	408.783	219.448	0.918	1.113	0.582	1.207	393.252	42.323
66	493.143	306.504	399.823	186.640	0.901	1.141	0.634	1.057	387.967	37.847
70	632.652	286.667	459.659	345.985	1.090	1.067	0.450	1.589	421.581	54.688
73	468.674	350.785	409.730	117.889	0.992	1.305	0.737	0.761	402.039	25.154

74	618.489	319.033	468.761	299.456	1.198	1.187	0.522	1.380	443.819	48.417
76	463.752	337.091	400.421	126.661	0.921	1.254	0.732	0.773	392.350	27.312
77	575.163	353.702	464.432	221.461	1.221	1.316	0.639	1.043	443.059	38.504
78	506.167	379.348	442.757	126.819	1.079	1.412	0.800	0.577	414.691	25.055
80	416.846	339.735	378.291	77.111	0.834	1.264	0.842	0.455	374.594	18.499
81	415.304	374.193	394.748	41.111	0.917	1.392	0.942	0.168	391.863	9.899
84	511.657	309.896	410.777	201.761	0.987	1.153	0.623	1.089	397.524	39.433
85	438.52	279.83	359.177	350.03	0.723	0.681	0.638	1.045	350.303	36.200
87	520.433	344.904	432.669	175.530	1.113	1.283	0.666	0.963	422.943	33.728
90	457.204	369.106	413.155	88.098	1.032	1.373	0.808	0.555	409.108	19.269
97	325.015	236.385	280.700	88.630	0.443	0.880	0.751	0.718	271.617	27.269
99	506.733	331.109	418.921	175.624	0.967	1.232	0.707	0.846	403.231	34.658
93[1]	430.481	436.484	433.483	-6.002	1.166	1.624	1.319	-0.922	422.901	-1.394
94[2]	492.188	271.889	382.039	220.299	0.812	1.012	0.628	1.074	360.769	44.759
98[3]	508.900	249.175	379.038	259.725	0.725	0.927	0.528	1.362	347.628	51.036
$\bar{x}$	411.137	268.737	339.937	142.4	0.669	1.000	0.739	0.753	325.2	32.008
σ	119.964	74.881	72.825	137.052	0.289	0.279	0.379	1.094	69.506	37.89
CV%	29.100	27.860	21.423	96.244	43.168	27.864	51.253	145.318	21.373	145.317
LSD 0.05	161.97	10.110	98.324	185.047	0.398	0.376	0.512	1.477	93.847	51.166

PY: mean paddy yield (g/m²); Co: control, no water stress condition; WS: water stress condition; TOL: stress tolerance index; MP: mean productivity; STI: stress tolerance index; YI: yield index; YSI: yield stability index; SSI: stress susceptibility index; GMP: geometric mean productivity; %YR: percentage yield reduction.

[1] Moroberekan and [2]NERICA4 are reproductive drought tolerant checks from the AfricaRice Center, Bouaké, Cote d'Ivoire.

[3] NERICA8 (M_0 Gγ) original mutant parental line.

FIG. 2. Field grown plants showing different morphological traits under water stress (WS) conditions. (a) Field grown plants — 5002 (F76) showing severe leaf rolling symptoms under WS condition; (b) field grown plants — 4081 (F71) showing few abortion symptoms under WS two days after water replenishment; (c) field plot — 5065 (NERICA4), drought tolerant check under WS; main spikelets are aborted, while tillers had seeds with no leaf firing; (d) field plot — 2089 (NERICA4) drought tolerant check under no water stress.

14. VALIDATION OF RELATIONSHIPS BETWEEN PADDY YIELD AND ITS CONTRIBUTING TRAITS

Biomass yield, number of fertile tillers, plant height, panicle length, and number of secondary branches bearing seeds, harvest index and hundred seed weight increased paddy yield, while later mutants decreased their yield under control conditions. Under water stress conditions, the relationships of increased paddy yield from these traits existed, but with lesser expression. The number of secondary branches bearing seeds under water stress did not increase paddy yield. This is probably due to spikelets abortion noted under water stress conditions compared to control conditions (Table 9). Selected physiological parameters, number of green leaves at physiological maturity (NGLPM), spikelet abortion, leaf rolling and stalk lodging also had no impact on paddy yield. Better performing mutants under control condition did perform well under water stress conditions. Biomass yield under water stress did not increase paddy yield (PY) under control conditions but did increase paddy yield under water stress conditions. More biomass production is linked to higher paddy yield under control conditions than water stress conditions. It is plausible that in mutants grown under non-water stress, assimilates are produced and are translocated to sink tissue more efficiently as opposed to water stress conditions where water management by mutant is critical for cell growth and survival.

Relationships are stronger between paddy yield (PY) and DTIs under control conditions than water stress conditions. Yield stability index (YSI) was negative under control conditions with PY. It is attributed to the fact that mutants show a higher yield gap among them under control conditions than water stress conditions, leading to loss of stability. Under water stress conditions, the YSI was involved with increased PY. Negative stress tolerance (TOL) and stress susceptibility index (SSI) relationships with PY are positive under water stress conditions because their lower indices are preferred as compared to their higher ones. But the behaviour of TOL, like SSI, with PY under control conditions is normally negative because mutants increase yield gap among them under control conditions compared to water stress conditions.

The relationships of DTIs with yield have been shown in previous studies. Under variable stress, water stress [1] and nitrogen deficiency conditions [34], researchers claimed that STI, GMP, MP are better indices for the selection and identification of better genotypes for genetic gain. The data from our study also suggest such indices are useful as selection criteria for the selection of superior genotypes.

TABLE 9. CORRELATION BETWEEN PADDY YIELD, BIOMASS, DTIs, MORPHOLOGICAL AND PHYSIOLOGICAL TRAITS OF MUTANTS, PEARSON LINEAR CORRELATION MATRICES

Yields and drought stress parameter correlations				Yields and morphological parameter correlations					Yields and physiological parameter correlations		
Variables	r/P	PY-Co	PY-WS	Variables	r/P	PY/Co	Variables	PY/WS	Variables	r/P	PY/WS
PY g/m²/WS	r	0.216		BY/Co	r	0.860	BY/WS	0.724	NGLPM/WS	r	-0.052
	P	0,000			P	0.000		0.000		P	0.399
BY g/m²/WS	r	0.083	0.724	NTLP/Co	r	-0.084	NTLP/WS	0.042	SA/WS	r	0.049
	P	0.180	0.000		P	0.173		0.501		P	0.449
BY g/m²/Co	r	0.858	0.183	NTFP/Co	r	0.430	NTFP/WS	0.337	LR/WS	r	0.000
	P	0.000	0.003		P	0.000		0.000		P	0.074
TOL	r	0.809	-0.399	NDPF/Co	r	-0.251	NDPF/WS	-0.053	STL/WS	r	-0.017
	P	0.000	0.000		P	0.000		0.389		P	0.778
MP	r	0.876	0.659	PH/Co	r	0.374	PH/WS	0.202			
	P	0.000	0.000		P	0.000		0.001			
STI	r	0.755	0.760	PL/Co	r	0.313	PL/WS	0.117			
	P	0.000	0.000		P	0.000		0.058			
YI	r	0.216	1.000	NSB/Co	r	0.220	NSB/WS	0.034			
	P	0.000	*		P	0.000		0.584			
YSI	r	-0.638	0.397	HI/Co	r	0.563	HI/WS	0.460			
	P	0.000	0.000		P	0.000		0.000			
SSI	r	0.638	-0.397	HSW/Co	r	0.202	HSW/WS	0.140			
	P	0.000	0.000		P	0.001		0.023			
GMP	r	0.789	0.757								
	P	0.000	0.000								
%YR	r	0.638	-0.397								
	P	0.000	0.000								

WS: water stress condition; Co: control, no water stress condition until maturity; PY: paddy yield (mean) (g/m²); BY: (mean) biomass yield (g/m²); TOL: stress tolerance index; MP: mean productivity; STI: stress tolerance index; YI: yield index; YSI: yield stability index; SSI: stress susceptibility index; GMP: geometric mean productivity; %YR: yield reduction percentage; WS: water stress condition; r: Pearson coefficient of correlation; P: probability level; *: Cells have same values; NTLP: number of total leaves produced by the main shoot; NTFP: number of total fertile tillers produced/m²; PH: plant height (cm); PL: panicle length (cm); NSB: number of secondary branches bearing seeds; HI: harvest index; HSW: hundred seed weight (g); NGLPM: number of green leaves recorded at physiological maturity; SA: spikelets abortion; LR: first leaf rolling reading after water withholding before irrigation; STL: stalk lodging recorded the last day of water and before water application.

15.   VALIDATION OF MUTANTS FOR DROUGHT TOLERANCE

There were 26 mutants used, including two known reproductive drought tolerant checks (Moroberekan and NERICA4) and the mutant parental line. These mutants were selected among 85 mutants. Number of observations for the multivariate analysis is 43 from the 26 mutants, including checks. Mutants were chosen based on their DTIs, high yielding under control and water stress condition and selection pressure (Table 10). Observations from morphological, physiological, and DTI estimates were used to build a dendrogram using four groups with Euclidian distance with MINITAB-18 software. This analysis is done when prior grouping information is not available among mutants.

TABLE 10. LIST OF M_4 MUTANTS SELECTED BASED ON DTI, SELECTION PRESSURE, HIGHER YIELD AND DROUGHT SUSCEPTIBILITY

| | 20% selection pressure and top DTI appearance | | | | | | | | NDPF | | Paddy yield g/m² | |
| | | | | | | | | | | | | Water |
Fam	MP	TOL	GMP	SSI	%YR	STI	YSI	YI	NWS	WS	Control	
2									75	78	576.963	231,985
5									83	77	514.825	231,625
7									81	73	530.327	236,02
20	MP		GMP			STI		YI	78	77	453.948	329.156
36	MP		GMP			STI			82	73	495.683	280.637
42	MP		GMP			STI			80	74	585.990	274.272
58		TOL						YI	87	75	390.519	315.733
62	MP		GMP			STI		YI	78	76	495.846	362.326
65	MP		GMP			STI			73	75	518.507	299.059
66	MP		GMP			STI			83	76	493.143	306.504
70	MP		GMP			STI			75	74	632.652	286.667
73	MP		GMP			STI		YI	84	77	468.674	350.785
74	MP		GMP			STI		YI	79	75	618.489	319.033
76	MP		GMP			STI		YI	81	75	463.752	337.091
77	MP		GMP			STI		YI	77	79	575.163	353.702
78	MP		GMP			STI		YI	79	77	506.167	379.348
80			GMP	SSI	YR	STI	YSI	YI	76	76	416.846	339.735
81	MP		GMP	SSI	YR	STI	YSI	YI	**75**	**75**	415.304	374.193
84	MP		GMP			STI			77	72	511.657	309.896
85									81	81	438,52	279,83
87	MP		GMP			STI		YI	79	74	520.433	344.904
90	MP		GMP			STI		YI	85	72	457.204	369.106
93[1]	MP	TOL	GMP	SSI	YR	STI	YSI	YI	**92**	**90**	430.481	436.484
94[2]									80	73	492.188	271.889
96		TOL		SSI	R		YSI	YI	82	76	287.407	313.667
98[3]									76	71	508.900	249.175
99	MP		GMP			STI		YI	77	78	506.733	331.109

MP: mean productivity; TOL: tolerance index; GMP: geometric mean productivity; SSI: stress susceptibility stress; STI: stress tolerance stress; YSI: yield stability index; YI: yield index; NWS: no water stress; WS: water stress; NDPF: number of days from planting to flowering. 20% selection pressure was applied on each drought tolerance index ranking basis. Only mutant listed on top of each selection index was retained. [1]Morobereka, drought tolerant check; [2]NERICA4, drought tolerant check; [3]NERICA8 (M0 Gγ) is M_4 parental line and was not identified by DTI considering 20% selection pressure.

15.1. Dendrogram of similarity among mutants

Four groups composed of several subgroups were constructed. Group 1 is made up of eight mutants (2, 42, 5, 7, 98 (NERICA48) 70, 74, 77). Group 2 has 13 mutants (20, 73, 62, 78, 87, 99, 36, 94 (NERICA4, drought resistant check), 65, 84, 66, 85). Group 3 contained four mutants (58, 80, 81, 90) and Group 1 with Moroberekan, a drought resistant lowland rice variety (Fig. 3).

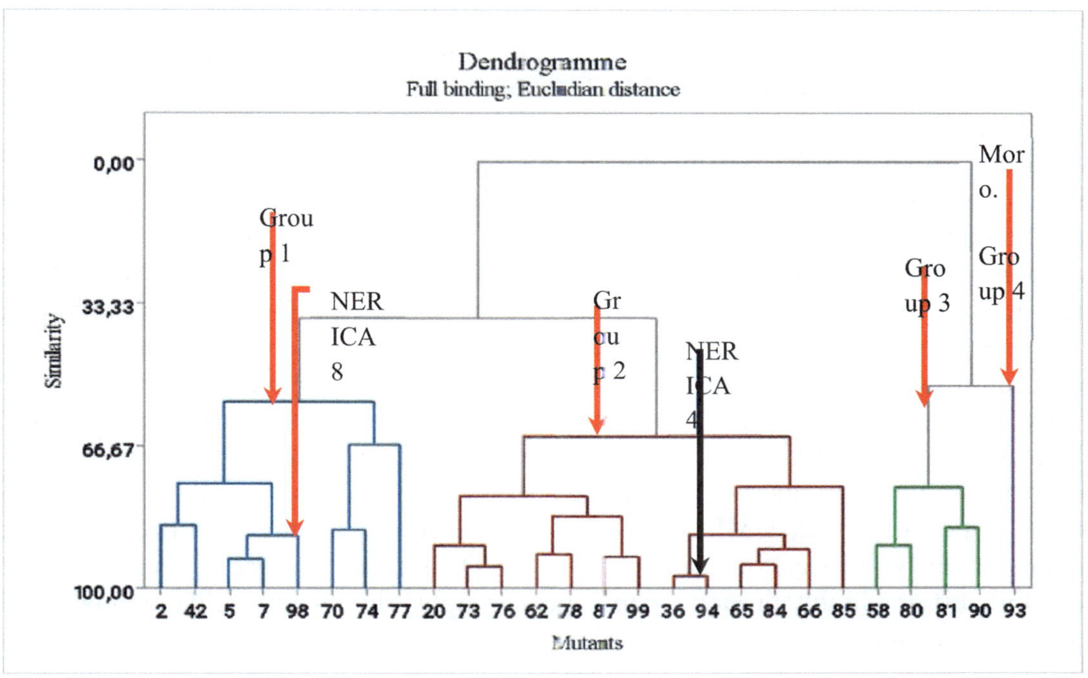

FIG. 3. Dendrogram of cluster analysis of 26 M_4 upland rice M_4 mutants, including three checks, two drought tolerant using morphological and physiological data and DTI estimates at Sotuba, Mali, during the 2020 normal cropping season.

16.   CONCLUSION AND PERSPECTIVES

Mutant responses varied in morphology (maturity, paddy yield, hundred seed weight, harvest index) and in physiology (spikelet abortion, senescence, and lodging) across irrigation. The first drought tolerant check, Moroberekan, was a little bit late, but no later than some mutants, was stable in yield across irrigation levels. The second tolerance check, NERICA4, was similar in maturity compared to most mutants, had average yield and was less stable than the first check. The two drought susceptible checks were too late and were not used in the data interpretation. The original parent NERICA8, earlier than the two controls, is in the same maturity group as most productive mutants across water levels. NERICA8 belongs to the second group of productivity under water stress and was among the first group in the control conditions. Average yield reduction of all mutants was 34.636%. Yield losses were 51% for NERICA8, 44.579% for NERICA4, −1.394 for Moroberekan. The highest yield loss was 59.80% for mutant (mutant 2) and the lowest was 36.5% (mutant 8). Differences between mutants in yield were greater under control conditions than in water stress, condition. Several very productive mutants were identified, but not statistically different from the parental NERICA8 under control conditions and more productive and different than the NERICA8 under water stress conditions. Broad sense heritability was higher in the combined analysis than in the single ones, while the water stress expressed more heritability compared to the water stress conditions. Productive mutants under control were under water stress conditions, with lower r^2 under water stress conditions. Multivariate analysis revealed mutants close to the two drought tolerant checks. Drought tolerance indices identify mutants with low and stable yields, productive in the control conditions and less productive in the water stressed conditions, in yield. The combination of indices, especially (MP, GMP and STI) with selection pressure identified very productive mutants under both irrigation conditions. Using DTIs alone is more of a yield protection tool than of identifying highly productive mutants. Drought tolerant and yield potential confirmation experiments (three on station, two on farm) were conducted during the 2021–2022 cropping season.

REFERENCES TO CHAPTER 2–5

[1]   MAU, S.Y., et al., Brown spot disease severity, yield and yield loss relationships in pigmented upland rice cultivars from East Nusa Tenggara, Indonesia, Biodiversitas **21** (2019) 1625–1634.

[2]   ZU, X., et al., A new method for evaluating the drought tolerance of upland rice cultivars, Crops J. **5** (2017) 488–498.

[3]   BADIGANNAVAR, A., et al., Physiological, genetic and molecular basis of drought resilience, Ind. J. Plant Physiol. **23** 4 (2018) 670–688.

[4]   PANDEY, V., SHUKLA, A., Acclimation and tolerance strategies of rice (*Oryza sativa* L.) under drought stress, Rice Sci. **22** 4 (2015).

[5]   SARSU, F., et al., "Screening protocols for heat tolerance in rice at the seedling and reproductive stages", Pre-Field Screening Protocols for Heat-Tolerant Mutants in Rice, Springer, Cham (2018) 9–24; https://doi.org/10.1007/978-3-319-77338-4_2

[6]   RAHMAN, M.T., ISLAM, M.T., ISLAM, M.O., Effect of water stress at different growth stages on yield and yield contributing characters of transplanted Aman rice, Pak. J. Biol. Sci. **5** 2 (2002) 169–172.

[7]   BING YUE, et al., Genetic basis of drought resistance at reproductive stage in rice: separation of drought tolerance from drought avoidance, Genetics **172** 2 (2006) 1213– 1228.

[8]   FAROOQ, M., WAHID, A., KOBAYASHI, N., FUJITA, D., BASRA, S.M.A., Plant drought stress: Effects, mechanisms and management, Agron. Sustain. Dev. **29** (2009) 185–212; doi: 10.1051/agro:2008021.

[9]   KUMAR, A., BASU, S., RAMEGOWDA, V., PEREIRA, A., Mechanisms of Drought Tolerance in Rice, Burleigh Dodds Science Publishing, Cambridge (2017) 131–163.

[10]   SINGH, B., REDDY, K.R., REDOÑA, E.D., WALKER, T., Developing a screening tool for osmotic stress tolerance classification of rice cultivars based on *in vitro* seed germination, Crop Sci. **57** 1 (2017) 387–394.

[11]   FISCHER, R.A., MAURER, R., Drought resistance in spring wheat cultivar: I. Grain yield response, Austr. J. Agric. Res. **29** (1973) 897–912.

[12]   ROSIELLE, A.A., HAMBLIN, J., Theoretical aspects of selection for yield in stress and non-stress environments, Crop Sci. **21** (1981) 943–946.

[13]   BIDINGER, F.R., MAHALAKSHMI, V., RAO, G.D.P., Assessment of drought resistance in pearl millet (*Pennisetum americanum* (L.) Leeke), I. Factors affecting yields under stress, Austr. J. Agr. Sci. **38** (1987) 37–48.

[14]   FERNANDEZ, G.C.J., "Effective selection criteria for assessing plant stress tolerance", Adaptation of Food Crops to Temperature and Water Stress (KUO, C.G., Ed.), World Vegetable Center, Taipei (1992) 257–270.

[15]   HOSSAIN, A.B.S., SEARS, A.G., COX, T.S., PAULSEN, G.M., Desiccation tolerance and its relationship to assimilate partitioning in winter wheat, Crop Sci. **30** (1999) 622– 627.

[16]   FARSHADFAR, E., SUTKA, J., Multivariate analysis of drought tolerance in wheat substitution lines, Cereal Res. Commun. **31** (2002) 33–40.

[17]   MOOSAVI, S.S., et al., Introduction of new indices to identify relative drought tolerance and resistance in wheat genotypes, Desert **12** (2008) 165–178.

[18]   DADBAKHSH, A., YAZDANSEPAS, A., AHMADIZADEH, M., Study drought stress on yield of wheat (*Triticum aestivum* L.) genotypes by drought tolerance indices, Adv. Environ. Bio. **5** 7 (2011) 1804–1810.

[19]   FARSHADFAR, E., ELYASI, P., Screening quantitative indicators of drought tolerance in bread wheat (*T. aestivum*) landraces, Eur. J. Exp. Biol. **2** 3 (2012) 577–584.

[20]   FISCHER, R.A., WOOD, J.T., Drought resistance in spring wheat cultivars. III. Yield association with morphological traits, Austr. J. Agr. Res. **30** (1979) 1001–1020.

[21]   GAVUZZI, P., et al., Evaluation of field and laboratory predictors of drought and heat tolerance in winter cereals, Can. J. Plant Sci. **77** (1997) 523–531.

[22]   BOUSLAMA, M., SCHAPAUGH, T., Stress tolerance in soybean Part I: Evaluation of three screening techniques for heat and drought tolerance, Crop Sci. **24** (1984) 933– 937.

[23]   CHOUKAN, R., TAHERKHANI, T., GHANNADHA, M.R., KHODARAHMI, M., Evaluation of drought tolerance in grain maize inbred lines using drought tolerance indices, Iran. J. Agric. Sci. **8** 1 (2006) 79–89.

[24]   DOUKORO, D., ABBEY, G., KALIFA, T., Drought monitoring and assessment of climate parameters variability in Koutiala and San Districts, Mali, Am. J. Climate Change **11** (2022) 230–249; doi: 10.4236/ajcc.2022.113011.

[25]   KANG, D.-J., FUTAKUCHI, K.,. Effect of moderate drought-stress on flowering time of interspecific hybrid progenies (*Oryza sativa* L. × *Oryza glaberrima* Steud.), Crop Sci. Biotech. **22** 1 (2019) 75–81.

[26]   O'TOOLE, J.C., "Adaptation of rice to drought prone environments", Drought Resistance in Crops with Emphasis on Rice, International Rice Research Institute, Los Baños (1982) 195–213.

[27]   TORRES, R.O., HENRY, A., Yield stability of selected rice breeding lines and donors across conditions of mild to moderately severe drought stress, Field Crops Res. **220** (2018) 37–45.

[28]   HSIAO, T.C., Plant responses to water stress, Annu. Rev. Plant Physiol. **24** (1973) 519– 570; http://dx.doi.org/10.1146/annurev.pp.24.060173.002511.

[29]   DEMBELE, J.S., et al., Plant density and nitrogen fertilization optimization on sorghum grain yield in Mali, Agron. J. **113** (2021) 4705–4720.

[30]   PANDA, D., MISHRA, S.S., BEHERA, P.K., Drought tolerance in rice: Focus on recent mechanisms and approaches, Rice Sci. **28** 2 (2021) 119–132.

[31]   BISWAL, A.K., KOHLI, A., Cereal flag leaf adaptations for grain yield under drought: Knowledge status and gaps, Mol. Breeding **31** (2013) 749–766; https://doi.org/10.1007/s11032-013-9847-7.

[32]   RAMAN, A., et al., Drought yield index to select high yielding rice lines under different drought stress severities, Rice **5** (2012) 31; https://doi.org/10.1186/1939-8433-5-31.

[33]   GARG, H.S., BHATTACHARYA, C., Drought tolerance indices for screening some of rice genotypes, Int. J. Adv. Biol. Res. **7** 4 (2017) 671–674.

[34] LESTARI, A.P., SOPANDIE, D., ASWIDINNOOR, H., Estimation for stress tolerance indices of rice genotypes in low nitrogen conditions, Thai J. Agric. Sci. **52** (2019) 180– 190

2–6. MUTATION BREEDING AND PRE-FIELD SCREENING FOR SCREENING FOR DROUGHT TOLERANT RICE

A.R. HARUN[1], N.Abd.A. SHAMSUDIN[2], F. AHMAD[1], Z.S. KAMARUDIN[3],
N.A. HASAN[4], N. HISHAM[2], S.N. YUSOF[2], S. HUSSEIN[1],
Y.O. OYEBAMIJI[2], S. SALLEH[1], S.A.A. RAHMAN[1],
A.N.Abd. WAHID[1], Md.A.N. JA'AFAR[4], F. SARSU[5]

[1]Agrotechnology and Biosciences Division,
Malaysian Nuclear Agency, Selangor, Malaysia

[2]Department of Biological Sciences and Biotechnology,
Faculty of Science and Technology, Universiti Kebangsaan Malaysia,
Selangor, Malaysia

[3]Department of Crop Science, Faculty of Agriculture,
Universiti Putra Malaysia, Selangor, Malaysia

[4]School of Biology, Faculty of Applied Sciences,
Universiti Teknologi MARA Cawangan Negeri Sembilan,
Kampus Kuala Pilah, Sembilan, Malaysia

[5]Joint FAO/IAEA Centre of Nuclear Techniques in Food and Agriculture,
Department of Nuclear Sciences and Applications, Vienna

Abstract

Drought is a significant factor limiting the rice industry worldwide, including in Malaysia. therefore, there is an imperative need to develop and validate a standard protocol for the efficient screening of drought tolerance rice lines developed through mutation breeding. In the paper the standard protocol for plant breeders and scientists was designed and validated using pre-field (greenhouse and growth chamber) and field conditions. Indicators such as morpho-physiological, biochemical and yield parameters were utilized to compare the candidate mutant rice lines with the check varieties. The morpho-physiological and biochemical measurements under drought conditions include MDA, antioxidant enzymes, plant height, number of tillers, photosynthetic rate, stomatal conductance, relative chlorophyll content, and relative water content in order to gain a comprehensive understanding of the mechanism underlying the drought tolerance traits of the selected mutant rice lines. Selected drought tolerant mutant rice lines exhibited an increase in proline accumulation, production of enzymatic antioxidants to scavenge reactive oxygen species (ROS), a decrease in MDA accumulation, and high photosynthetic and stomatal conductance values. This indicates that the evaluated candidate mutant rice lines performed better under drought conditions. Selected mutant lines were then subjected to multilocation trials (MLTs) across Malaysia to ensure improved performance of mutant lines in major rice fields throughout the entire country.

Key words: Drought, morpho-physiological traits, mutant lines, rice.

1. INTRODUCTION

Rice (*Oryza sativa* L.) is an important staple food crop; more than half of the world population consume rice as their primary source of carbohydrate [1]. Located in the equatorial region, and exposed to a relatively high temperature, ranging from 27° to 32°C with high humidity and rainfall throughout the year, Malaysia is suitable for rice cultivation [2]. As the staple food of

Malaysia, the production of rice is deeply integrated into the country's economy, politics and social structure. Rice is a main focus in the National Agrofood Policy 2021–2030 (NAP 1.0 and NAP 2.0), as well as in the 12th Malaysia Plan that is set to reach 75% self-sufficiency level (SSL) in rice. However, for more than three decades, the rice SSL for Malaysia remains at 60–70% [3].

Currently, Malaysia is dependent on surplus stock from other rice producing countries to fill a 30% SSL gap [4]. This dependency is at risk whenever the rice exporting countries face unexpected issues involving adverse climatic situations. One of the critical impacts faced by Malaysia is drought, which is essentially a freshwater deficit (Fig. 1). It is one of the major abiotic stresses that can cause rice yields to plummet [5]. This is due to the paddy rice plants, the type of rice most cultivated in Malaysia, that require waterlogged conditions to ensure high productivity.

FIG. 1. A rice field in a main granary area of Malaysia affected by drought.

There are about 700 000 ha of rice land in Malaysia, with 43% of the total planting area being fully rainfed dependent [3]. The annual average rainfall in Malaysia is more than 2500 mm and is strongly influenced by monsoons from the southwest and northwest. According to Firdaus et al. [2], inconsistent rainfalls had affected non-granary areas, where high temperatures and low precipitations were observed in Alor Setar, leading to yield losses in Muda Agricultural Development Authority (MADA) areas. To overcome this problem, the development of high yielding rice varieties with drought tolerant traits has become the main focus of rice breeders in Malaysia.

The rice breeding programme in Malaysia started in 1915 at the Department of Agriculture's Titi Serong Research Station, Perak. Since then, a total of 56 rice varieties, including MR219, MR220, MR297, MR303, MR307, UKMRC2 and UKMRC8, have been released in Malaysia. Most of these rice cultivars were high yielding, but highly susceptible to abiotic stresses such as drought, submergence and salinity. Malaysia, through the Malaysian Agriculture Research and Development Institute (MARDI), had released a variety called Aeron 1, claimed to perform well under low water conditions. However, this variety was not favourable among farmers due to its low yield.

Breeders and agricultural scientists have embraced the mutation breeding approach, since new mutant traits can be created using this tool. Mutation breeding is widely used to improve plant varieties by changing their genetic make-up in a shorter period compared with conventional and molecular breeding methods. In addition, selection of desired traits without changing other characteristics of a certain variety with economic importance, can be achieved. Mutation breeding is one of the breeding methods to create variation in crops [6]. In Malaysia, lack of a new superior rice variety adaptable to climate change such as drought is the main problem in rice production [2]. Thus, improving the rice genotype via induced mutation is crucial to broaden the genetic diversity of rice in Malaysia. Furthermore, this technique is suitable for developing new rice varieties since stable populations can be achieved faster compared with the conventional technique. The Malaysian Nuclear Agency (MNA) has successfully developed, commercialized and disseminated the new mutant rice seeds to farmers in Malaysia. The new mutant rice variety, named NMR152, was derived from radiation mutation of the mega variety MR219. Despite cultivation under water stress conditions, this variety produced high yield and showed good agronomic traits [7]. The yields remain high, stable and constant across the granaries and non-granary areas in Malaysia. The approval of the NMR152 genotype as the national certified seed by the authority is a highly significant contribution of nuclear technology in the agriculture sector and can improve the livelihood of the rice farmers.

To improve drought tolerance, the assessments must involve pre-field screening using polyethylene glycol (PEG) and field screening at the reproductive stage under normal and drought stress conditions. Along with the assessment, the selected positive (tolerant) and negative (highly susceptible) checks must be included for validation of the drought tolerant ability. The common positive checks include AdaySel, Aeron 1, Apo, Kuku Balam and Nagina22, while common negative checks include IR64, MR219 and MR297.

There are many effects of drought stress, for example, morpho-physiological changes, biochemical, and molecular, that affect various cellular and whole plant processes. To investigate the biological changes mechanism under drought condition, physiological traits (photosynthetic rate, stomatal conductance, chlorophyll fluorescence, relative chlorophyll content and relative water content), morphological traits (plant height, number of tillers per plant, flag leaf area and plant biomass), biochemical parameters (proline content and antioxidant enzymes) and yield component (number of spikelets per plant, 1000 grain weight and weight per plant) are recorded. Under glasshouse conditions, only two traits are measured, namely the photosynthetic rate and stomatal conductance, whereas under field condition, all the traits are measured on ten plants per genotype in both treatments at peak flowering stage (85 days after transplanting (DAT)). This chapter will discuss mutation breeding experiments, isolation and advancement of rice mutant lines for resilience to drought.

2. MUTAGENESIS, SCREENING AND SELECTION

The prime objective has been to screen and select potential mutant lines suitable for growth and development under minimal water conditions, along with improved agronomic traits such as high yield and resistance to major blast diseases of rice. Towards this end, the popular mega rice variety, MR219 was used for gamma irradiation at a dose of 300 Gy. Screening and selection was initiated at M_2 generation with 20 000 mutant populations under simulated non-flooded water regime in the glasshouse. In addition, 500 panicles with a high percentage of grain filling were selected for M_3 population screening (Fig. 2). Further screening and selection was done in the field under water stress conditions for selecting potential mutants in M_4 generations (Fig. 3).

FIG. 2 Screening and selection of M₂ generations in the glasshouse. (a) vegetative stage, (b) reproductive stage, and (c) harvesting the panicles.

FIG. 3. Screening and selection on M₃ and M₄ generations in the rice field.

Yield evaluation (M₅–M₈ generations) in field conditions was done to determine the yield and other agro-morphological traits under water stress and non-stress (well-watered, control) conditions. From the drought scoring and yield evaluation data, two potential mutants showed good performance under water stress conditions and good agro-morphological traits were selected, and named as NMR151 and NMR152 (Tables 1 and 2). These two potential mutants have been registered under the Department of Agriculture (DOA) for Plant Variety Protection (PVP) and have been certified as new varieties in 2020. One mutant genotype, NMR152, has been chosen for further evaluation for multilocation trial (MLT), local verification trial (LVT), and pest and disease trials. This mutant genotype has been consistent with good drought scoring data and yield component data under water-stressed conditions.

TABLE 1. LEAF ROLLING AND LEAF DRYING SCORING
FOR NMR152, NMR151 AND MR219

Genotype	Leaf rolling	Leaf drying
NMR152	1	1
NMR151	3	1
MR219	5	3

TABLE 2. YIELD AND YIELD COMPONENT FOR NMR152, NMR151 AND MR219 IN TWO SEASONS

Genotype	Yield and yield component				
	Panicle numbers/m^2	Percentage filled grain (%)	Number of grain/panicles	1000 grain weight (g)	Yield (t/ha)
Season 1 (Off-season)					
NMR152	299.00[a]	90.50[a]	182.00[a]	30.20[a]	10.00[a]
NMR151	283.00[a]	89.50[a]	175.00[a]	29.00[a]	9.30[a]
MR219	242.00[b]	88.00[a]	150.00[b]	26.90[b]	7.50[b]
Average	274.70	89.30	169.00	28.50	8.93
Season 2 (Main season)					
NMR152	310.00[a]	88.00[a]	192.00[a]	30.80[a]	10.60[a]
NMR151	296.00[a]	84.50[a]	180.00[a]	29.30[a]	10.00[a]
MR219	275.00[b]	84.00[a]	169.00[b]	28.90[b]	8.00[b]
Average	293.70	85.50	180.30	29.70	9.53

Means followed by the different letters within a column are significantly different at the p$\leq$0.05.

3.   MULTILOCATION TRIALS AND LOCAL VERIFICATION TRIALS

Multilocation trials (MLTs) have been conducted at major rice granary and non-granary areas in Malaysia (Fig. 4). NMR152 consistently gave high yields compared with positive and negative checks IR77298-14-1-2-10 and MR219, respectively (Table 3).

FIG. 4. Multilocation trials at several major granary and non-granary areas in Malaysia.

TABLE 3. YIELD (t/ha) MEAN DATA OF NMR152, MR219 AND IR77298-14-1-2-10 AT MLT PLOTS UNDER DROUGHT STRESS AND NON-STRESS CONDITIONS

Location	Genotype		
	NMR152	IR77298-14-1-2-10	MR219
Bukit Merah, Perak	5.82[a]	3.93[b]	5.68[a b]
Bumbung Lima, Pulau Pinang	8.08[a]	8.18[a]	7.49[a]
Teluk Cengai, Kedah (main season)	9.31[a]	8.85[a]	8.56[a]
Teluk Cengai, Kedah (off-season)	6.04[a]	4.49[b]	5.48[b]
Titi Serong, Perak	7.17[a]	4.07[b]	7.03[a]

Means followed by different letters within a column are significantly different at p$\leq$0.05.

Local verification trials (LVTs) were conducted in several rice growing areas, with rice farmers participating in agronomy practices. Genotype NMR152 showed stable and high yield across all locations, with a yield range of 6.0 to 10.0 t/ha. The evaluation of significant pests and diseases in rice crops, for example, foliar blast, panicle blast, bacterial leaf blight (BLB), sheath blight, brown planthopper, and tungro virus diseases, have been conducted (Fig. 5). Mutant NMR152 demonstrated moderate resistance to foliar blast, BLB and sheath blight, and showed resistance to panicle blast. On the other hand, NMR152 showed moderate susceptibility to brown planthopper and tungro viruses.

FIG. 5. Pest and disease screening of (a) foliar blast, and (b) bacterial leaf blight.

Rice mutant variety NMR152 is the first rice mutant successfully commercialized by the Malaysian Nuclear Agency and is presently available to be distributed to rice farmers in Malaysia. More new varieties with high yield and drought tolerant traits are in great need of Malaysia in response to the negative effects of climate change, which is expected to be more severe in the near future.

3.1.   Pre-field screening protocol for drought tolerance in rice mutants at seedling stage

A robust, cost effective and reliable laboratory based drought screening tool at the early stages of plant growth is required to speed up the selected breeding programme for maximum genetic gain. The phenotyping protocol for seedling and early vegetative stage screening should be highly consistent and repeatable. Hydroponics is a simple culturing method; it maintains a consistent stress condition, contains sufficient nutrients, and allows genotypic variances to be attributed to innate tolerance differences [8]. The Yoshida culture based method proposed by Gregorio et al. [9] has been extensively used as a rapid method for screening large numbers of genotypes or populations. While there are several ways to screen in hydroponics, the use of perforated Styrofoam™ sealed with net worked effectively as the Styrofoam™ platforms rested on culture solution at the International Rice Research Institute (IRRI), Manila. Before scoring, four day old germinated seeds are cultivated for three additional days on nutrient solution under stress (typically 3–5% of PEG, i.e. 10–12 dS m^{-1}). Every tray must have a tolerant genotype and a sensitivity check to validate the screening. After two weeks, the seedling injury score is calculated based on the damage to the entry as described in Standard Evaluation System (SES) for rice [10, 11].

Polyethylene glycol (PEG) is a non-ionic, inert substance with a wide range of molecular weights that is commonly employed in drought experiments to create water stress, constant water potential and induces drought (osmotic) stress without the confounding environmental conditions that are generally associated with field trials [12–14]. PEG as a drought inducer has been studied in maize, barley, wheat, potato, sunflower, soybean, and rice [15–19].

Drought is a complex trait, and it depends on the interaction of different morphological, biochemical and physiological responses [19]. Morphological characteristics, for example, early maturity, early vigour and rapid growth, and physiological characteristics, including diffusive resistance of stomata, osmotic adjustment, leaf rolling, stomata closing and opening, the position of stomata, leaf water retention, and leaf senescence were linked with drought tolerance. In a large scale screening, leaf rolling character and leaf death are good parameters for measuring degrees of drought resistance [20]. Leaf rolling can be done in the morning or in the middle of the day to score visually. Since the flag leaf in a rice crop plays a vital role in grain filling and growth, a plant with the characteristics of delayed leaf rolling under water stress and faster recovery rate after eliminating the water stress was regarded as a favourable trait [21]. As a result, the selection process continues to discover genotypes with nearly erect flag leaves, allowing for extended photosynthesis.

Physiological characteristics of rice are affected by scarcity of water, for example in decreases in net photosynthetic rate, transpiration rate, stomatal conductance, water use efficiency, internal CO_2 concentration, photosystem II (PSII) activity, relative water content and membrane stability index [22]. In biochemical parameters, drought stress causes cell harm by increasing cellular temperature, protein deposition and denaturation due to lower water intake, faster transpiration rate, tighter stomata, and excessive production of reactive oxygen species (ROS) [23]. Drought increased the formation of metabolites, such as proline, glycine betaine, soluble polysaccharides, and gamma-aminobutyric acid (GABA) [24]. This results in toxicity and reduced enzyme activity and reduced photosynthesis, bleaching, curling, wilting, and eventual plant death.

3.2. Screening for drought tolerance in rice at the seedling stage

Experiments were performed using controlled environment facilities and test plant materials were compared to standard genotypes of known drought stress tolerance. Rice controls used in this study were:

(1) Towuti: A moderate, drought tolerant genotype, Indonesian origin.
(2) NMR151: An intermediate, drought tolerant mutant, Malaysian origin.
(3) NMR152: An intermediate, drought tolerant mutant, Malaysian origin.
(4) MR219: A drought susceptible genotype from Malaysia.

3.3. Screening for drought tolerance using hydroponics

The hydroponics method, the preparation of nutrient stock solutions and the setting up of the hydroponics system are described in Ref. [25]. Glass jars were filled with PEG-Yoshida working solution until the water level was about 1 mm above the mesh of the hydroponic set. Pre-germinated, seven day old young healthy seedlings (odd-looking or diseased seedlings should be removed at this stage) were transferred into a glass jar filled with PEG-Yoshida working solution [26], which mimicked the specific drought condition (unirrigated water stress). Dirty or contaminated seeds were discarded since they may rot during germination. If the symptoms were severe, start over with a fresh batch of seedlings and hydroponic materials. The hydroponic set experiment was then incubated at room temperature (24–27°C) and was allowed to grow for seven days. The water was replaced with Yoshida working solution, since vigorous seedlings require nutrients for germination. On day seven after being transferred to PEG- working solution, the seedlings show signs of leaf rolling (Fig. 6). The experiment was laid out in a complete randomized design (CRD) with three levels of drought stress and three replications. Seedling height and seedling dry weights were measured on day 14.

FIG. 6. Leaf rolling observed after seven days of drought treatment in susceptible and tolerant rice genotypes.

3.4.   Precautionary steps during plant growth in hydroponics

There will be a gradual decrease in solution volume over time due to evaporation from glass jar and transpiration from plants. Hence, there is a need for this to be brought back to the level of full capacity (volume of PEG working solution touching the glass jar) every two days. The pH of the PEG working solution needs to be adjusted to pH5.0. The nutrient solution is completely replaced if algal or microbial contamination is detected.

3.5.   Evaluation of rice genotype performance

At the 7th and 14th days, the seedlings from the glass jar were taken out using a pair of forceps. The number of germinated seeds was recorded at 24 h intervals. Seeds were considered germinated when both plumule and radicle extended to more than 2 mm from the seeds. The germination index was calculated after final germination using the following equation:

$$\text{Germination \% (GI)} = \frac{\text{Germination \% in each treatment}}{\text{Germination \% in control}} \times 100$$

3.6.   Seedling height, shoot length, root length and dry weight

Ten seedlings were chosen randomly, and seedling height was measured. Shoot length, root length, were measured in centimetres (cm) by a graduated scale, and total length was calculated from the recorded data. Dry weight was determined after drying the seedlings at 70°C for 48 h. Shoot and root were weighed in grams (g) using the electrical balance in fresh, dry and turgid conditions. Data on germination and seedling characteristics for each treatment were compared with the control to determine the drought tolerant rice genotypes.

3.6.1. *Relative seedling height*

The relative seedling height (RSH) was calculated using the following equation:

$$\text{Relative Seedling height (RSH)} = \frac{\text{Seedling length in treatment}}{\text{Seedling length in control}} \times 100$$

3.6.2. *Relative dry weight*

The relative dry weight (RDW) was calculated using the following equation:

$$\text{Relative dry weight (RDW)} = \frac{\text{Dry weight in each treatment}}{\text{Dry weight in control}} \times 100$$

The Standard Evaluation System (SES) [27] for rice from the International Rice Research in 1996 was used for screening of drought tolerant rice genotypes (Table 4). Visual scores for stress symptoms were on a scale of 0–9, where a lower score denotes tolerance and a higher score denotes susceptibility.

TABLE 4. STANDARD EVALUATION SYSTEM FOR RICE [27]

Trait	Score	Description
Drought resistance	0	Highly resistant: No symptoms
	1	Resistant: Light tip drying
	3	Tip drying to ¼ length in most leaves
	5	Moderately susceptible: ¼ to ½ of leaves fully dried
	7	Susceptible: More than 2/3 of all leaves fully dried
	9	Highly susceptible: All plants apparently dead
Drought recovery score	1	90–100% of plants recovered
	3	70–89% of plants recovered
	5	40–69% of plants recovered
	7	20–39% of plants recovered
	9	0–19% of plants recovered
Leaf rolling score	0	Leaves are healthy
	1	Leaves start to fold
	3	Leaves are folded (deep V shaped)
	5	Leaves are fully cupped (U shaped)
	7	Leaf margins touching (O shaped)
	9	Leaves are tightly rolled

The collected data were analysed to assess their statistical significance. The Statistix 10 program was used to perform statistical analysis. Means were separated by the least significant difference (LSD).

4. RESULTS

The data collected are: (1) germination rate (%); (2) drought score; (3) shoot length reduction (%); (4) root length reduction (%); and (5) relative dry weight (g). The results for all data collected are shown in Tables 5–9, and in Fig. 7).

TABLE 5. GERMINATION RATE (%) FOR TESTED GENOTYPES

Genotype	Germination rate (%)		
	0% PEG	5% PEG	10% PEG
Towuti (drought tolerant)	100.00	100.00	90.00
NMR151	100.00	100.00	76.60
NMR152	100.00	100.00	90.00
MR219 (drought susceptible)	100.00	93.30	53.33

The maximum seed germination percentage was observed under control conditions. However, the results showed negative correlation between the seed germination rate and water stress severity. At 5% concentration of PEG-6000, Towuti, NMR151 and NMR152 showed maximum (100%) germination rate, while MR219 exhibited the minimum germination rate of 93.30% (Table 5). At 10% concentration of PEG-6000, Towuti and NMR152 obtained the highest germination rate of 90.00% followed by MNR151 (76.60%) and MR219 (53.33%). It was observed that germination percentage with decreasing water potential of the environment probably was triggered by the low hydraulic conductivity of the environment, where PEG 6000 makes water unavailable to seeds, affecting the imbibition process of the seed, which is fundamental for germination.

TABLE 6. DROUGHT SCORES IN TESTED GENOTYPES

Genotype	Drought score	
	5% PEG	10% PEG
Towuti (drought tolerant)	1.00	1.20
NMR151	1.20	1.30
NMR152	1.80	2.30
MR219 (drought susceptible)	4.00	6.50

Note: 1 = Highly tolerant, 2–3 = tolerant; 4–5 = moderately tolerant; 6–9 = susceptible.

In rice, drying of tip and rolling of leaf was studied as one of the best criteria in estimating levels of drought tolerance in a large scale screening [28]. The SES for rice was used to score two mutant genotypes together with control genotypes during early stage. Based on the results presented in Table 6, mutant genotype NMR151 was found to be highly tolerant to water stress together with Towuti at both 5% and 10% of PEG concentrations. Another mutant genotype MNR152 was classified as highly tolerant at 5% PEG concentration, but tolerant at 10% PEG concentration. The drought susceptible control genotype MR219 was classified as moderately tolerant and susceptible to drought at 5% and 10% PEG concentrations, respectively.

TABLE 7. SHOOT LENGTH REDUCTION PERCENTAGE IN TESTED GENOTYPES: CASE STUDY IN MALAYSIA

Genotype	Shoot length reduction (%)		
	0% PEG	5% PEG	10% PEG
Towuti (drought tolerant)	0.00 (9.41 cm)	11.26 (8.35 cm)	48.35 (4.86 cm)
NMR151	0.00 (12.01 cm)	4.24 (11.50 cm)	27.31 (8.73 cm)
NMR152	0.00 (13.71 cm)	9.55 (12.40 cm)	19.76 (11.00 cm)
MR219 (drought susceptible)	0.00 (11.08 cm)	28.52 (7.92 cm)	64.30 (3.96 cm)

It has been confirmed that drought stress leads to growth reduction, which is reflected in plant height, biomass and other growth functions [29]. Due to water stress environment, the decrease in the water potential gradient between the external environment and seeds mainly causes the inhibition of radicle emergence and at the same time impairs seedling height [30]. As root length is more affected by drought than shoot length, the effect of drought is exhibited mostly on the shoot as well as aerial parts of the plant, which will bear the most economical parts of the crops. Hence, the shoot parameters will also help the breeder while selecting the superior genotypes against drought. In the present study, the shoot lengths significantly vary with increased external water potential and all of the treatments caused a decrease in root elongation in most genotypes compared to the control. Significant differences were observed for shoot length between the genotypes and different PEG-6000 concentrations. Furthermore, mutant genotypes NMR151 and NMR152 also showed the lowest reduction in shoot length compared with drought tolerant genotype, Towuti while susceptible genotypes, MR219 showed a reduction of more than 60% under a 10% concentration of PEG-6000 (Table 7).

TABLE 8. ROOT LENGTH REDUCTION (%) IN TESTED GENOTYPES

Genotype	Root length reduction (%)		
	0% PEG	5% PEG	10% PEG
Towuti (drought tolerant)	0 (13.69 cm)	9.93 (12.33 cm)	20.01 (10.95 cm)
NMR151	0 (15.62 cm)	13.25 (13.55 cm)	33.03 (10.46 cm)
NMR152	0 (13.78 cm)	14.37 (11.80 cm)	42.45 (7.93 cm)
MR219 (drought susceptible)	0 (11.75 cm)	25.87 (8.71 cm)	57.56 (5.00 cm)

Early and rapid elongation of roots is an important indication of drought tolerance. The structure and development of the rice root system determines crop function under drought stress. Ability for deep root growth and large xylem diameters in deep roots may increase root acquisition of water when sufficient water at depth is available. In the present study, it has been found that root lengths significantly vary with the increase of external water potential and consequently, all of the treatments caused a decrease in root elongation in most genotypes compared to the control. Root plays a major role in plant survival during drought and also drought tolerant can be characterized by extensive root growth. Both mutant genotypes, NMR151 and NMR152 showed lower root length reduction percentage compared to their parent MR219 (Table 8). Long roots were reported as a component trait for drought tolerance as they play a direct role with high penetration ability and have large xylem vessel radius and lower axial resistance to water flux aiding in greater water acquisition.

TABLE 9. RELATIVE DRY WEIGHT (g) IN TESTED GENOTYPES

Genotype	Relative dry weight (g)		
	0% PEG	5% PEG	10% PEG
Towuti (drought tolerance)	0.020	0.019	0.017
NMR151	0.037	0.026	0.022
NMR152	0.022	0.020	0.015
MR219 (Drought susceptible)	0.017	0.013	0.010

For relative dry weight, a decreasing pattern can be seen with the increased concentration of PEG in Table 9. At 10% PEG, all varieties tested exhibited lowest relative dry weight as compared with 0% PEG. Water stress that was induced by the PEG appears to reduce the absorption and utilization of water by forcing water to exit from the plant's cell to such an extent that the tolerance mechanisms employed by these plants in a drought are insufficient to maintain normal growth. Among varieties tested, the highest relative dry weight was found in the mutant variety which is NMR151 with 0.037 g, 0.026 g and 0.0022 g for 0%, 5% and 10% of PEG, respectively. The lowest relative dry weight was found in drought susceptible control genotype, MR219 in all PEG concentrations.

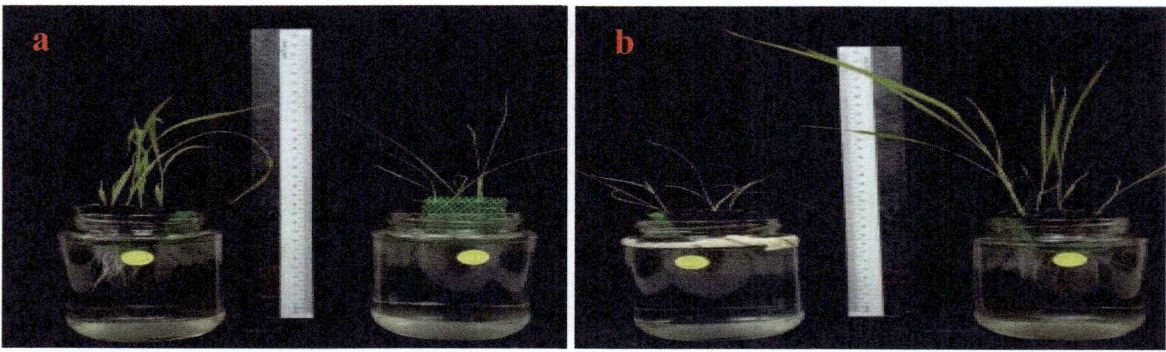

FIG. 7. Comparison of tolerant (NMR151) and susceptible (MR219) genotypes at (a) 5% and (b) 10% PEG.

5. CONCLUSIONS

It can be concluded that as the level of PEG or water stress increased, early seedling growth was strongly affected in all rice genotypes. The overall results showed that Towuti (positive check) was the most drought tolerant variety (highly tolerant), while mutants NMR151 and NMR152 were found to be drought tolerant. This finding was confirmed with data obtained from field screening conditions. As for the concentration of PEG, 10% was the optimum level to induce water stress in the study, since 5% PEG was not sufficient to create tension for the seedlings to grow. In vitro drought screening methods facilitate progress in our understanding of drought resistant traits and our selection of drought resistant genotypes. Exposure to polyethylene glycol (PEG-6000) solutions has been effectively used to mimic drought stress with limited metabolic interferences as those associated with the use of low molecular weight osmolytes that the plant can take up. Hence, PEG based in vitro screening used in this study is suitable as a simple, rapid and cost effective method for screening seedling traits of a large set of germplasms for drought tolerance with good accuracy. Since drought field based screening is labour intensive and sometimes problematic due to rainfall can eliminate water deficits, an early stage screening with PEG will help breeders to eliminate the susceptible lines and choose only the resistant plant for the next mutant generation.

5.1. Field screening protocol for drought tolerance in rice plant at reproductive stage

A majority of hands-on drought breeding programmes highlighted direct selection for grain yield under water stress. However, indirect selection for carefully selected secondary traits can be an alternative way in improving selection response. This chapter covered further validation of identified lines through morphological, physiological, and biochemical indicators that are reported to be correlated with plant stress response to classify the mechanism of drought tolerance of screened lines. Yield data were used to evaluate tested rice genotypes and comparisons made to drought tolerant and drought susceptible standards. Identified drought tolerant genotypes lines were tested under field conditions for the main season and off-season to evaluate real performance of selected promising lines in particular hotspot area at the northwest of Peninsular Malaysia (Kedah), known to experience severe drought periods. Identified lines were subjected to physiological and biochemical tests, including photosynthetic rate, stomatal conductance, total chlorophyll content, relative water content, proline content, and antioxidant enzymes (catalase, ascorbate peroxidase, and guaiacol peroxidase) activity at rice plant reproductive stage. The laboratory work was conducted at the Food Crop Molecular Laboratory, Universiti Putra Malaysia, Malaysia.

5.2. Screening protocol for drought tolerance in rice at reproductive stage

5.2.1. Plant materials, experimental site and experimental layout

Two drought tolerant mutants (NMR151 and NMR152), a popular local variety (MR219), a drought susceptible control variety (IR64), and a drought tolerant control variety (Aeron 1) were screened in this experiment. The field trial was conducted at Muda Agricultural Development Authority (MADA), Kota Sarang Semut, Kedah, Malaysia (latitude 6°13'10"N, longitude 100°14'18"E). Twenty-one day old seedlings of all genotypes were transplanted at one seedling per hill on the basis of randomized complete block design (RCBD) with four replications. The area of each main plot was 100 m^2. In one plot, plants were grown under favourable water conditions with supplementary surface irrigation (control) and in another plot water was drained off and irrigation was withheld to induce drought stress (Fig. 8). All rice genotypes were randomly assigned to the 4 m^2 subplots with spacing of 20 cm × 20 cm between and within rows.

5.2.2. Drought treatment at the reproductive stage

Two treatments were applied to test genotypes: (1) plants with no drought stress exposure (control); and (2) plants subjected to drought stress at reproductive stage. The plot with drought stress treatment was drained at 25 days after transplanting (DAT). Five tensiometers were used in this experiment to determine the soil moisture tension. All tensiometers were inserted randomly at 30 cm depth in the soil of drought stress treatment. Re-irrigation was done periodically when soil water tension fell below −30 kPa.

FIG. 8. Experiment under field conditions showing: (a) control treatment during main season, (b) drought stress treatment during main season, (c) control treatment during off-season, and (d) drought stress treatment during off-season.

5.2.3. Screening morpho-physiological traits for drought tolerance at the reproductive stage

During drought stress treatment, the leaf rolling and leaf drying scores of each genotype were taken as 1 to 9 scales, by referring to the IRRI SES for rice (Table 10). Ten representative plants for each genotype in each treatment were randomly selected at peak flowering stage (85 DAT) to record observations. The morpho-physiological traits, yield and yield component traits were recorded for plant height, number of tillers, flag leaf area, total dry weight per plant, photosynthetic rate, stomatal conductance, relative chlorophyll content, relative water content, grain weight, grain yield, 1000 grain weight, spikelets per panicle and harvest index as described in Table 11.

TABLE 10. DESCRIPTION OF LEAF ROLLING AND LEAF DRYING SCORE [31]

Scale	Description		Rate
	Leaf rolling score	Leaf drying score	
0	Leaves healthy	No symptoms	Highly resistant
1	Leaves start to fold	Slight tip drying	Resistant
3	Leaves folding (deep V shaped)	Tip drying extended to ¼ length in most leaves	Moderately resistant
5	Leaves fully cupped (U shaped)	¼ to ½ of the leaves fully dried	Moderately susceptible
7	Leaves margins touching (O shaped)	More than 2/3 of all leaves fully dried	Susceptible
9	Leaves tightly rolled	All plants apparently dead	Highly susceptible

TABLE 11. QUANTITATIVE TRAITS FOR RICE

Trait	Description
Plant height	The average height from ground level to the tip of the tallest tiller after flowering stage using a measuring tape
Number of tillers	Count the number of tillers per plant after tertiary tiller arose which included the tillers bearing panicles and not bearing panicles
Flag leaf area	Leaf length × leaf width × calibration factor (calibration factor used for all cereal crops is 0.75)
Total dry weight	Culms and leaves dry weight + panicle dry weight
Photosynthetic rate Stomatal conductance	The measurements were recorded on the fully expanded and exposed leaves (3rd or 4th leaf from the tip) using a portable photosynthesis system (Li-6400xt, Li-cor, Inc., Lincoln, NE, USA) in the morning (0900–1000 h) at peak flowering stage (85 DAT). The CO_2 flow rate was 400 μ mol m^{-2}s^{-1} and saturating photosynthetic photon flux density (PPFD) was 900 mmol m^{-2}s^{-1}.
Relative chlorophyll content	Fresh leaf sample (0.2 g) was cut into 0.5 cm pieces following the method by Ashraf et al. (2005).
Relative water content	Flag leaf of the main culm was excised and immediately soaked into liquid nitrogen and kept in ice box. Relative water content was measured based on the method of Turner and Begg (1981).
Grain weight	Weigh total grains produced per plant.
Grain yield	Weigh filled grains per plant.
1000 grain weight	Weigh any 1000 filled grains.
Spikelets per panicle	Count the number of filled grains per panicle.
Harvest index	Divide the value of grain weight by biomass(Biomass = Grain weight + Culms and leaves dry weight).

Drought tolerance indices are used for drought tolerance analysis on the basis of yield performance. Indices were calculated as follows:

$$\text{Relative yield index, REI} = [(Y_i)_s/(Y_s)] \times [(Y_i)_{ns}/(Y_{ns})] \quad [32]$$

$$\text{Mean productivity index, MPI} = [(Y_i)_{ns} + (Y_i)_s]/2 \quad [32]$$

$$\text{Mean relative performance, MRP} = [(Y_i)_s/(Y_s)] - [(Y_i)_{ns}/(Y_{ns})] \quad [33]$$

$$\text{Stress tolerance level, TOL} = (Y_i)_{ns} - (Y_i)_s \quad [34]$$

$$\text{Stress tolerance index, STI} = [(Y_i)_{ns} \times (Y_i)_s]/(Y_{ns})^2 \quad [35]$$

$$\text{Stress intensity, SI} = 1 - (Y_s/Y_{ns}) \quad [34, 35]$$

$$\text{Stress susceptibility index, SSI} = [1 - ((Y_i)_s/(Y_i)_{ns})]/SI \quad [36]$$

$$\text{Drought tolerance efficiency, DTE} = [(Y_i)_s/(Y)_{ns}] \times 100 \quad [35]$$

where Y_{ns} is the mean yield of all selected genotypes evaluated under control; Y_s is the mean yield of all selected genotypes under drought stress; $(Y_i)_{ns}$ and $(Y_i)_s$ denote with respect to the yield of the ith genotype under control and drought stress.

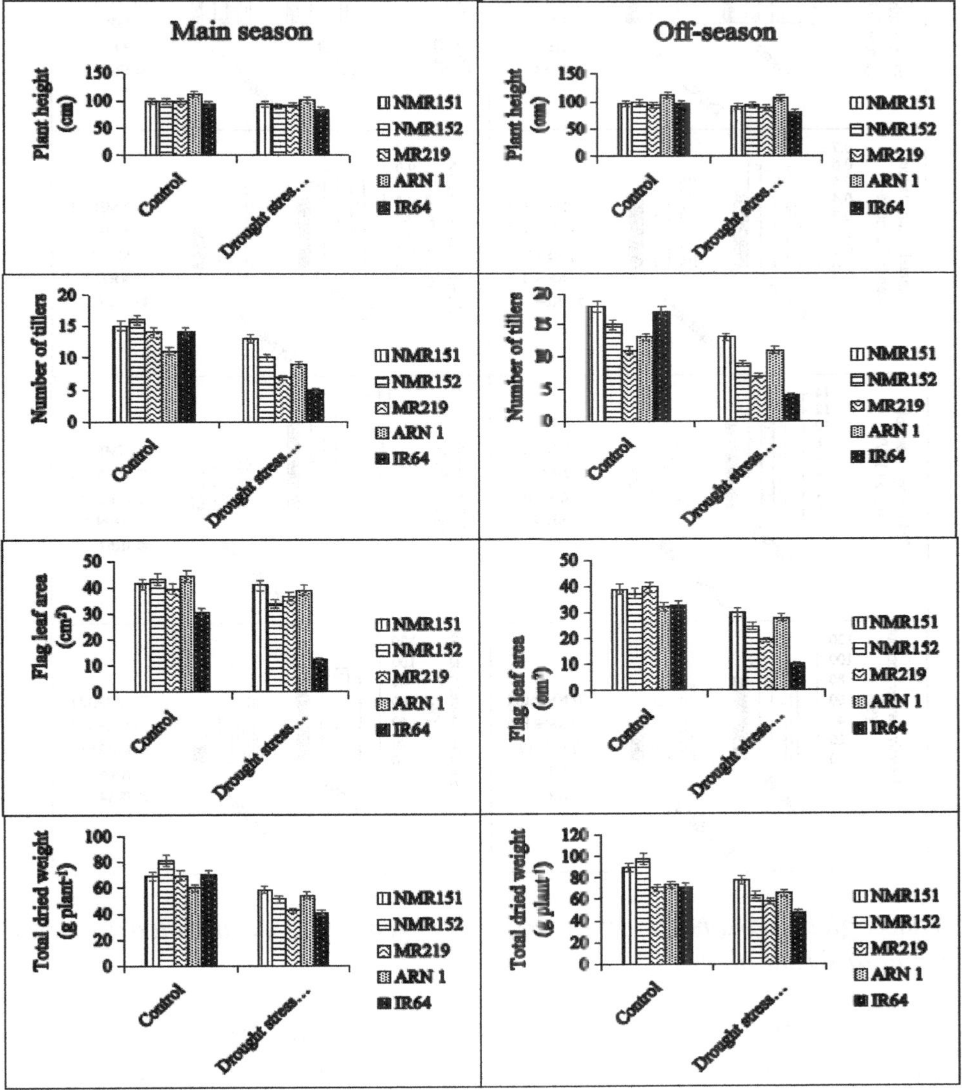

FIG. 9. Morphological parameter analysis in rice genotypes as affected by drought stress treatment during different planting seasons.

Generally, all studied morpho-agronomical traits were higher in control than in drought stress treatment (Fig. 9). Mutant genotype MNR151 recorded the highest values of number of tillers, flag leaf area and total dried weight under drought stress, while the lowest values of these traits were observed in drought susceptible control rice genotype IR64 (Fig. 9). From the morphological data recorded, it can be concluded that both mutant rice genotypes NMR151 and NMR152 performed well under drought stress compared with all control genotypes (Fig. 10). These two mutant rice genotypes also had higher vegetative and reproductive growth under control treatment, suggesting that the genetic controls were constitutive. Similar results were observed for physiological traits where the water stress has greater impacts on photosynthetic rate, stomatal conductance, relative chlorophyll content, and relative water content where all of these traits were reduced under drought stress compared to control treatment (Fig. 10). The advanced rice mutant genotypes also performed better in terms of physiological traits than their wild parent

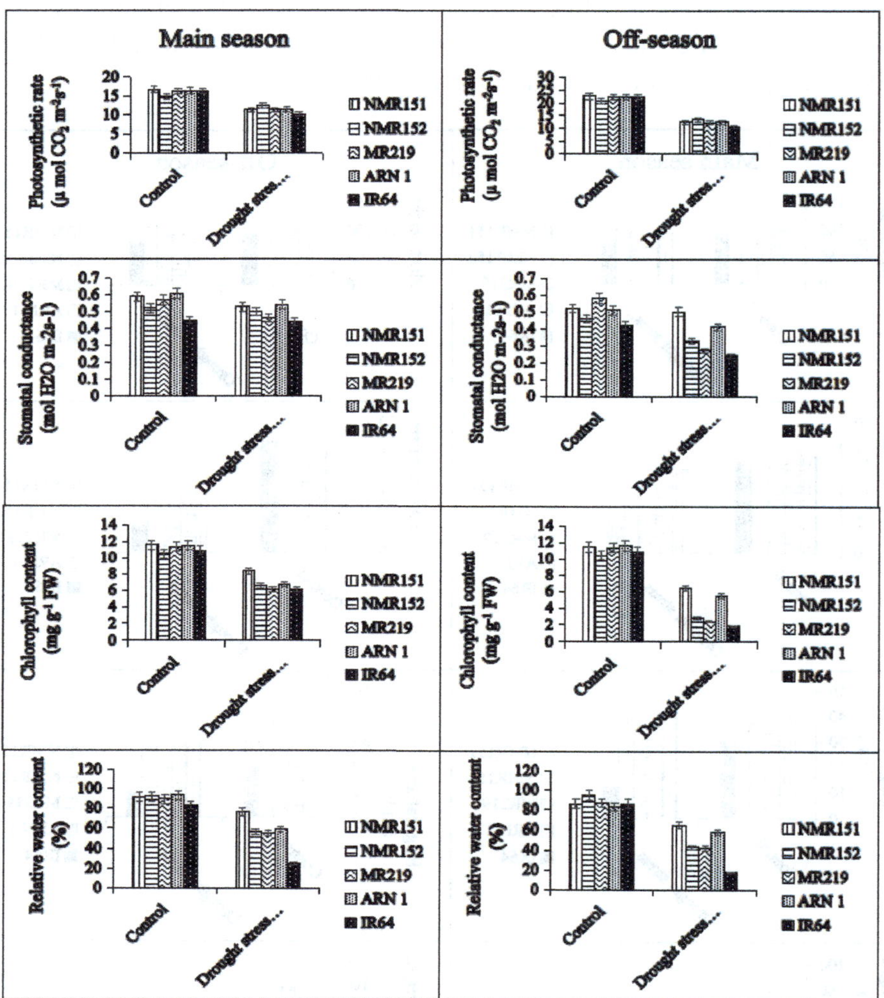

FIG. 10. Physiological parameter analysis in rice genotypes as affected by drought stress treatment during different planting seasons.

MR219 under drought stress treatment (Fig. 10). Under drought stress treatment, NMR151 and NMR152 recorded the highest values of photosynthetic rate, stomatal conductance, relative chlorophyll content, and relative water content while the greatest reduction of these traits were observed in IR64. The role of these traits in minimizing the impacts of drought stress in rice has been extensively discussed in past studies.

NMR151 and NMR152 were identified based on their yield and yield attribute performance under drought stress treatment (Table 12). Analysis of the drought tolerance indices indicates that the ability of drought tolerance for each rice genotype was different in response to drought stress intensity of each season. MNR152 showed the highest yield under drought stress treatment in both seasons. This mutant genotype also outperformed other rice genotypes under control treatment during the off season (Table 13).

The role of morpho-physiological and agronomic traits in minimizing the impacts of drought stress in rice has been extensively discussed in past studies. Water stress brings about osmotic stress which affects the turgor pressure, causing a decline in cell expansion and growth which finally affects plant productivity [37, 38]. In addition, photosynthesis is another factor that plays a major role in determining plant growth and productivity under drought stress conditions. Lack of stomatal conductance, which decreases under drought stress, could also be associated with a reduction in the photosynthesis rate since stomatal conductance is a primary driver of hydrological changes that regulate plant response to environmental stresses [39]. Therefore, a reduction in the photosynthesis rate could be attributed to a decrease in the transpiration rate and stomatal conductance as observed in this experiment (Fig. 11).

TABLE 12. GENOTYPE MEAN YIELD, RELATIVE YIELD PERFORMANCE AND MEAN RELATIVE YIELD IN RESPONSE TO DROUGHT STRESS TREATMENT IN RICE GENOTYPES DURING DIFFERENT PLANTING SEASONS

Treatment	NMR151	NMR152	MR219	Aeron 1	IR64
Control plant at main season					
Mean yield (g)	32.58[c]	34.38[b]	35.03[b]	25.36[d]	47.72[a]
RY (g)	0.68	0.72	0.72	0.53	1.00
Mean RY (g)	0.73				
Drought stress plant at main season					
Mean yield (g)	34.00[bc]	59.81[a]	36.62[b]	30.52[c]	24.73[d]
RY (g)	0.57	1.00	0.61	0.51	0.41
Mean RY	0.62				
Control plant at off-season					
Mean yield (g)	53.15[b]	92.17[a]	45.69[c]	46.82[c]	46.22[c]
RY (g)	0.58	1.00	0.50	0.51	0.50
Mean RY (g)	0.62				
Drought stress plant at off-season					
Mean yield (g)	50.88[b]	65.49[a]	45.87[bc]	44.83[c]	29.89[d]
RY (g)	0.78	1.00	0.70	0.68	0.46
Mean RY (g)	0.72				

RY = relative yield. Means followed by different letters within a row are significantly different from each other according to the DNMRT at $p \leq 0.05$.

TABLE 13. DROUGHT TOLERANCE INDICES OF GENOTYPES UNDER DROUGHT STRESS TREATMENT DURING TWO PLANTING SEASONS

Genotype	Drought tolerance indices						
	REI	MPI	MRP	TOL	STI	SSI	DTE (%)
Main season							
NMR151	0.87 (4)	33.29 (3)	1.87 (3)	1.43 (1)	1.27 (1)	0.20 (1)	95.79 (1)
NMR152	1.00 (3)	35.83 (2)	2.01 (2)	1.59 (2)	0.69 (3)	0.19 (2)	95.71 (2)
MR219	1.42 (2)	29.37 (4)	1.68 (4)	9.27 (4)	0.56 (4)	1.21 (4)	72.68 (4)
Aeron 1	0.61 (5)	27.94 (5)	1.55 (5)	5.15 (3)	0.80 (2)	0.83 (3)	83.13 (3)
IR64	1.60 (1)	47.10 (1)	2.57 (1)	25.41 (5)	0.48 (5)	2.04 (5)	57.91 (5)
Off-season							
NMR151	0.68 (3)	45.78 (3)	1.66 (4)	0.69 (1)	1.71 (1)	0.12 (1)	98.51 (1)
NMR152	0.88 (2)	52.02 (2)	1.88 (2)	2.28 (3)	0.60 (3)	0.35 (3)	95.79 (3)
MR219	1.96 (1)	78.83 (1)	2.82 (1)	16.68 (4)	0.59 (4)	2.07 (4)	71.19 (4)
Aeron 1	0.66 (4)	45.83 (4)	1.67 (3)	2.00 (2)	0.77 (2)	0.31 (2)	95.80 (2)
IR64	0.51 (5)	38.05 (5)	1.44 (5)	26.33 (5)	0.43 (5)	2.23 (5)	64.32 (5)

REI = relative yield index; MPI = mean productivity index; MRP = mean; relative performance; TOL = stress tolerance level; STI = stress tolerance index; SSI = stress susceptibility index; DTE = drought tolerance efficiency; ( ) = ranking for each index.

6.   CONCLUSIONS

For over three decades, the SSL for rice in Malaysia has remained at 60–71%, while the remaining is imported from Thailand, Viet Nam and Pakistan. This high dependency on imports can threaten the stability of the country's staple food supply. Therefore, increasing the country's rice production is a key target of the government. However, abiotic stress such as drought, submergence and salinity are major constraints to rice production in Malaysia. Despite the green revolution's goal of reducing food insecurity and malnutrition problems encountered in many regions across the globe, the increasing drought frequency and incidences especially at critical stages, viz. seedling and reproductive stages have become major hurdles and serious threats in achieving sustainable rice production. This has brought about the urgent need to produce rice cultivars tolerant to drought stress particularly at the seedling and reproductive stages. The use of induced mutation breeding for the development of drought tolerance rice cultivars has gathered pace and is becoming more popular among rice breeders. However, a standard protocol, which has undergone rigorous testing and stringent validation is required to facilitate efficient screening of tolerant cultivars via mutation breeding. A standard protocol was designed for plant breeders and scientists and validated via pre-field (glasshouse and growth chamber) and field conditions. The candidate mutant rice lines were compared with the check varieties using morphological, physiological, biochemical and agronomical traits which have indirect and direct links to drought stress. Selected drought tolerant lines experienced increase in proline accumulation and production of enzymatic antioxidants to scavenge ROS, reduced MDA accumulation and high photosynthetic and stomatal conductance values. This signifies that evaluated candidate mutant lines exhibited improved performance under drought conditions.

Based on the trials conducted and the results obtained, it was concluded that screening for drought tolerance rice cultivars can be critically evaluated at the seedlings, vegetative and reproductive stages. Pre-field seedling evaluation of candidate mutant lines with PEG in a controlled environment showed that the lines tested were tolerant to drought stress with a high germination rate as well as seedling growth-traits. The pre-field and field screenings results of advanced mutant lines at the tillering and reproductive stage using the physiological and biochemical traits as indicators are highly correlated, demonstrating that the designed protocol can be consistently used to determine drought resilient genotypes in a mutation breeding programme. Furthermore, considerable numbers of identified drought-tolerant advanced mutant rice genotypes tested under greenhouse conditions exhibited improved performance under drought stress in the field, thus reducing the numbers of genotypes to be subjected to stringent evaluation in the field for further validation. The present studies showed that evaluation and selection of drought tolerant genotypes can be conducted in the greenhouse at first prior to field trials with few hundred genotypes as compared to thousands due to greenhouse screening.

However, the processes in the development of tolerant drought genotypes are considered tedious and complicated due to the complexity of the drought tolerance mechanism. These protocols are mainly designed for and by plant breeders and scientists in order to facilitate rice breeding development programmes in the screening of large populations, under pre-field and field conditions, and to identify useful plant genetics resources for drought tolerance. To date, little work has ensued on the use of molecular approaches and next generation sequencing in identifying important candidate genes and quantitative traits loci (QTLs) in mutant rice genotypes under drought conditions. In the future, more emphasis will be given to the application of molecular approaches and advanced biotechnology for the identification of important mutated regions and candidate genes, which can be useful in improvement of yield and associated traits, and in conferring drought tolerance or other abiotic stress and to hasten breeding process of rice mutant varieties.

REFERENCES TO CHAPTER 2–6

[1] BIN RAHMAN, A.N M., ZHANG, J., Trends in rice research: 2030 and beyond, Food Energy Security **12** (2023) e390. https://doi.org/10.1002/fes3.390.

[2] FIRDAUS, R., TAN, M., RAHMAT, S., GUNARATNE, M.,. Paddy, rice and food security in Malaysia: A review of climate change impacts, Cogent. Soc. Sci. **6** (2020) 1– 18.

[3] CHE OMAR, S., SHAHARUDIN, A., TUMIN, A., The status of the paddy and rice industry in Malaysia, Khazanah Research Institute, Kuala Lumpur, (2019).

[4] RAHIM, F.H.A., NURULNAZIHAHH, N.Z., Supply and demand of rice in Malaysia: A system dynamics approach, Int. J. Sup. Chain Mgt **6** (2017) 1–7.

[5] SANDHU, N., ARVIND, K., Bridging the rice yield gaps under drought: QTLs, genes and their use in breeding programs, Agronomy **7** (2017) 27.

[6] MIZUNO, H., Mutant rice and agricultural modernization in Asia, Hist. Technol. (2021) 360–381.

[7] KAMARUDIN, Z.S., YUSOP, M.R., TENGKUMUDA, M.M., ISMAIL, M.R., HARUN, A.R., Growth performance and antioxidant enzyme activities of advanced mutant rice genotypes under drought stress condition, Agronomy **8** (2018) 279.

[8] AKTER, S., JAHAN, I., HOSSAIN, M.A., HOSSAIN, M.A. Laboratory- and field-phenotyping for drought stress tolerance and diversity study in lentil (*Lens culinaris* Medik) Phyton — Int. J. Exp. Bot. **90** (2020) 950–970.

[9] GREGORIO, G.B., SENADHIRA, D., MENDOZA, R.T., Screening Rice for salinity Tolerance, IRRI Discussion Paper Series, International Rice Research Institute, Manila (1997) 1–30

[10] SINGH, H., MACKILL, K.T., Sensitivity of rice to water deficit at different growth stages, Phil. Crop Sci. **16** (1991) S11.

[11] INTERNATIONAL RICE RESEARCH INSTITUTE, Standard Evaluation System (SES) for Rice, International Rice Research Institute, Manila (2013).

[12] ALMANSOURI, M., KINET, J.M., LUTTS, S., Effect of salt and osmotic stresses on germination in durum wheat (*Triticum durum* Desf.), Plant Soil **231** (2001) 243–254.

[13] TURKAN, I., BOR, M., OZDEMIR, F., KOCA, H., Differential responses of lipid peroxidation and antioxidants in the leaves of drought tolerant *P. acutifolius* Gray and drought sensitive *P. vulgaris* L. subjected to polyethylene glycol mediated water stress, Plant Sci. **168** (2005) 223–231.

[14] LANDJEVA, S., NEUMANN, K., LOHWASSER, U., BORNER, M., Molecular mapping of genomic regions associated with wheat seedling growth under osmotic stress, Biol. Plant **52** (2008) 259–266.

[15] AHMAD, R., AHMAD, M.R., ASHRAF, M., ASHRAF, E.A., WARAICH, E.A., Sunflower (*Helianthus annuus* L.) response to drought stress at germination and seedling growth stages, Pak. J. Bot. **41** (2009) 647–654.

[16] JAIN, M., MITTAL, M., GADRE, R., Effect of PEG-6000 imposed water deficit on chlorophyll metabolism in maize leaves,. J. Stress Physiol. Biochem. **9** (2013) 262–271.

[17]   MIRBAHAR, A.A., SAEED, R., MARKHAND, G.S., Effect of polyethylene glycol-6000 on wheat (*Triticum aestivum* L.) seed germination, Int. J. Biol. Biotech. **10** (2013) 401–405.

[18]   ZHANG, D.J., et al., A novel soybean intrinsic protein gene, GMTIP2;3, involved in responding to osmotic stress, Front. Plant Sci. **6** (2016) 1237.

[19]   OGUZ, M.C., AYCAN, M., OGUZ, E., POYRAZ, I., YILDIZ, M., Drought stress tolerance in plants: Interplay of molecular, biochemical and physiological responses in important development stages, Physiologia **2** 4 (2022) 180–197 https://doi.org/10.3390/physiologia2040015.

[20]   KIM, Y., et al., Root response to drought stress in rice (*Oryza sativa* L.), Int. J. Mol. Sci. **21** (2020) 1513.

[21]   EVANS, L.T., WARDLAW, I.F., FISCHER, R.A. "Wheat", Crop Physiology (EVANS, L.T., Ed.), Cambridge University Press, Cambridge (1975) 101–149.

[22]   ZHU, R., et al., Cumulative effects of drought-flood abrupt alternation on the photosynthetic characteristics of rice, Environ. Exp. Bot. **169** (2020) 103901.

[23]   GILL, S.S., TUTEJA, N., Reactive oxygen species and antioxidant machinery in abiotic stress tolerance in crop plants, Plant Physiol. Biochem. **48** (2010) 909–930

[24]   SINHA, R., IRULAPPANV, MOHAN-RAJU, B., SUGANTHI, SENTHIL-KUMAR, M., Impact of drought stress on simultaneously occurring pathogen infection in field-grown chickpea, Sci. Rep. **9** (2019) 5577.

[25]   OTHMAN, T.A., et al., Transcriptomic profiling of Malaysia rice cultivars (*Oryza sativa*) under effect of osmotic stress induced by PEG 6000, Int. J. Agric. Biol. **21** (2019) 375–384.

[26]   CHUTIA, J., BORAH, S.P., Water stress effects on leaf growth and chlorophyll content but not the grain yield in traditional rice (*Oryza sativa* Linn.) genotypes of Assam, India II. Protein and proline status in seedlings under PEG induced water stress, Am. J. Plant Sci. **3** (2012) 971–980.

[27]   International Network for Genetic Evaluation of Rice. Standard Evaluation System for Rice, International Rice Research Institute, Los Baños (1996).

[28]   PANDEY, V., SHUKLA, A., Acclimation and tolerance strategies of rice under drought stress, Rice Sci. **22** (2015) 147–161; https://doi.org/10.1016/j.rsci.2015.04.001.

[29]   SWAPNA, S., SHYLARAJ, K.S., Screening for osmotic stress responses in rice varieties under drought condition, Rice Sci. **24** 5 (2017) 253–263.

[30]   SOKOTO, M.B., MUHAMMAD, A., Response of rice varieties to water stress in Sokoto, Sudan Savannah Nigeria, J. Biosci. Med. **2** 1 (2014) 68–74.

[31]   INTERNATIONAL RICE RESEARCH INSTITUTE, International Network for Genetic Evaluation of Rice, Standard Evaluation System for Rice, IRRI, Manila (2014).

[32]   FISCHER, R.A., MAURER, R., Drought resistance in spring wheat cultivars: I. Grain yield responses, Austr. J. Agric. Res. **29** (1978) 897–912.

[33]   ROSIELLE, A.A., HAMBLIN, J., Theoretical aspects of selection for yield in stress and non-stress environment, Crop Sci. **21** (1981) 943–946

[34] FERNANDEZ, G., "Effective selection criteria for assessing stress tolerance", Adaptation of Vegetables and Other Food Crops in Temperature and Water Stress Tolerance (Proc. Int. Symp. Taipei, 1992), Asian Vegetable Research and Development Centre, Taipei (1992) 257–270.

[35] HOSSAIN, A.B.S, SEARS, A.G., COX T.S., PAULSEN, G.M.,. Desiccation tolerance and its relationship to assimilate partitioning in winter wheat, Crop Sci. **30** (1999) 622–627.

[36] KUMAR A, BASU S, RAMEGOWDA V, PEREIRA A. Mechanisms of Drought Tolerance in Rice, Burleigh Dodds Science Publishing, Cambridge, UK (2016) 1–34.

[37] GOCHE, T., et al., Comparative physiological and root proteome analyses of two sorghum varieties responding to water limitation, Sci. Rep. **10** 1 (2020)1–18.

[38] HAQUE, K.M.S., KARIM, M.A., BARI, M.N., ISLAM, M.R.,. Genotypic variation in the effect of drought stress on phenology, morphology and yield of aus rice, Int. J. Biosci. **8** 6 (2016) 73–82.

[39] SHARIF, P., et al., Effect of drought and salinity stresses on morphological and physiological characteristics of canola, Int. J. Environ. Sci. Technol. **15** 9 (2017) 1–8.

2–7. EVALUATION OF GAMMA RAY INDUCED COARSE RICE MUTANTS FOR YIELD AND DROUGHT TOLERANCE

Z.U. QAMAR[1], M. ASHRAF[1], S. BIBI[1], F. SARSU[2], A. KHAN[1]

[1]Nuclear Institute for Agriculture and Biology, Faisalabad, Pakistan

[2]Joint FAO/IAEA Centre of Nuclear Techniques in Food and Agriculture,
Department of Nuclear Sciences and Applications,
International Atomic Energy Agency,
Vienna

Abstract

Climate resilient cultivars offer a cost effective solution to mitigate the drought stress. During the current study, 24 coarse rice mutants (Nos 1–24), along with tolerant (Nagina-22, BRRI Dhan-56), susceptible (WAB-56-104) and parent (IR-6) checks, were evaluated to assess their performance under optimum field conditions and water stressed conditions maintained by withholding irrigation water 15 days prior to heading and terminating at grain formation. Data on days to maturity (DTM), plant height (PH), productive tillers per plant (PT), panicle length (PL), 1000 GW (TGW), tiller fertility percentage (TF%), panicle fertility percentage (PF%) and paddy yield per plant (PY/P), were used to compute the variation, pair wise comparisons, Pearson's correlation coefficients, stress tolerance index (STI) and principal components (PCA). Mutants exhibited significant variation under normal, stressed, and pooled analysis of variance. Two water regimes interacted significantly with each other and genotypes. Fisher's pairwise comparisons indicated significant differences among mutants and susceptible check; WAB-56-104. Generally agro-morphological traits exhibited highly significant positive associations with yield. The PCA also indicated significant positive influence of these traits under stress (TFS, PHS, PTS, PFS and PLS) and optimum (PTN, PLN, 1000GW, PHN and PFN) conditions, to improve the yield. The PCA indicated that under stress conditions TFS, PHS, PTS, PFS and PLS have much influence on selection efficiency for yield and can be selected together. Similarly, under optimum conditions traits like PTN, PLN, 1000GW, PHN and PFN have significant positive contribution during selection for yield. These results further highlight the importance of selecting genotypes based on yield components, which could result in simultaneous selection for complementary genes adding up to yield under stressed and optimum conditions. Highly significant positive association between STI-Y/P (stress tolerance index of yield per plant) and Y/PS (yield per plant under stress) authenticated the usefulness of the index. Eleven mutants with relatively higher Y/PS and stress adaptive traits were selected.

Key words: Rice, induced mutation, drought stress, tolerance index, principal component

1. INTRODUCTION

Rice (*Oryza sativa* L.) is the leading cereal for about 3.5 billion people as it fulfils 35–75% of their daily caloric requirement. About 95% of rice is grown and consumed in Asia, the region with more than half of the world's population [1]. Geographically, rice production extends from 50° N to 35° S [2]. An ever-increasing human population will reach 10 billion by 2050, which suggests a 2.5% per year growth rate in rice production is needed to meet the demand [3]. Global warming is threatening water availability with recurrent episodes of drought [4]. Climate change predictions show remarkable increases in drought and heat spells in semi-arid areas of the globe [5]. Drought stress, during the reproductive stage of rice, is among the major challenges facing rice productivity and quality [6–9]. Globally, about 50% of the cultivated rice area is anticipated to suffer from water stress, causing substantial yield losses in rice production

[10]. Predictions indicate that almost half the world's population will be living in areas of high water stress by 2030 [11].

In Pakistan, rice (*Oryza sativa* L.) is the second major cereal, followed by wheat, and accounts for 7.4 million tonnes of production, an 0.6% share of GDP and exports worth US $2 billion per year [12]. The climate of Pakistan is mostly arid to semi-arid, with less than 254 mm rainfall per year. Rainfall patterns have remarkable variations in terms of intensity and distribution, leading to serious incidents of drought every four to five years. Additionally, about 75% of the mean annual rainfall occurs during a relatively short, hot, span of time, July–September, which makes it less efficient due to the high evapotranspiration rate during these months. The Indus Basin Irrigation System (IBIS), a major source of surface water in Pakistan, has shown significant reduction from 233 128 million cubic metres to 113 480 million cubic metres since 1962 and a further 31% reduction is predicted by the year 2025. Additionally, underground water in the country is mostly substandard and costly due to high pumping costs. It places the country in the world's 'high water stress' category.

Breeding drought tolerant rice lines carrying associated agronomic and adaptive characters is important to enhance productivity and food security among rice producing countries of the world. Induced mutation is a powerful tool to create novel genetic variation which can be used for the selection of superior genotypes under changing climatic conditions [13]. Genetic variations created through physical or chemical mutagens induce changes in agronomic traits and the development of mutant varieties [14]. Phenotyping has shown promise for screening of breeding lines based on drought adaptive morphological characteristics, including yield and its components [15–17]. Selection for such traits through the classical breeding tools has improved rice yields remarkably under both optimum and limited water conditions. Among important agronomic traits, reduced plant height (PH) is strongly related to harvest index in dry land cereal crops, especially in water limited environments [18]. Yield and associated traits that can be used for the selection of drought responsive lines include plant height (PH)t, number of productive tillers per plant (PT/P), tiller fertility percentage (TF%), panicle fertility (PF), yield per plant (Y/P) and thousand seed weight (TSW). Reduced number of days to heading (DTH) and days to maturity (DTM) are also important when breeding for terminal drought stress tolerance since they allow for drought escape [19]. Among various growth stages, drought stress, triggered by climate change at the reproductive growth stage of rice, is being felt throughout the rice-growing world [20–22]. Reports suggest that water stress, during this period leads to increased panicle sterility, eventually posing a serious threat to rice production [23]. Classically, selection should focus genotypes with relatively high yields under both stressed and optimum conditions for their improved adaptation to changing climatic conditions, which emphasizes the significance of the stress tolerance index (STI) of test genotypes. Thus, there is a need to select genotypes with a useful combination of agronomic traits contributing positively towards grain yield under water stressed conditions [24]. Thus, selection under water stressed conditions, using drought stress indices is an effective strategy to identify high yielding drought tolerant rice lines. Principal component analysis (PCA) is a powerful tool for quantifying genetic divergence among germplasm collections with respect to characters. In this study, we have evaluated the mutant populations (M_2–M_6) under water stress at reproductive and seedling stages, and selected 24 stable mutant lines in the M_6 generation, for their characterization to water stress at the reproductive stage.

2. METHODOLOGY

2.1. Creation of genetic variability and generation advancement

A coarse rice cultivar, IR-6 was mutagenized at a 250 Gy dose of gamma ray irradiation using ^{137}Cs. The M_1 generation was raised in the field and a single panicle was harvested from each of the M_1 plants. A single panicle was picked up from each of the M_1 plants and an M_2 generation, comprising about 2000 plants, was raised in the field.

2.2. Screening of segregating generations of mutants under limited water conditions at pre and post flowering stages

About 2000 putative mutants were transplanted, in the augmented design, in the field as the M_3 generation. Normal irrigation and other agronomic practices, for example the application of fertilizer and necessary plant protection measures were continued up to two weeks before the flowering of the mutants; afterward, irrigation was stopped to impose the stress up to two weeks after flowering (withholding irrigation two-week pre- and post-flowering stage). Data were recorded on yield and its associated traits and selections of drought tolerant mutants were performed based on high yield, panicle fertility, and a large number of fertile tillers per plant.

2.3. Screening of mutants under limited water conditions at seedling stage

These 2000 putative mutants were screened at seedling stage in the tunnel following the modified protocol of Standard Evaluation System for Rice (SES, 2013) of the International Rice Research Institute. For screening of drought tolerant mutants in the tunnel, paddy seeds were sown in natural field conditions and covered with farmyard manure and wheat straw. Irrigation was done twice a day by using a sprinkler irrigation method. The seed bed was kept moist until the seedlings emerged. After germination, normal agronomic practices were continued for up to 15 days. Then water stress was maintained by withholding irrigation water and continued until the seedlings were 36 days old. Then recovery irrigation was applied, and data were recorded on seedling height, seedling vigour, leaf rolling and leaf drying to assess the drought tolerance level of the mutants. After data recording re-irrigation was performed for up to two weeks for seedling recovery. The seedling recovery was recorded based on the number of mutants recovered.

2.4. Selection of mutants based on seedling stage and reproductive stage water stress

Final selections of stable mutant lines were made based on the performance of the mutants under seedling stage and reproductive stage water stress conditions. This screening process was repeated up to the M_5 generation and finally, 24 productive stable mutant progenies were selected for their characterization under normal and water stressed conditions using statistical analysis of phonological data.

2.5. Replicated yield trial under drought stressed and normal field conditions

(i) Selection of experimental site

Yield trials were conducted in the Faisalabad district (31°2′ N, 73°05′ E) of Punjab Province which is in the rice production Zone II of the country. This zone is situated in the broad strip of land between the rivers Ravi and Chenab where both canal and subsoil water is used for irrigation. The climate is subhumid, subtropical with 400–700 mm of rainfall; mostly in July–

August. The average monthly temperature is 5–18°C during winter and 20–49°C during summer in this rice production zone.

(ii)   Preparation of seed bed, raising of nursery seedlings and transplanting

Nursery seedlings were raised in the field area of the Nuclear Institute for Agriculture (NIAB), Faisalabad, Punjab (183 m above mean sea level) (31°2′ N, 73°05′ E); using the traditional dry method of sowing. About 2000 seeds, of each mutant and standard cultivars were sown under the field conditions in the nursery beds. Seeds were covered with a thin layer of farmyard manure. Wheat straw was spread over the farmyard manure to save the seeds from birds and direct hitting of sprinkled water. Sprinkling of water continued for three days. Regular plant protection measures were adopted during the nursery period of the seedlings. The quality of irrigation water was monitored and mostly showed EC: 0.76-0.8 dS m-1, pH7–7.5 and SAR 2–2.5. Standard cultural practices and plant protection measures used in the irrigated ecosystem of rice in Punjab province were observed.

(iii)   Transplanting for yield trials

Two sets of a yield trial, each one comprising 24 mutant lines, parent variety IR-6, two drought tolerant, BRRI Dhan-56 and Nagina-22, and a drought susceptible standard cultivar, WAB56-104 were used for these field trials. About 35 day old seedlings of the mutants and standard cultivars were transplanted. Each plot size was 15 m^2 and a single seedling per hill was transplanted by maintaining a 20 cm plant to row spacing in a triplicated randomized complete block design (RCBD) in natural open field conditions.

(iv)   Application of water stress and data recording

One set of seedlings was allowed to grow under normal irrigation throughout the growing season, while the irrigation of the other set was stopped two weeks before flowering and continued up to two weeks after flowering (pre- and post-flowering water stress). All the fertilizers except nitrogen were applied at the time of transplanting of the seedlings. Nitrogen was applied in three equal splits: a basal dose, 25 days after transplanting (DAT) and 40 DAT. Zinc sulphate (25%) was applied 10 DAT at 5 kg/acre. In order to control the stem borer and leaf folder, aneriistoxin analogue insecticide, Padan®, was applied 45 DAT at 9 kg/acre. Minor, intercultural operations, and pest control measures were done as and when necessary. At maturity, five plants were harvested within each replication of the normal and water stressed set, and data were recorded on yield and its associated traits like plant height (cm), productive tillers, tiller fertility (%), panicle length (cm), panicle fertility (%), thousand seed weight (g), and grain yield per plant (g). Data were also recorded on maximum and minimum temperatures, relative humidity, and total sunshine hours to observe any extreme variation in climatic conditions during the crop cycle. Grain yield obtained on per plant basis was adjusted to 14% grain moisture content.

(v)   Meteorological data recording

In order to monitor any sudden fluctuation in the weather during 2021, the meteorological data were recorded at the weather observatory of the Ayub Agriculture Research Institute, Faisalabad, at latitude 31°– 44' N, longitude = 73°–6' E, altitude = 184.4 m. The distance from this institute is 500 m to the experimental site at the Nuclear Institute of Agriculture and Biology (NIAB).

(vi)   Soil moisture monitoring of the experimental site

The soil moisture was measured with a neutron probe. The soil moisture contents in the root zone were monitored throughout the experiment at the seedling stage in the tunnel as well as in the open field. The measurements were recorded at five soil depths (0–15, 15–30, 30–50, 50–70 and 70–100 cm) using an on-site calibrated neutron moisture probe. On an average basis, the distribution of volumetric moisture content (θv; mm^{-1}) in different soil layers was measured at 7–10 day intervals and converted into total soil moisture (mm) in the root zone. The moisture movement following the irrigation was recorded at different soil depths. The temporal changes in soil moisture content during different irrigation events were also recorded. The weather data of the experimental site were also included after harvesting the crop.

(vii)   Statistical analysis

Analysis of variance, comparison of means, Pearson's correlation, and principal component analysis (PCA) were performed using Minitab Version 18.1 computer software for Windows. The significance of correlation between yield and other agronomic/seedling growth traits was determined at 0.01 and 0.05 levels of probability. For PCA, procedures as mentioned [25–28] were applied. The variable loadings (correlation coefficients) and the variances (eigenvalues) regarding the components were computed for all the characters at the first step following a correlation matrix as all the traits had equal importance with different scales. The proportion of the total variance explained by each principal component was additive, with each new component contributing less than the preceding one to the explained variance. According to Ref. [29], data were considered in each component with an eigenvalue >1 which determined at least 10% of the variation. The higher eigenvalues were considered as best descriptive of traits in principal components. Successively, the components were designated whose eigenvalue was >1, and varimax rotations were performed until all the communalities were ~0.7. The values of yield and its associated agro-morphological traits (plant height, total number of tillers per plant, tiller fertility (%), panicle length, thousand grain weight, days to maturity and paddy yield per plant) were included in the PCA. The eigenvalues generated by PCA were used to grade mutants for their water stress tolerance. The first two PC scores (PC1 and PC2), accounting for maximum variability of the parameters tested, were used to classify the mutants. The mutants that had +PC1 and +PC2 scores were classified as tolerant, those with +PC1 and −PC2 scores as moderately tolerant, those with −PC1 and +PC2 scores as moderately susceptible, and those with −PC1 and −PC2 scores as susceptible following Kakani et al. (2005). To identify high yielding genotypes, under limited water and normal water conditions, stress tolerance indices (STIs) were computed using the following formula [30]:

$$STI = (Yn * Ys)/(Xp)^2$$

where Ys = paddy yield of a test genotype under limited water conditions; Yn = paddy yield of a test genotype under normal water conditions, and Xn = mean yield of test genotypes under normal water conditions.

3.   RESULTS

The data depicted in tables and figures is presented in Annex at the end of the reference section.

3.1. Analysis of variance under normal field conditions

The mutants and standards exhibited highly significant differences for various agro-morphological traits, such as PH (cm), PT/P, PL (cm), PF%, Y/P (g), 1000 GWand DTM (Table A–1). Under the water stressed field conditions, the tested genotypes (mutants and check cultivars) showed highly significant differences for all the agro-morphological traits such as PH, PT/P, PL, total spikelet, TF% (%), Y/P, and 1000 grains weight (Table A-2).

Notably, under normal field conditions, the genotypes and standards showed 100% TF as compared to water stressed conditions; hence this parameter was not included in the statistical analysis of the agro-morphological traits under normal field conditions (Table A–1). However, under water stressed field conditions, mutants and standards showed highly significant diversity for TF% (Tables A–1 and A–2); hence this parameter was subjected to statistical analysis. In order to study the interaction of various plant traits with two water regimes, a two way analysis of variance was performed (Table A–3). Significant to highly significant differences were observed among the genotypes, moisture levels and the interaction between genotypes and moisture levels for all the agro-morphological traits. Days to maturity showed significant differences for replicates suggesting that replications have played a significant role to identify the experimental error for this trait, DTM.

3.2. Correlation under normal field conditions

Significant differences among the genotypes for various agro-morphological traits emphasized the importance of further statistical analysis to study the association of these traits among themselves and yield. Therefore, correlation analysis was performed to observe the trait association under normal and stressed field conditions. Paddy Y/P showed significant to highly significant positive association with PT(0.37), PL (0.72), PH(0.55), PF%(0.55) and 1000 GW (0.45), while a non-significant negative association was noted with DTM –0.03). In addition to association with Y/P, agro-morphological traits showed a varying degree of association among themselves: significant association between PH and PT (0.37), PL and PF% (0.41); and highly significant association between PL and PT (0.73), PF% and PH (0.50). Significant negative association, of any trait, was not recorded under normal water conditions (Table A– 4). All the mutants showed 100% tiller-fertility under optimum conditions and no variation. Therefore, this trait exhibited no correlation with other agro-morphological traits (upper diagonal of Table A–4).

3.3. Correlation under stressed field conditions

In under water stress field conditions, Y/P showed a highly significant positive association with PT (0.63), TF% (0.59), PL (0.43), and PF% (0.40). Contrary to normal water conditions, PH showed a non-significant association with Y/P (0.35). All these traits showed a highly significant association with each other (Table A–5). Similar to normal water conditions, 1000 GW and DTM showed non-significant associations with all other traits except PH, which showed a significantly negative association with 1000 GW. The highest degree of positive and significant associations was observed between PF and PL (0.83); PT and PF (0.78); PT and PLS (0.72). In general, PH, 1000 GW and DTM showed non-significant associations with Y/P and other traits.

3.4. Pairwise comparison of mean performance

Pairwise comparison of the mean performance, under normal and water stress condition, of the mutants and standard cultivars are presented in Table A–6. These comparisons were performed for those agro-morphological traits which exhibited significant association with paddy Y/P. DTM and 1000 GW showed non-significant association with yield, therefore they were not considered for this comparison.

3.5. Pairwise comparison of mean performance under normal water field conditions

Under normal water conditions, drought tolerant check cultivar Nagina-22 was the tallest with 128 cm PH, while drought susceptible check cultivar WAB-56-104 was the shortest with PH 72 cm. Among the newly induced mutant lines, mutant DT11 was the shortest (83 cm) and mutant DT24 was the tallest (103 cm), as compared with the parent cultivar IR-6 (95 cm). With respect to PT, the minimum number of PT (8) was produced by WAB-56-104, while the maximum number of PT (19) was produced by the mutant DT3 followed by mutant Nos DT2, DT4, DT113, DT15 and BRRI Dhan-56 (18). Under normal water conditions, the mutants and standard cultivars exhibited 100% TF% and no variation was observed. Regarding PF%, the minimum PF% (78%) was exhibited by the susceptible cultivar WAB-56-104, while maximum PF% was shown by mutant No. 2 (94%). Among the mutant lines, mutant DT5 produced minimum PF% (78%), which is non-significantly different from that of the susceptible check-WAB-56-104 (82%). The rest of the mutants and standard cultivars exhibited non-significant variation for this trait. Regarding paddy Y/P, the lowest yield was produced by the susceptible check cultivar WAB-56-104 (8.0 g), while the highest Y/P was produced by mutant No. 3 (35.0 g), followed by mutant DT6 (33 g), DT2 (32 g), DT15 (32 g), DT4 (32 g) and Nagina-22 (32 g). Since these mutants were selected from the M_2–M_5 generations under drought stress, relatively less diversity was observed among the mutants for various traits under irrigated conditions, as indicated by pairwise comparisons using LSD tests.

3.6. Pairwise comparison of mean performance under water stressed field conditions

Under water stress conditions, drought resistant check cultivar, BRRI Dhan, was the tallest at 97 cm PH followed by Nagina-22 (85 cm). The drought susceptible check cultivar WAB-56-104 was the shortest with a pH of 58 cm. For PH, BRRI Dhan and Nagina showed significant difference with each other. Among the induced mutants, PH ranged between 79 (mutant Nos DT7 and DT16) to 84 cm (mutant No. DT21, followed by mutant No. DT4). The rest of the mutants showed almost non-significant differences for PH under stress condition. The minimum number of PTs was produced by the susceptible check cultivar WAB-56-104 (five tillers), followed by Nagina-22 (8). Among the mutants, the maximum number of PTs was produced by mutant Nos DT1, DT7 and DT16, each one producing 17 tillers non-significantly different from the tolerant cultivar BRRI Dhan (16 tillers per plant). The minimum number of PT was produced by mutant No. 11, followed by mutant Nos DT14 (12 tillers per plant) and DT24 (12 tillers per plant). The rest of the mutants showed almost a non-significant difference with one another. The maximum TF% was exhibited by mutant No. DT4 (91%), non-significantly different from the tolerant check BRRI Dhan and majority of the mutant lines. The minimum TF% (40%) was produced by the susceptible check cultivar WAB-56-104, followed by Nagina-22 and mutant No. DT18, each having 58% TF%. The maximum PF% was exhibited by mutant No. DT16 (93%), which was non-significantly different from the tolerant check and majority of the mutant lines. The minimum PF% (25%) was exhibited by the susceptible check WAB-56-104, which was significantly different from all other mutants, standards and parent cultivar-IR-6 (77%). The highest Y/P (19 g) was produced by mutant Nos DT1 and DT8, not significantly different from mutant Nos. DT4, DT6 and DT12, which produced 16 g, 16 g and 18 g, respectively. The minimum yield was produced by the susceptible check cultivar WAB 56-104, non-significantly different from mutant Nos DT3, DT11, DT18, DT19, DT20, IR-6 and Nagina-22. Overall, seven mutants showed better yield over the tolerant check cultivar BRRI-Dhan. Regarding the mean performance of the mutants, for various agro-morphological traits, a declining trend was observed under stress conditions as compared with the performance under normal water conditions.

3.7. Correlation of stress tolerance indices of various agro-morphological traits with yield per plant under stressed field conditions.

Stress tolerance indices (STIs) of various agro-morphological traits were computed (Table A-7). Correlation analyses were performed among various STIs and Y/P under stress. Paddy Y/P, under stress conditions, showed highly significant association (0.625) with an STI of PT/P (STI-PT), STI of TF% (0.585), STI of PF (0.339) and STI of paddy Y/P (0.925). Y/P, under stress, showed non-significant association with STI-PH, STI-PL, STI-1000GW and STI-DTM. Significant to highly significant correlations were recorded among other STIs. Strong association of STIs with the paddy Y/P, under stress conditions, indicated the efficiency of this index in identifying high yielding mutant lines having better tolerance to water stress conditions. Stress tolerance indices of various agro-morphological traits and paddy Y/P, under stress condition, are presented in Table A–7. Eight DT mutants, 1, 8, 12, 6,4,13,16 and 5, produced higher yield and better tolerance as compared with the tolerant check cultivar BRRI Dhan. Mutant Nos 10, 21 and 2 showed stress yield performance comparable to BRRI Dhan, tolerant check. However, mutant No. 10 showed slightly better resistance over this check cultivar. Mutant No. DT14 and DT24 and DT7 produced an average PY and tolerance to the water stress. Three mutants, Nos DT23, DT17, DT22, showed the PY and stress tolerance below the average ranging: PY ranged from 10.67 to 11 and STI ranged from 0.33 to 0.38. Mutant Nos DT9, DT15 and DT30 exhibited the PY below average, but stress tolerance comparable to average of the population (0.45). Mutant Nos DT7, DT8, DT9 and DT10 exhibited poor yield response (9.00–9.33 g) and low values STIs (0.31–0.34). The average grain Y/P was reduced by 55.54% under stress as compared with the control. The minimum and maximum STI was 0.06 and 0.77 observed on the genotypes WAB-56-104 and mutant No. 8, respectively. Mean STI was 0.45 with 46% of the genotypes having above average STI for yield. Mean STIs for PH, PT, TF%, PL, PF%, 1000GW and DTM, were 0.85, 0.84, 0.76, 0.98, 0.93, 0.85 and 0.97, respectively.

3.8. Principal component analysis (PCA)

Under normal and stressed field conditions, associations among various genotypes and variables, with respective principal components are further demonstrated by the score plot and loading plots in Figs A–1 and A–2 (under normal field) and Figs A–3 and A–4 (under stressed field). The narrow angles between dimension vectors (Figs A–2 and A–4) in the same bearing indicated a high association of the variables in terms of differentiating genotypes. Genotypes topping up in a specific trait were plotted closer to the vector line and further in the direction of that particular vector, often on the peaks of the curving body. Under normal water conditions, three principal components were important for contributing 79.3% of the total variance, of which 64.5% was contributed by the first two components. The first component contributed 47.7% to the total variance, while the second component contributed 16.7% to the total variance. The communality values, for all the variables under normal water field conditions were found to be within permissible limits, ranging from 0.7 to 0.95. Traits such as PT, PL, YP, 1000GW contributed significantly towards the first principal component while PH and DTM had negative factor loadings (Table A–8) into the first component. PH had the highest positive factor loading into the second component, followed by PF and YP and PT. DTM showed significant association with the third component, but this component describes a very small proportion (14.9%) of the total variance (79.3%). Score plot under normal water conditions categorized the mutants and standard cultivars into four groups. Fifteen DT mutants, (Nos 1, 2, 3, 4, 5, 6, 7, 8, 9, 14, 15, 18, 20, 22, 25), including the parent cultivar IR-6 (No. 25) were concentrated on the positive side of the first principal component. Being located on the positive side these mutants and parent cultivar were labelled as tolerant (quadrant I) to moderately tolerant (quadrant II). The quadrant-1 represents the genotypes with high yield; due to the positive association of PF, PT and PL, while the third quadrant represents the genotypes with short PH, low yield and number of PT per plant and short PL. Out of seven, six genotypes in the third quadrant (Nos 11, 16, 17, 19, 21 and 23) were concentrated near the origin indicating a non-significant difference among each other and the population mean. The susceptible check cultivar WAB-56-104 (No. 28) describes a distinct behaviour having the longest vector in the negative sides of both principal components (PC1 and PC2). The drought tolerant check cultivars Nagina-22, BRRI Dhan along with newly induced mutant Nos 10 and 24 were categorized in the fourth quadrant, which describes the tall growing genotypes with relatively long DTM and low grain weight.

Likewise, under stressed field conditions, three principal components were important, contributing 81% of the total variation detected (Table A–9). The first two principal components were the most important, with a cumulative contribution of 68.4% to the total variation. The first principal component contributed 49% to the total variance, while the second principal component contributed 19% to the total variance. Factor analysis of the first three components, having Eigen vector value ≥ 1, described the communality values within acceptable limits, ranging from 0.749 to 0.859. All the variables had high positive loading into the first principal component except DTM (–0.054). Significant positive trait associations were shown by TFS, YPS, PF, PT and PH with the first principal component. The variable DTM had the lowest negative loading (–0.054) and 1000GW (0.06) had the lowest positive loading to the first principal component. TF% had a positive factor loading into the second component, while all other components had a negative factor loading. The variables PLS, PT and DTM had the highest negative association to the second principal component. Traits like PH had the maximum positive contribution (0.581) and 1000 GWS showed the highest negative contribution to the third principal component (–0.903).

The score plot, under stressed field conditions, categorized the mutants and standard cultivars into four groups. About 71.4% of the mutants were scattered in the positive side of the first component. This site represents the genotypes with higher TF%, PF%, number of PT per plant, longer panicles which are positively contributing towards Y/P. The tolerant check cultivar, BRRI Dhan, was also scattered into this side of the first component. The third quadrant represented the mutants having early maturing genotypes with low PT, describing a negative contribution towards the Y/P under stressed conditions. The PL, TF% and PF% were strongly correlated and were parallel to the reference line. Under the stressed conditions the check cultivars, Nagina-22 and mutant No. DT11, were categorized in the third quadrant, which was designated as susceptible due to –PC1 and +PC2 scores. The susceptible check cultivar WAB-56-104 was scattered in the fourth quadrant, which describes the low PF and short PH with higher grain weight and was designated as moderately susceptible due to –PC1 and +PC2 scores. Eleven DT mutants (Nos 1, 2, 4, 5, 6, 8, 12, 13, 16, 10, 21) with relatively higher PY under water stressed conditions and favourable adaptive agro-morphological traits were selected for future water stress tolerance breeding.

4. DISCUSSION

Physical mutagens are efficient and cost effective tools to create new allelic combinations in an arbitrary fashion. These combinations can be selected for targeted objectives by applying appropriate screening tools. During the current study, 24 coarse rice mutants, induced by the irradiation of cultivar IR6 (gamma rays ^{137}Cs) were evaluated under stressed and optimum field conditions. Stress was managed by withholding irrigation water 15 days prior to heading and terminating at grain formation.

Drought tolerance is a complex trait and is governed by the interaction of various morphological responses. It can occur at any plant growth stage and has detrimental effects on crop yield [31]; however, the reproductive stage is considered as the most sensitive as major yield loss occurs at this stage [32]. Terminal water stress is being felt throughout the rice growing areas [21, 22] and is characterized by increased sterility, leaving unfilled grains and heavy yield penalties [23]. Up till now, very few drought tolerant rice cultivars have been documented [33, 34]. Advanced breeding programmes emphasize selection of novel allelic combinations under managed water stress in the field for high throughput and robust screening. Selection under managed water stress, in the field, is cost effective and provides real time estimates of yield and stress performance [34]. Escape, tolerance, and avoidance are different adaptive mechanisms that plants confer to tackle the harsh climate; however, lack of efficient screening methods to select drought tolerant rice cultivars is still a major challenge for breeders. Biochemical and molecular screening tools often lack effective correspondence with the farmer's field because of the environmental interactions with the genetics of the selected gene pool. Phenotyping remains a mainstay for screening germplasm resources based on drought adaptive and constitutive agro-morphological characteristics, including yield and its components [15, 16]. A selection of genotypes that produce a harmonious combination of significant agro-morphological traits, cumulatively contributing towards yields under target stressed environments, offers an effective strategy to combat the issue sustainably [24]. Application of stress tolerance indices under the managed water stress in the field offers effective yield based screening strategy and PCA can help to identify the genotypes conditioning stress resistance and high productivity.

The current study was based upon the selection of gamma ray induced high yielding water stress tolerant lines having resistance to drought stress at terminal stage. A diverse array of stable mutants, induced by gamma rays, ^{137}Cs source, was used to select the stress tolerant lines with high productivity. Significant to highly significant genotypic differences were observed among all the traits recorded describing the worth of the current mutant gene pool for water stress tolerance. These mutants can be used to detect genotypes with high levels of tolerance to water stress, as indicated by differential genotypic responses to different water regimes. The observed significant effects of the mutants, water regimes and mutant by water regime interactions were anticipated as all the evaluated mutants were selected from diverse mutagenized populations and most of the agro-morphological traits were quantitatively inherited, hence differentially respond to the environment.

4.1. Paddy yield response of genotypes to normal and stressed water conditions

From crop breeding viewpoint, drought tolerance is defined as the capacity of plants to grow and produce economic yields under water limited conditions [35]. Stress indices are quantitative measures that assess water stress response by yield performance from one or several environments based on timing, frequency and severity of stress. Such an index is advantageous to assess the performance of a genotype compared to the use of raw yield data [36]. Selecting for better paddy yield, under stressed conditions, allows the mutants to sustain the productivity since the same genotypes will be expected to perform well in either stress or normal water conditions [37]. However, genotypes that maintain high yield under stress and non-stress conditions are very rare in a population [38]. During the current study, no such mutant was observed. However, mutant Nos DT1, DT8, DT12, DT6, DT4, DT13, DT16 and DT5 maintained relatively higher paddy yield and stress tolerance as compared with the tolerant check BRRI Dhan (Table A–7). Maintaining a higher paddy yield and stress tolerance by these mutants authenticates the proficiency of the index, proposed by Fernandez [30] and used in the current study, to select genotypes with relatively higher yield under stressed and optimum environment. Additionally, paddy yield under stress (Y/PS) displayed a highly significant positive and strong association (0.925) with its STI (STI-Y/PS), which reflects the reliability of this index for selecting stress tolerant genotypes with optimum yield (Table A–7). Overall, 46% of the newly induced mutants exhibited better paddy yield, under stressed conditions, as compared to the population mean yield (12.10 g). Under normal water conditions, the highest paddy yield was produced by mutant No. 3 (35 g). However, under water stress conditions this mutant produced very poor yield per plant (10 g). Conventionally, breeders accept that a genotype with high yield potential will perform well under most environments. However, the performance of mutant No. 3 clearly deviated from this classical concept. Thiry et al. [36] described that this concept does not consider the significance of yield stability and adaptation to a stress environment, perhaps a reason for slow progress in crop breeding for stress tolerance. Generally, under a stress environment, most of the mutants, tested in the current study showed better performance over the parent cultivar IR-6, tolerant check Nagina-22 and susceptible check variety WAB-56-104.

4.2. Association of agro-morphological traits under different water regimes and testing environments

Under both stressed and optimum conditions, Y/P exhibited a significant to highly significant association with PT, PL, PF% and TF%, which describes the direct contribution of these traits (Tables A–4 and A–5). Therefore, these traits need to be targeted during selection to high productivity and water stress tolerance. Similar findings have been reported by Sareen et al. [39] and Mwadzingeni et al. [37]. These findings are reflected in the high yielding and stress tolerant mutants of the current study: Nos DT1, DT4, DT5, DT6, DT 8, DT12, DT13 and DT16 (Figs A–3 and A–4). Moreover, it was found that late maturing genotypes maintaining long fertile panicles and higher number of fertile tillers contribute relatively higher yields under stressed conditions compared to other yield components (Fig. A–4). Slafer et al. [40] described that, under limited water conditions the number of grains produced per plant compensates well for the decline in paddy seed weight. However, under normal water conditions (Fig. A–2), all the yield components have considerable contribution to grain yield, suggesting that selection for any of the yield associated traits, in the current study, could significantly enhance the paddy yields. Under optimum water conditions, late maturing and tall growing genotypes have enough time and capacity to accumulate photo-synthates to produce higher paddy yields while under stress conditions; genotypes outperforming in these traits suffered from yield penalties due to high evapo-transpiration losses and tendency of plants to reduce their life cycle for survival, availing less time to accumulate the photo-synthates. This could be the reason for the decline in ranks, under water stress conditions, of the majority of the mutants in the current study. However, late maturing and tall growing genotypes tend to lodge, under optimum water conditions, and pay yield penalties, whereas early maturing genotypes have an advantage of drought escape, permitting the genotype to harvest available water resources efficiently and utilize them during critical and sensitive seedling and reproductive growth stages [18].

Moreover, late maturing genotypes put an extra strain on the input resources like irrigation, weedicides, pesticides and fertilizer inputs due to the extra time spent in the harsh field conditions. The current study suggests that the life cycle of the cultivar should neither be too short nor too long and plant height should also be rational according to the prevailing field conditions.

Principal component analysis (PCA) is very useful to give the dimensionality to large data sets for quantification of genetic divergence, among germplasm collections, based on the traits under study. The principal component analysis indicated that under stress; PHS, PTS, TFS and PFS have the highest contribution to selection for yield improvement (Table A–9). This further signifies the selection of genotypes based on yield associated agro-morphological traits that could result in real time selection for complementary genes, acting in additive fashion, to improve the yield. Under such scenarios, focusing on a few major genes may result in increased survival rate at the expense of grain yield [16]. Based on the score plots, genotypes were classified as: tolerant having +PC1 and +PC2 scores; moderately tolerant with +PC1 and –PC2 scores; susceptible having –PC1 and –PC2 scores, moderately susceptible with –PC1 and +PC2 scores (Fig. A-3). In general, all the yield associated traits, except 1000 grain weight, were located on the positive side of the first principal component with major contribution to paddy yield. Vector of 1000 GW was located on the negative side of the first component, which covers a major proportion of the variation under stress conditions (49%). Plant height showed a major contribution to yield under stress conditions, but as compared with the other traits it showed a declining trend towards the negative side of the first principal component (Fig. A–4).

Under the optimum conditions, the high positive loading of PFN, PTN, PLN and 1000GWN into the first principal component indicates that they have a strong influence and can be concurrently selected for because of their direct influence on each other (Fig. A–2). This could be explained by the fact that tall growing genotypes with increased panicle fertility, productive tillers and panicle length have high yielding ability under normal conditions. Days to maturity and 1000 grain weight exhibited a non-significant contribution towards yield improvement (Fig. A–2, Table A–9).

5.   CONCLUSIONS

The current study highlights the use of gamma rays in the selection of high yielding water stress tolerant coarse rice lines having resistance to drought stress at the terminal stage. A diverse array of stable mutants, induced by gamma rays from a ^{137}Cs source, was used to select the stress tolerant lines with high productivity. Statistical analysis indicated significant differences among the mutants. Under optimum conditions, all the agro-morphological traits, except DTM, exhibited significant to highly significant positive association with Y/P. Under stressed conditions, an almost similar trait association was found except that PHS and 1000GW showed non-significant positive associations with Y/P. Association analysis suggested the direct contribution of PT, TF, PL and PF to yield both under stressed and optimum field conditions. The positive and strong association between STI-Y/PS and Y/PS (0.925) authenticated the proficiency of the tolerance index used in the current study. Three mutants (Nos DT1, DT8 and DT12) exhibited relatively higher yield and STI-Y/PS, which further verified the proficiency of this index. Under stressed conditions, the traits like TFS, PHS, PTS and PFS should be given high preference, followed by PLS, during selection for higher yield. Eleven mutants with relatively higher Y/PS and STI-Y/PS have been selected to further screen these mutants against simultaneous occurrence of terminal drought and heat stress to mitigate the increasingly unpredictable changing climate scenarios.

ACKNOWLEDGEMENTS

The authors are thankful to the IAEA for providing technical and financial support through Coordinated Research Project Pak-22224, "Breeding High Yielding Rice for Limited-Water Conditions Through Mutagenesis and Related Approaches". The authors are further thankful to the Pakistan Atomic Energy Commission and the Nuclear Institute for Agriculture and Biology Faisalabad for their active cooperation and encouragement throughout the project.

ANNEX

Source	df	Mean square (under normal irrigation conditions)						
		PHN (cm)	PTN	PLN (cm)	PFN	Y/PN	1000WTN	DTMN
Genotypes	27	251.78**	13.7**	12.53**	58.95**	73.54**	2.36**	12.20**
Replicates	2	2.94	2.6	1.3	24.3	2.8	0.00	2.08
Error	54	1.70	3.05	0.80	9.50	6.830	0.35	0.32

TABLE A–1. ANALYSIS OF VARIANCE OF EIGHT PHENOTYPIC TRAITS OF 24 COARSE RICE MUTANTS ALONG WITH A PARENT. IR6 AND CHECKS EVALUATED UNDER NORMAL FIELD WATER CONDITIONS

PHN: Plant height under normal field; PTN: plant tillers under normal field; PLN: panicle length under normal field; PFN: panicle fertility under normal field; Y/PN: yield per plant under normal field; 1000WTN: 1000 grains weight under normal field; DTMN: days to maturity under normal field *: significant at 5%, **: highly significant at 1%, NS: non-significant, N: normal field condition.

TABLE A–2. MEAN SQUARES AND TEST OF SIGNIFICANCE AFTER ANALYSIS OF VARIANCE OF EIGHT PHENOTYPIC TRAITS OF 24 COARSE RICE MUTANTS, ALONG WITH A PARENT VARIETY. IR6 AND FOUR STANDARD CULTIVARS EVALUATED UNDER WATER STRESSED FIELD CONDITIONS

SOV	df	Mean square (under stressed)							
		PHS	PTS	PLS	PFS	TFS	Y/PS	1000GWS	DTMS
Genotypes	27	92.8**	16.95**	11**	549.8**	490.54**	33.18**	2.9**	16.93
Replicates	2	0.16[ns]	0.47[ns]	0.08[ns]	3.12[rs]	111.20[ns]	3.06[ns]	0.92[ns]	24.53
Error	54	0.73	1.82	0.66	20.50	43.46	3.75	0.24	0.45

PHS: plant height under stress; PT: plant tillers under stress; TF: tillers fertility under stress; PL: panicle length under stress; ET: empty tiller under stress; PF: panicle fertility under stress; Y/P: yield per plant under stress; 1000GWS: 1000 grain weight under stress; DTMN: days to maturity under stress; *: Significant; **: Highly significant; Prob.: probability; S: stressed field condition.

TABLE A–3. POOLED ANALYSIS OF VARIANCE OF EIGHT PHENOTYPIC TRAITS OF 24 COARSE RICE MUTANTS ALONG WITH PARENT. IR6 AND CHECK CULTIVARS EVALUATED ACROSS NORMAL AND WATER STRESSED FIELD CONDITIONS

SOV	df	Mean Square (under stressed open field)							
		PH	PT	PL	PF	TF	Y/P	1000GW	DTM
Genotypes	27	273.8**	26.53**	22.41**	351.60**	245.3**	57.13**	2.19**	13.69**
Environment	1	9713**	293.3**	17.36**	1755.6**	24875**	9971.5**	453.4**	164.02**
Genotype × environment	27	70.73**	4.10*	1.20*	237.23**	245.3**	49.59**	3.00**	15.44**
Error	110	1.21	2.405	0.714	14.94	22.9	5.21	0.298	0.748

PH: plant height, PTS: plant tillers, TFS: tillers fertility, PLS: panicle length, PF: panicle fertility, Y/P: yield per plant, 1000GW: 1000 grains weight. DTM: days to maturity; *: Significant. **: highly significant; probability at 1% and 5% level of significance.

TABLE A–4. PEARSON'S CORRELATION COEFFICIENTS DESCRIBING ASSOCIATION OF EIGHT AGRO-MORPHOLOGICAL TRAITS EVALUATED UNDER OPTIMUM (UPPER DIAGONAL) AND TWO WATER STRESSED (LOWER DIAGONAL) FIELD CONDITIONS

		Optimum conditions							
	Traits	PH	PT	TF%	PL	PF	Y/P	1000GW	DTM
Stressed conditions	PH	1	0.37^*	-	0.15	0.5^{**}	0.55^{**}	-0.01	0
	PT	0.53^{**}	1	0	0.73^{**}	0.50^{**}	0.77^{**}	0.33	-0.06
	TF%	0.58^{**}	0.45	1	-	-	-	-	
	PL	0.48^{**}	0.72^{**}	0.36	1	0.41^*	0.72^{**}	0.43^{**}	-0.07
	PF%	0.63^{**}	0.78^{**}	0.57^{**}	0.83^{**}	1	0.55	0.18	0.1
	Y/PS	0.35	0.63^{**}	0.59^{**}	0.43^*	0.40^*	1	0.45	-0.03
	1000GW	-0.44^*	-0.04	-0.15	-0.32	-0.33	0.21	1	0.06
	DTM	-0.02	0.36	0.04	0.31	0.16	0.29	0.16	1

PH: plant height; PT: plant; PL: panicle length, PF: panicle fertility, Y/P: yield per plant; 1000GW: 1000 grain weight; *: significant at 5%; **: highly significant at 1%; Prob.: probability; (–): under optimum conditions tiller fertility was 100% and showed no correlation with other traits.

TABLE A–5. PAIRWISE COMPARISON OF MUTANTS BASED ON SELECTED TRAITS SHOWING SIGNIFICANT ASSOCIATION WITH YIELD PER PLANT UNDER STRESSED AND NORMAL FIELD

	Plant height (cm)		Productive tillers/plant		Tiller fertility %		Panicle fertility (%)		Yield per plant (g)	
Drought tolerant mutants	Stress	Normal	Stress	Normal	Stress	Normal	Stress	Normal	Stress	Normal
DT1	81^{efgh}	96^{fg}	17^a	17^{abcd}	89^{ab}	100	84^{cde}	93^{abc}	19^a	26^{efghi}
DT2	81^{efgh}	98^{ef}	14^{cdefg}	18^{abc}	89^{abc}	100	82^{def}	94^a	12^{defgh}	32^{abc}
DT3	80^{ijkl}	101^d	14^{cdefg}	19^a	49^{jk}	100	81^{ef}	91^{abcde}	10^{ghij}	35^a
DT4	83^{bcd}	96^{ghi}	15^{abcd}	18^{ab}	91^a	100	$76f$	$90^{abcdefgh}$	16^{abc}	32^{abc}
DT5	79^{jkl}	$93jk$	15^{abcde}	17^{abcd}	72^{fg}	100	83^{def}	82^{ij}	14^{bcde}	30^{bcde}
DT6	81^{fghi}	99^{de}	15^{abcde}	17^{abcd}	80^{abcdef}	100	88^{abcde}	92^{abc}	16^{abc}	33^{ab}
DT7	$79l$	97^{efg}	17^{ab}	17^{abcd}	62^{ghi}	100	83^{def}	91^{abcdef}	12^{defgh}	31^{bcd}
DT8	81^{ghij}	99^{de}	15^{abcd}	17^{abcd}	84^{abcde}	100	85^{abcde}	$91^{abcdefg}$	19^a	31^{bcd}
DT9	81^{fghij}	94^{ijk}	13^{efgh}	17^{abcd}	68^{fghi}	100	88^{abcde}	$90^{abcdefgh}$	11^{defgh}	28^{cdefgh}

DT10	82[defg]	97[efg]	14[cdefg]	16[abcde]	75[ef]	100	85[abcde]	87[defghi]	14[cdef]	29[bcdef]
DT11	80[hijk]	83[m]	11[h]	16[abcde]	86[abcde]	100	89[abcde]	92[abcd]	9[hij]	27[def]
DT12	82[de]	98[ef]	14[cdefg]	17[abcd]	86[abcde]	100	92[ab]	88[bcdefgh]	18[ab]	29[ghi]
DT13	81[efgh]	96[ghi]	15[abcde]	18[ab]	73[bcdef]	100	91[abc]	86[ghi]	15[bcd]	30[cdefg]
DT14	81[efgh]	96[gh]	14[cdefg]	17[abcd]	77[cdef]	100	89[abcd]	90[abcdefgh]	12[defgh]	30[bcdef]
DT15	82[def]	98[ef]	12[fgh]	18[a]	80[abcdef]	100	88[abcde]	88[cdefgh]	11[defgh]	32[bcdef]
DT16	79[kl]	92[k]	17[ab]	17[abcd]	75[ef]	100	93[a]	86[fghi]	15[bcd]	25[abc]
DT17	81[efghi]	88[l]	15[abcde]	17[abcd]	70[fgh]	100	88[abcde]	91[abcde]	11[fghi]	24[ghi]
DT18	81[efghi]	94[hia]	13[defgh]	16[abcde]	53[ij]	100	89[abcde]	93[ab]	9[hij]	28[hi]
DT19	79[kl]	89[l]	14[bcdef]	15[bcdef]	63[ghi]	100	85[bcde]	89[abcdefgh]	9[hij]	26[cdefgh]
DT20	81[fghi]	96[fg]	14[cdefg]	16[abcde]	71[fg]	100	88[abcde]	87[efghi]	9[hij]	27[efghi]
DT21	84[bc]	94[hij]	13[defgh]	14[ef]	87[abcd]	100	87[abcde]	91[abcde]	13[cdefg]	26[efghi]
DT22	81[efghi]	89[l]	13[defgh]	15[bcdef]	89[ab]	100	86[abcde]	85[hi]	11[fghi]	26[fghi]
DT23	81[fghi]	97[efg]	13[defgh]	15[def]	85[abcde]	100	83[def]	89[bcdefgh]	11[efghi]	27[efghi]
DT24	81[efghi]	103[c]	12[gh]	13[f]	89[ab]	100	86[abcde]	89[abcdefgh]	12[defgh]	25[defghi]
IR6	80[ghijk]	95[ghij]	13[defgh]	15[bcdef]	77[def]	100	84[cde]	86[efghi]	8[ij]	29[ghi]
Nagina - 22	85[b]	128[a]	8[i]	15[cdef]	58[hij]	100	52[g]	93[abc]	9[hij]	32[cdefg]
BRI Dhan	97[a]	110[b]	16[abc]	18[a]	89[ab]	100	86[abcde]	91[abcde]	14[cdef]	23i
WAB-56	58[m]	72[n]	5[j]	8[g]	40[k]	100	25[h]	78[J]	6[j]	8[j]
Mean	*81*	*96*	*14*	*16*	*76*	*100*	*83*	*89*	*12*	*29*
Maximum	*99*	*128*	*18*	*22*	*100*	*100*	*94.0*	*95*	*19*	*35*
Minimum	*56*	*71*	*4*	*6*	*31*	*100*	*9.52*	*68*	*5*	*7.20*
SE Mean	*0.604*	*0.99*	*0.28*	*0.28*	*1.56*	*0.00*	*1.51*	*0.48*	*0.4*	*0.52*

Mutants and standard cultivars sharing a common letter do not different from each other under stressed and normal water conditions; N: normal field condition.

TABLE A–6. CORRELATIONS AMONG DIFFERENT STRESS TOLERANCE INDICES AND YIELD PER PLANT UNDER STRESS CONDITIONS

	STI-PH	STI-PT	STI-TF%	STI-PL	STI-PF	STI-Y/P	STI-1000GW	**Y/PS**
STI-PT	0.316							
Prob.	0.101							
STI-TF%	0.326	0.378*						
Prob.	0.091	0.047						
STI-PL	0.252	0.719**	0.411*					
Prob.	0.196	0	0.03					
STI-PF	0.325	0.725**	0.562**	0.853**				
Prob.	0.091	0	0.002	0				
STI-Y/P	0.358	0.684**	0.521**	0.594**	0.475*			
Prob.	0.062	0	0.005	0.001	0.011			
STI-1000GW	-0.351	0.238	-0.082	0.078	-0.082	0.295		
Prob.	0.067	0.222	0.678	0.692	0.68	0.128		
Y/PS	0.27	0.625**	0.585**	0.429	0.399*	0.925**	0.328	
Prob.	0.165	0	0.001	0.023	0.035	0	0.088	
STI-DTM	-0.062	0.195	-0.023	0.21	0.012	0.023	0.292	0.05
Prob.	0.755	0.319	0.909	0.285	0.953	0.908	0.131	0.801

Significance at * (5%) and ** (1%).

TABLE A–7. STRESS TOLERANCE INDICES OF YIELD AND ITS ASSOCIATED TRAITS

Genotypes	STI-PH	STI-PT	STI-TF%	STI-PL	STI-PF	STI-1000GW	STI-DTM	STI-Y/P	Y/PN	Y/PS
DT1	0.85	1.08	0.89	0.99	0.98	0.94	0.97	0.65	26	19.33
DT 8	0.86	0.98	0.84	1.09	0.97	0.96	0.99	0.77	31	19.33
DT 12	0.88	0.92	0.86	1.02	1.02	0.84	0.88	0.66	29	17.67
DT 6	0.87	0.96	0.80	1.06	1.02	0.89	0.97	0.70	33	16.33

DT 4	0.86	1.04	0.91	1.02	0.86	0.89	0.99	0.65	32	16.00
DT 13	0.84	1.02	0.78	1.06	0.98	0.81	0.98	0.56	30	14.67
DT 16	0.79	1.07	0.75	0.96	1.00	0.81	0.97	0.47	25	14.67
DT 5	0.80	0.94	0.72	1.07	0.85	0.86	0.97	0.55	30	14.33
BRRI.Dhan	1.15	1.08	0.89	0.89	0.99	0.84	0.99	0.42	23	14.13
DT 10	0.86	0.85	0.75	1.02	0.94	0.81	0.97	0.53	29	14.00
DT 21	0.86	0.68	0.87	1.02	1.00	0.77	0.98	0.43	26	13.00
DT 2	0.87	0.94	0.89	0.99	0.98	0.86	0.97	0.51	32	12.33
DT 14	0.85	0.88	0.77	1.06	1.01	0.86	0.98	0.47	30	12.33
DT 24	0.91	0.58	0.89	1.00	0.97	0.82	0.98	0.40	25	12.33
Average	0.85	0.84	0.76	0.98	0.93	0.85	0.97	0.45	35	12.10
DT 7	0.83	1.05	0.62	1.05	0.96	0.89	0.99	0.46	31	11.67
DT 9	0.82	0.81	0.68	1.05	1.00	0.89	0.98	0.41	28	11.33
DT 15	0.87	0.85	0.80	1.05	0.97	0.82	0.99	0.47	32	11.33
DT 23	0.85	0.74	0.85	0.98	0.93	0.86	0.98	0.38	27	11.00
DT 17	0.77	0.96	0.70	0.96	1.01	0.84	0.98	0.33	24	10.67
DT 22	0.79	0.78	0.89	0.92	0.92	0.79	0.99	0.36	26	10.67
DT 3	0.87	1.01	0.49	1.04	0.94	0.89	0.97	0.45	35	10.00
DT 18	0.83	0.82	0.58	1.06	1.04	0.86	0.98	0.34	28	9.33
DT 19	0.76	0.84	0.63	1.02	0.95	0.78	0.97	0.32	26	9.33
DT 20	0.85	0.84	0.71	1.01	0.96	0.86	0.96	0.32	27	9.33
DT 11	0.72	0.69	0.86	0.89	1.03	0.81	0.88	0.31	27	9.00
Nagina 22	1.17	0.46	0.58	0.72	0.61	0.72	0.92	0.36	32	8.91
IR6	0.83	0.78	0.77	1.07	0.92	0.85	0.98	0.29	29	7.83
WAB-56-104	0.45	0.16	0.40	0.46	0.24	0.93	0.98	0.06	8	6.50

STI–PH: stress tolerance index of plant height; STI–PT: stress tolerance index of productive tillers per plant; STI–TF%: stress tolerance index of tiller fertility; STI–PL: stress tolerance index of panicle length; STI– PF: stress tolerance index of panicle fertility; STI-1000GW: stress tolerance index of grain weight; STI–DTM: stress tolerance index of days to maturity; STI-Y/P: stress tolerance index of paddy yield per plant; Y/PN: yield per plant under normal water conditions; Y/PS: yield per plant under stress conditions.

TABLE A–8. PRINCIPAL COMPONENT ANALYSIS OF COARSE RICE MUTANTS BASED ON YIELD AND ASSOCIATED COMPONENTS UNDER NORMAL FIELD CONDITIONS

Statistical variables	*Eigen analysis of the correlation matrix*		
	PC1	PC2	PC3
Eigenvalue	3.3416	1.1721	1.0404
Percent of variation	0.477	0.167	0.149
Cumulative	0.477	0.645	0.793

Rotated factor loadings and communalities (Varimax rotation)			
Traits	Factor 1	Factor 2	Factor 3	Communality
PH-N	-0.035	0.898	-0.033	0.808
PT-N	0.708	0.504	-0.142	0.775
PL-N	0.838	0.260	-0.133	0.787
PF-N	0.300	0.743	0.149	0.664
YP-N	0.699	0.613	-0.065	0.869
1000GW-N	0.801	-0.152	0.188	0.700
DTM-S	-0.021	0.055	0.973	0.950
Variance	2.4257	2.0807	1.0476	5.5541
% variance	0.347	0.297	0.150	0.793

PHN: plant height under normal field; PTN: plant tillers under normal field; PL N: panicle length under normal field; PFN: panicle fertility under normal field; Y/PN: yield per plant under normal field; 1000WTN: 1000 grain weight under normal field; DTMN: days to maturity under normal field; *: significant; **: highly significant, NS: non-significant.

TABLE A–9. PRINCIPAL COMPONENT ANALYSIS OF COARSE RICE MUTANTS BASED ON YIELD AND ASSOCIATED COMPONENTS UNDER WATER STRESS FIELD CONDITIONS

	Eigen analysis of the correlation matrix		
Statistical variables	PC1	PC2	PC3
Eigenvalue	3.9370	1.5336	1.0097
Percent variability	0.492	0.192	0.126
Cumulative	0.492	0.684	0.810

Rotated factor loadings and communalities Varimax Rotation			
Traits	Factor 1	Factor 2	Factor 3	Communality
PHS	0.640	-0.046	0.581	0.749
PTS	0.599	-0.653	0.203	0.826
TFS	0.875	0.024	0.159	0.791
PLS	0.368	-0.669	0.506	0.838
PFS	0.560	-0.480	0.561	0.859
YPS	0.806	-0.346	-0.261	0.837
1000GWS	0.063	-0.095	-0.903	0.829
DTM-S	-0.054	-0.838	-0.216	0.752
Variance	2.6381	1.9370	1.9051	6.4802
% Variance	0.330	0.242	0.238	0.810

PHS: plant height under stress; PT: plant tillers under stress; TF: tillers fertility under stress; PL: panicle length under stress; ET: empty tiller under stress; PF: panicle fertility under stress; Y/P: yield per plant under stress; 1000WT: 1000 grain weight under stress; DTMN: days to maturity under stress.

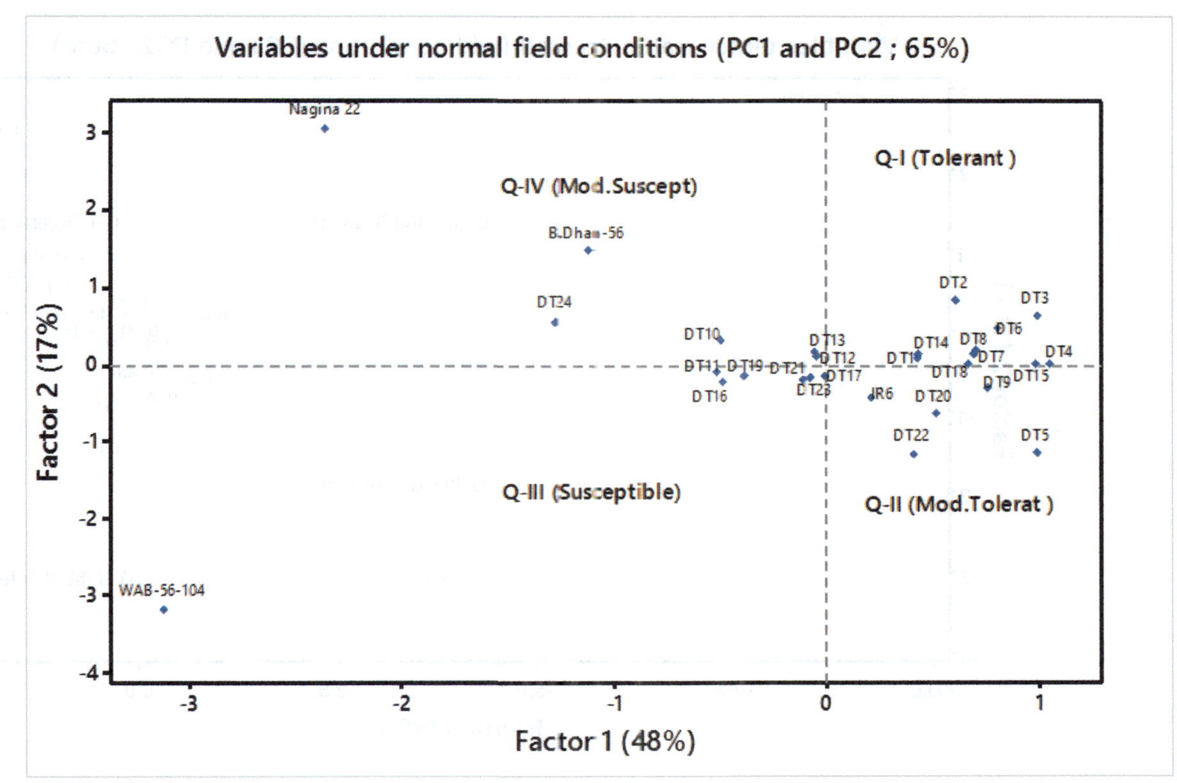

FIG. A–1. *Classification of coarse rice mutants based on yield and its associated traits under normal irrigation field conditions through PCA. Mutants that had +PC1 and +PC2, +PC1 and −PC2, −PC1 and +PC2, −PC1 and −PC2 scores were classified as tolerant, moderately tolerant, moderately susceptible, and susceptible, respectively.*

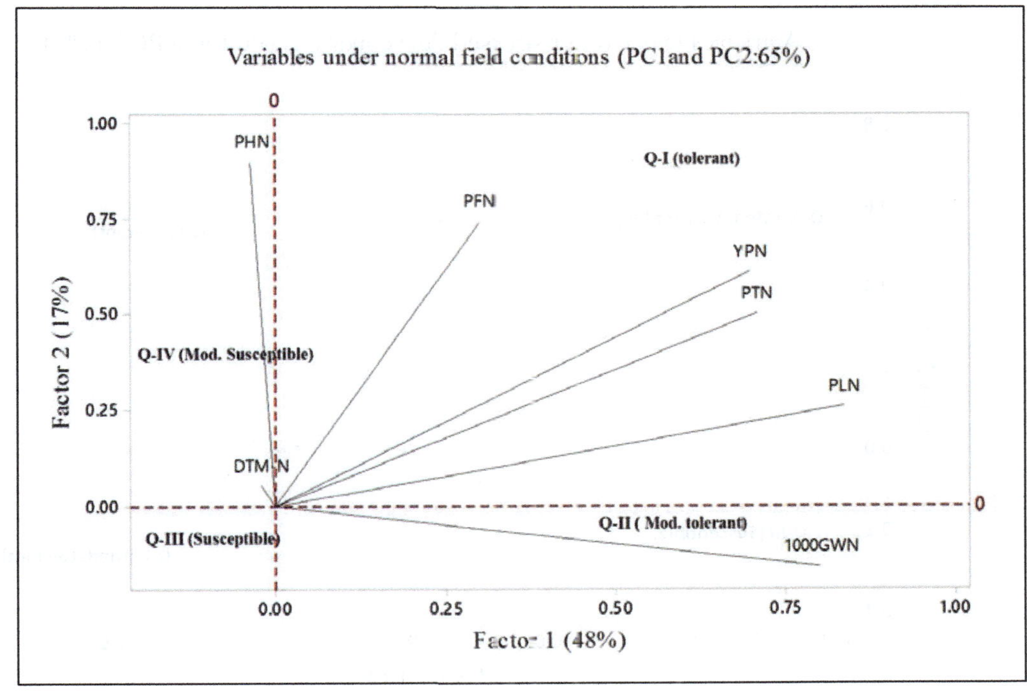

FIG. A–2. *Loading plot for various traits under normal field conditions.*

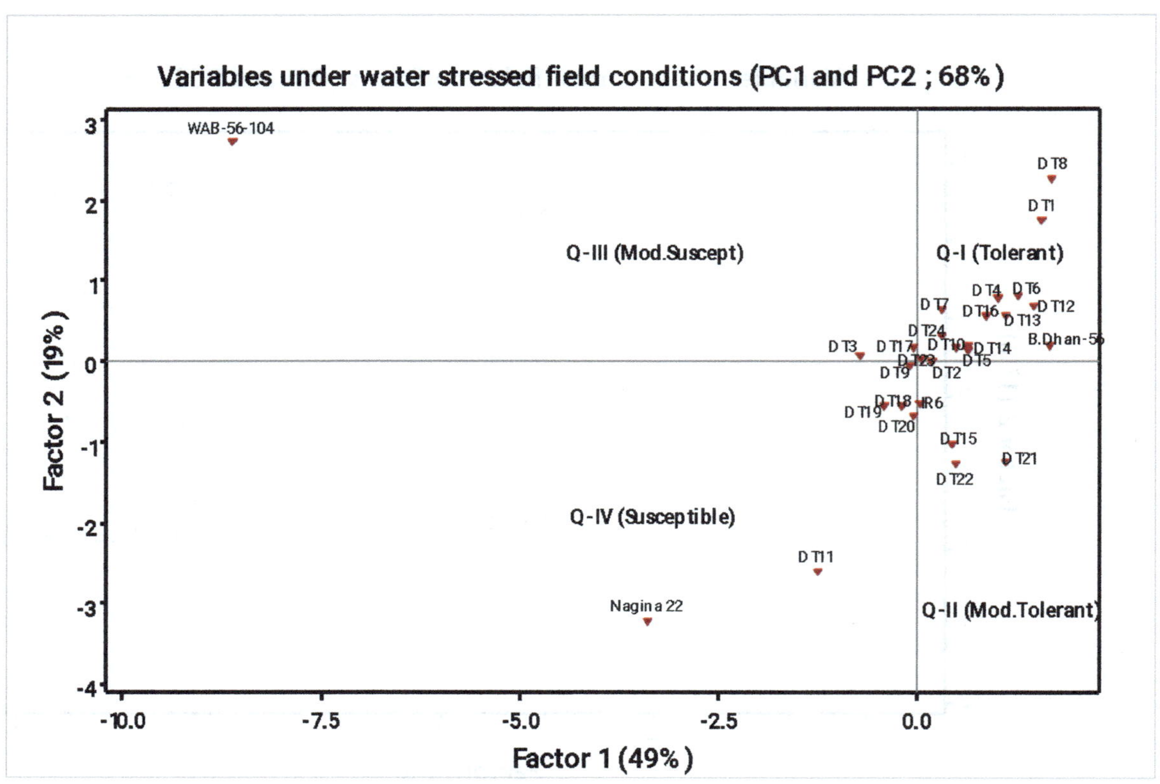

FIG. A–3. Classification of coarse rice mutants based on yield and its associated traits under water stressed field conditions through PCA. Mutants that had +PC1 and +PC2, +PC1 and −PC2, −PC1 and +PC2, −PC1 and −PC2 scores were classified as tolerant, moderately tolerant, moderately susceptible, and susceptible, respectively.

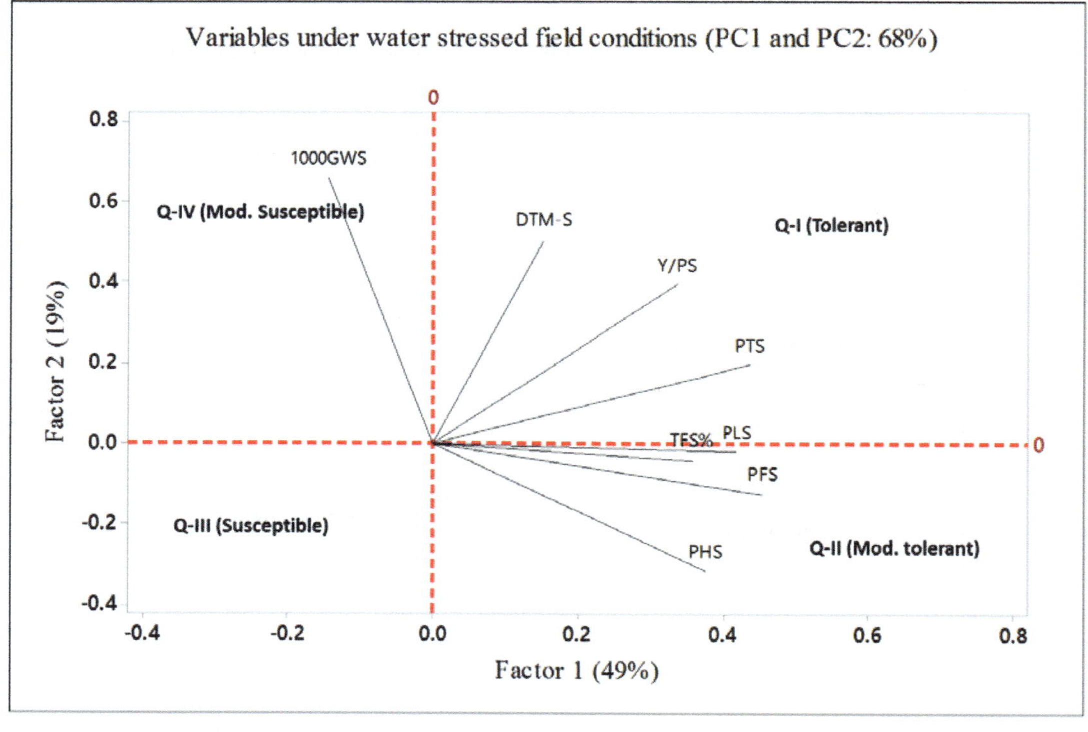

FIG.A–4. Loading plot for various traits under stressed field conditions.

REFERENCES TO CHAPTER 2–7

[1]   PANDA, D., BEHERA, P.K., MISHRA, S., MISHRA, B.S., Differential drought tolerance responses in short-grain aromatic rice germplasms from Koraput valley of Eastern Ghats of India, Plant Physiol. Rep. (2022) 1–13.

[2]   PATHAK, T.B., et al., Climate change trends and impacts on California agriculture: A detailed review, Agronomy **8** 3 (2018) 25; https://doi.org/10.3390/agronomy8030025.

[3]   BEENA, R., Studies on physio-morphological traits and genetic markers associated with drought responses in rice (*Oryza sativa* L), PhD thesis submitted to Tamil Nadu Agricultural University, Coimbatore, Tamil Nadu (2005).

[4]   KANG, J., et al., PDR-type ABC transporter mediates cellular uptake of the phytohormone abscisic acid, Proc. Natl Acad. Sci. USA **107** (2010) 2355–2360.

[5]   LAMAOUI, M., JEMO, M., DATLA, R., BEKKAOUI, F., Heat and drought stress in crops and approaches for their mitigation, Front. Chem. 19 February (2018) https://doi.org/10.3389/fchem.2018.00026.

[6]   HUSSAIN, T., HUSSAIN, N., AHMED, M., NUALSRI, C., DUANGPAN, S., Responses of lowland rice genotypes under terminal water stress and identification of drought tolerance to stabilize rice productivity in southern Thailand, Plants **10** 12 (2021) 2565; https://doi.org/10.3390/plants10122565.

[7]   RAYEE, R., XUAN, T.D., KHANH, T.D., TRAN, H.-D., KIFAYATULLAH, K., efficacy of irrigation interval after anthesis on grain quality, alkali digestion, and gel consistency of rice, Agriculture **11** (2021) 325.

[8]   PANDEY, A., KUMAR, A., PANDEY, D.S., THONGBAM, P.D., Rice quality under water stress, Indian J. Adv. Plant Res., **1** (2014) 23–26.

[9]   MATSUURA, A., TSUJI, W., AN, P., INANAGA, S., MURATA, K., Effect of pre- and post-heading water deficit on growth and grain yield of four millets, Plant Prod. Sci. **15** (2012) 323–331.

[10]   ABDELRAHMAN, M., et al., Developing novel rice genotypes harboring specific QTL alleles associated with high grain yield under water shortage stress, Plants **10** 10 (2021) 2219; https://doi.org/10.3390/plants10102219

[11]   AHUJA, S., "Water quality worldwide", Handbook of Water Purity and Quality, Elsevier, Amsterdam (2021) 19–33.

[12]   Anonymous, 2019.

[13]   MAHAGAN, G, CHAUHAN, B.S., TIMSINA, J., SINGH, P.P., SINGH, K., Crop performance and water- and nitrogen-use efficiencies in dry-seeded rice in response to irrigation and fertilizer amounts in northwest India, Field Crops Res. **134** (2011) 59–70; https://bit.ly/3eaIVPW

[14]   SUPRASANNA, P., MIRAJKAR, S.J., BHAGWAT, S.G., "Induced mutations and crop improvement", Plant Biology and Biotechnogy (BAHADUR, B., VENKAT RAJAM, M., SAHIJRAM, L., KRISHNAMURTHY, K., Eds), Springer, New Delhi (2015); https://doi.org/10.1007/978-81-322-2286-6_23.

[15]   MONNEVEUX, P., JING, R., MISRA, S., Phenotyping for drought adaptation in wheat using physiological traits, Front. Physiol. **3** (2012) 429.

[16] PASSIOURA, J.B., Phenotyping for drought tolerance in grain crops: When is it useful to breeders? Funct. Plant Biol. **39** 11 (2012) 851–859.

[17] AMRUTHA, S., et al., Characterization of Eucalyptus camaldulensis clones with contrasting response to short-term water stress response, Acta Physiol. Plant. **43** (2021) 1–13; doi: 10.1007/s11738-020-03175-0.

[18] BLUM, A., Plant Breeding for Water-Limited Environments, Springer, London (2010).

[19] LOPES, M.S., The yield correlations of selectable physiological traits in a population of advanced spring wheat lines grown in warm and drought environments, Field Crops Res. **128** (2012) 129–136.

[20] RAMAN, A., et al., Drought yield index to select high-yielding rice lines under different drought stress severities, Rice **5** 1 (2012) 1–12.

[21] ADEBOYE, K.A., ODUWAYE, O.A., DANIEL, I.O., FOFANA, M., SEMON, M.,. Characterization of flowering time response among recombinant inbred lines of WAB638-1/PRIMAVERA rice under reproductive stage drought stress, Plant Genetic Res. **19** 1 (2021) 1–8.

[22] KAFY, A.A., et al., Assessment and prediction of seasonal land surface temperature change using multi-temporal Landsat images and their impacts on agricultural yields in Rajshahi, Bangladesh, Environ. Chall. **4** (2021) 100147.

[23] LATA, C., SHIVHARE, R., Engineering cereal crops for enhanced abiotic stress tolerance, Proc. Indian Natl Sci. Acad. (2021) 1–21.

[24] TARDIEU, F., Any trait or trait-related allele can confer drought tolerance: Just design the right drought scenario, J. Exp. Bot. **63** 1 (2012) 25–31.

[25] CHATFIELD, C., COLLINS, A.J., Introduction to Multivariate Analysis, Chapman and Hall, London (1980); http://dx.doi.org/10.1007/978-1-4899-3184-9.

[26] MAHLOCH, J.L., Multivariate technique for water quality analysis, J. Environ. Eng. Div, Am. Soc. Civil Eng. **100** (EE5) (1974) 1119–1124.

[27] MAZLUM, N., Multivariate analysis of water quality data. MSc thesis submitted to Dokuz Eylul University, Graduate School of National and Applied Sciences, Izmir (1994).

[28] MAZLUM, N., OZER, A., MAZLUM, S., Interpretation of water quality data by principal component analysis, Tr .J. Eng. Environ. Sci. 23 (1999) 19–26.

[29] BREJDA, J.J., et al., Distribution and variability of surface soil properties at a regional scale. Soil Sci. Soc. Am. J. **64** 3 (2000) 974– 982.

[30] FERANDEZ, G.C.J., "Effective selection criteria for assessing stress tolerance", Adaptation of Vegetables and Other Food Crops in Temperature and Water Stress (Proc. Int. Symp. Taipei) (KUO, C.G., Ed), World Vegetable Center, Taipei (1992).

[31] PANDA, D., MISHRA, S.S., BEHERA, P.K., Drought tolerance in rice: Focus on recent mechanisms and approaches, Rice Sci. **28** 2 (2021) 119–132.

[32] JAGADISH, S.V.K., et al., Genetic analysis of heat tolerance at anthesis in rice (*Oryza sativa* L.), Crop Sci. **50** (2010) 1633–1641.

[33] SINGH, R., et al., From QTL to variety-harnessing the benefits of QTLs for drought, flood, and salt tolerance in mega rice varieties of India through a multi-institutional network, Plant Sci. **242** (2016) 278–287.

[34] PANDEY, V., SHUKLA, A., Acclimation and tolerance strategies of rice under drought stress, Rice Sci. **22** 4 (2015) 147–161.

[35] FLEURY, D., JEFFERIES, S., KUCHEL, H., LANGRIDGE, P., Genetic and genomic tools to improve drought tolerance in wheat, J. Exp. Bot. **61** (2010) 3211–3222.

[36] THIRY, A.A., CHAVEZ DULANTO, P.N., REYNOLDS, M.P., DAVIES, W.J., How can we improve crop genotypes to increase stress resilience and productivity in a future climate? A new crop screening method based on productivity and resistance to abiotic stress, J. Exp. Bot. **67** 19 (2016) 5593–5603.

[37] MWADZINGENI, L., SHIMELIS, H., DUBE, E., LAING, M.D., TSILO, T.J., Breeding wheat for drought tolerance: Progress and technologies, J. Integr. Agr. **15** (2016) 935–943; doi: 10.1016/S2095-3119(15)61102-9

[38] SOFI, P.A., REHMAN, K., MUSHARIB, G. ASMATARA, Phenology based biomass accumulation and partitioning indices in relation to water stress in common bean (*Phaseolus vulgaris* L.), Int. J. Pure Appl. Biosci. **5** 6 (2017) 1441–1449.

[39] SAREEN, S., TYAGI, B.S., SARIAL, A.K., TIWARI, V., SHARMA, I., Trait analysis, diversity, and genotype × environment interaction in some wheat landraces evaluated under drought and heat stress conditions, Chil. J. Agric. Res. **74** (2014) 135–142; doi: 10.4067/S0718-58392014000200002.

[40] SLAFER, G.A., SAVIN, R., SADRAS, V.O., Coarse and fine regulation of wheat yield components in response to genotype and environment, Field Crops Res. **157** (2014) 71– 83.

2–8. DROUGHT TOLERANCE IN RICE: VIET NAM

N.T. HONG[1], V.T.M. TUYEN[1], N.T.M. NGUYET[1],
L.D. THAO[1], P.X. HOI[1], L.H. HAM[1,2]

[1] Agricultural Genetics Institute, Hanoi, Viet Nam

[2] University of Science and Technology,
Viet Nam National University, Hanoi, Viet Nam

Abstract

The utilization of drought tolerant varieties is an effective solution in the climate change situation. Seng Cu, a local and quality variety, was irradiated by gamma rays (cobalt-60) with the goal to develop a new drought mutant variety. The drought tolerance of rice is a complex trait and hard to screen, hence robust screening protocols play a key role in the efficiency of drought tolerance selection. The hydroponic method with PEG 6000 was applied at an early stage to screen mutant rice lines. It showed the clear separation of drought symptoms among mutant lines to provide strong evidence for selection. In the field condition, the drought stress was experimented at the reproductive stage; then primary and secondary traits were evaluated to assist in screening. The severe drought at the reproductive stage impacted critically on yield through grain number, fill grain number, grain weight of a sensitive variety, but less significantly on the tolerant variety. The evaluation of root traits at an early stage proved that if a plant owning the root system with: higher dried weight; more density of hair root; lower water content in the root; and shorter and more lateral root, will have better drought tolerance. When studying further indicators, the results showed that in water stress, a plant with the better tolerance presents a more significant reduction of stomata density and the higher accumulating of proline content. Through experiments in this project, the protocol combining methods to screen drought tolerance at different stages (seedling and reproductive stages) and at different levels (morphological-physiological and biochemical traits) was established to enhance the efficiency of drought screening and contribute to the establishment of two mutant promising rice lines with resilience to water stress.

Key words: drought screening, Seng Cu rice variety, mutation breeding, stomata density, proline content.

1. INTRODUCTION

Drought is one of the most destructive challenges faced by global agriculture [1]. The general definition refers to drought as a condition with an extended period of deficient precipitation compared with the statistical multi-year average for a region that results in water shortage for some activity, group, or environmental sector. In particular conditions, drought is a normal and cycling feature of the climate. For some others, drought is considered an unusually long dry period. For example, in Tay Nguyen Viet Nam, if precipitation is less than 100 mm in the rainy season, it is reported as a drought, but is a normal condition in the dry season. Thus, drought is defined as an unusual condition caused by insufficient rainfall over a long period of time, leading to insufficient water for production and daily life as well as the ecological environment.

For crop improvement, breeding varieties to have better resilience to water deficit is a suitable method to adapt to agricultural drought and climate change. Drought tolerance can be broadly achieved through three major mechanisms: (1) Drought escape due to early completion of the life cycle. This is particularly useful in a scenario where drought is a recurrent phenomenon at the end of the growing season. In such cases, early flowering varieties can escape terminal drought); (2) Dehydration avoidance, which enables the plant to uptake or conserve more water to avoid dehydration: this is achieved through traits related to root architecture, stomatal control, and transpiration efficiency; (3) Dehydration tolerance, which is achieved through traits such as cell membrane stability, osmotic adjustment, stem reserve mobilization, and stability of the flowering process. Conventional and molecular breeding approaches targeting these traits have been used in the past to develop drought tolerant rice varieties. Like other abiotic stresses, drought tolerance in a plant is a complex trait regulated by multi genes and is highly sensitive to environmental factors.

Designing protocol(s) for screening the target traits is the first step and forms a baseline for further studies on improving of traits of interest. Selecting drought tolerant mutant varieties requires screening of large populations to improve the chance of finding the target trait expression in mutant populations. Thus, a technique for effective drought screening can help to downsize the selected population in both early stage and early generation.

In the past 20 years, Viet Nam has faced ten drought events with large impacts. Recently, two severe drought events with a cycle of about five years were recorded. The event in 2015–2016 was reported as the most destructive drought in a 100 year historical record that affected the Mekong River Delta. The economic loss was severe, estimated at US $669 million (VND 15 000 billion) and the cost of recovery was reported at US $1500 billion (VND 34 trillion). In agriculture, a total of 330 000 ha was affected by the drought and salt water (mard.gov.vn). The next event was a severe drought in the 2019–2020 summer season; the total destruction was reported to be 450 000 ha of rice in the Central Highlands and the South Central Coast, representing a decrease of 34 600 t compared with 2018. It was considered as the most severe salinity drought in the history of the Mekong Delta.

## 2.	RICE BREEDING FOR DROUGHT TOLERANCE IN VIET NAM

A drought tolerance breeding programme for rice in Viet Nam before 2010 used the traditional approach to establish the final goal of tolerant varieties for farmers. Drought tolerant rice varieties were selected by Vietnamese breeders after screening local varieties (Khau Don, Khau Lac, Khau Lep Trong, Khau Te Lau, Khau Lanh, Khau Luong), or breeding between local and improved varieties (CH133, CH135, CH5, CH7, CH 207, CH 208). In the last decade, molecular markers have been applied to assist selection for drought tolerance. Results of some quantitative trait loci (QTLs), such as: qLR-4, qLR-8, qLD-8-b, qlD-8-a, qPR-8, and qPR-11 related to drought tolerance, have been published.

A comprehensive and thorough protocol is the key for the success of any drought tolerance breeding programme. Thus, the IAEA/Viet Nam Coordinated Research Project 'Improving Drought Tolerant Rice through Mutation Breeding in Viet Nam' focused on a robust screening protocol for selecting drought tolerant mutants and at the same time to promptly develop the results obtained of elite drought tolerant rice lines for further studies and applications.

## 3.	METHODOLOGY

The local lowland variety Seng Cu was used as material for the study. This variety is grown widely and provides the best quality in Lao Cai, the province located in the mountainous northern region with higher terrain and cool climate. It also presents a narrow adaptation due to the strong variation of both quality and yield at different ecozone.

Seeds of Seng Cu variety were irradiated with gamma ray in a range of doses: 200, 250 300, 350 and 400 Gy to identify the optimal dose through the survival rate of irradiated individuals.

To evaluate separately the effect of gamma irradiation on plants, the survival rate of the control (0 Gy) was set at 100%. Data were collected on the experiment with 100 seeds of each dose. The survival reduced from 80% at 200 Gy to 20% at 400 Gy. The dose with around 50% survival was 300 Gy => LD50 was 300 Gy. Figure 1(a) shows that the survival rates were distributed as the polynomial regression type with the highest $R^2 = 0.9862$. Figure 1(b) shows the effect of radiation on sterility in var. Seng Cu.

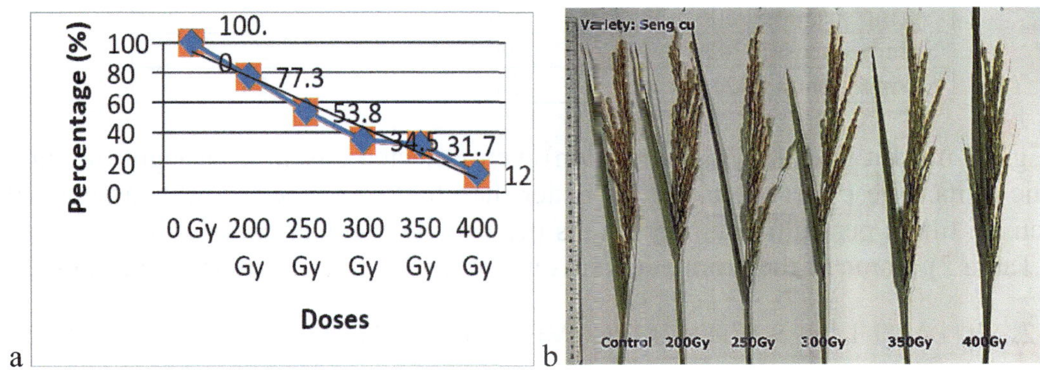

a   b

FIG. 1. (a) Effect of different doses on survival rate of M₁ plants; (b) effect of different doses on sterility in var. Seng Cu.

In the early generations, mutant lines were conserved to evaluate in field conditions for different morphological and agronomical characteristics, such as duration, plant height, culm angle, leaf angle, length of panicle, etc., required to establish the potential yield of lines. The mutants were selected and harvested as separate individuals till their performances were uniform. Then the good uniform mutant lines were planted in larger areas and in different locations for testing stability and adaptation. In the M_1 generation, there was significant variation in sterility, suggesting that the higher frequency of sterility at the higher dose (Fig. 1(b)). The data ranged from 39.3% to 88.8%, especially high at doses 350 Gy (72.5%) and 400 Gy (88.8%). It was shown from the chart that sterility data were the logarithmic regression type with $R^2 = 0.9629$.

The M_2 generation had strong segregation of phenotypes among plants. For example, common phenotypes such as plant height, growth duration, sterility, tillering ability, culm angle were observed. The mutants with improved traits compared with control were collected separately and planted as a new line in the next generation. The original parent Seng Cu is a good quality variety. However, it has negative traits such as less tillers, open flag leaf, small panicle, low seed density, long awn, besides low tillering, low grain density, low grain number, and therefore its potential yield is low (Table 1).

TABLE 1. OVERVIEW OF THE MUTANT BREEDING EXPERIMENTS

Season	Generation	No. of selected mutant lines
Spring season 2018	M_2	106
Summer season 2018	M_3	64
Spring season 2019	M_4	77
Summer season 2019	M_5	64
Spring season 2020	M_6	17
Summer season 2020	M_7	28
Spring season 2021	M_8	13

Thirteen good mutant lines (in M_8 generation) were selected based on drought tolerance and agronomic traits. These were found to be better than the wild type for traits associated with yield such as: tillers per culm, number grains per panicle, number spikelet per panicle, grain density (Table 2). Some of the promising mutants for different traits are shown in Fig. 2.

TABLE 2. MUTANT LINES WITH IMPROVED YIELD IN THE M_8 GENERATION

Name	Tillers per culm	Panicle length (cm)	Spikelet per panicle	Grain per panicle	Full grain per panicle	Grain density (grain/cm)
Seng Cu	5	26.50 ± 0.32	7.82 ± 0.13	98.56 ± 3.89	88.93 ± 4.11	3.70 ± 0.12
99R-3	5	28.68 ± 0.76	12.40 ± 0.24	234.20 ± 17.31	208.20 ± 16.50	7.93 ± 0.56
93-2-1-1h-1	4	31.97 ± 0.74	12.00 ± 0.00	297.00 ± 12.15	267.75 ± 11.70	9.37 ± 0.47
93-2-1-1h-2	9	34.98 ± 0.48	10.00 ± 0.28	155.66 ± 13.18	123.22 ± 11.96	5.65 ± 0.14
99R-2-1	11	33.70 ± 0.90	12.27 ± 0.35	255.18 ± 19.85	235.09 ± 20.05	5.08 ± 0.56
99R-2-2	10	33.36 ± 0.63	12.80 ± 0.24	220.50 ± 11.32	175.40 ± 13.32	7.26 ± 0.28
99R-2-3	12	32.84 ± 0.85	12.91 ± 0.33	218.83 ± 10.8	210.08 ± 11.31	7.65 ± 0.34
99R-2-4	10	31.83 ± 0.48	12.70 ± 0.26	203.50 ± 9.91	181.60 ± 9.11	6.83 ± 0.36
68-3R-1-1	7	31.14 ± 0.94	12.00 ± 0.21	276.85 ± 20.23	255.14 ± 20.58	10.97 ± 0.69
68-3R-1-2	5	34.34 ± 0.80	13.60 ± 0.40	335.20 ± 8.73	302.20 ± 8.08	11.29 ± 0.28
68-3R-1-4	7	29.10 ± 0.88	12.57 ± 0.29	223.14 ± 10.31	215.71 ± 10.64	8.33 ± 0.37
68-3R-1-6	6	29.30 ± 0.97	12.16 ± 0.30	289.66 ± 23.42	265.50 ± 21.94	10.34 ± 0.71
68-3R-1-8	9	31.07 ± 0.84	12.88 ± 0.20	276.55 ± 17.24	259.88 ± 18.30	9.47 ± 0.44
68-3R-1-10	8	32.67 ± 0.51	11.37 ± 0.46	279.25 ± 17.12	259.00 ± 14.22	10.18 ± 0.38

FIG. 2. Promising mutants observed for different traits. (a) Mutants for plant height (1: wild type; 2–6: mutant types); (b) sterility and awn mutants (1: wild type; 2 and 3: mutant types); (c) seed coat colour mutant (yellow: wild type; brown: mutant type); (d) longer duration and higher plant mutant; (e) open culm angle mutant.

4. EVALUATION OF PROMISING MUTANT LINES

Two lines with better stability and overall phenotype were found to be promising . In the spring season 2021, the field trial was done for assessing yield and adaptation in different regions. Two promising lines Drought Mutant 1 (99R-2) and Drought Mutant 2 (68-3R-1), and the original control variety Seng Cu were planted in Soc Son, Hanoi, Bat Xat-Lao Cai and Lak-Dak Lak. Soc Son, Hanoi represents the Red River Delta with two separate seasons by temperature: low temperature in spring season and high temperature in summer season. Bat Xat-Lao Cai is in the northern mountainous zone with higher terrain and cool weather almost every year. Lak-Dak Lak is a province in the highland zone with stable high temperature and two separate seasons by precipitation. Season schedules of each experiment sites are described below:

— Location 1 (Soc Son, Hanoi): Seedling 1 Feb. 2021; harvest 1 June 2021;
— Location 2 (Bat Xat-Lao Cai): Seedling 10 Jan. 2021; harvest 12 May 2021;
— Location 2 (Lak-Dak Lak): Seedling 15 Dec. 2020; harvest 12 April 2021.

TABLE 3. YIELDS OF PROMISING MUTANT LINES AT DIFFERENT LOCATIONS

Locations	Yield (t/ha)			
	Seng Cu (control)	Drought Mutant 1	Drought Mutant 2	Average
Hanoi	5.06	6.44	6.8	5.6
Lao Cai	5.48	7.14	6.64	6.42
Dak Lak	4.3	6.28	6.22	5.60
Average	4.95	6.62	6.41	—
CV (%)	0.115	0.065	0.035	—

The data in Table 3 show that the CV(%) of yield of two promising lines were lower than that of the original type. This value of Seng Cu was 0.115; of Drought Mutant 1 it was 0.065; and of Drought Mutant 2 it was 0.035. This indicated that the adaptation of two promising lines was improved and wider than that of Seng Cu. The average yield of the original variety Seng Cu was 4.95 t/ha; of Drought Mutant 1 it was 6.62 t/ha; of Drought Mutant 2 it was 6.41 t/ha. Yields of all types were established highest at Lao Cai location (average of three 6.42 t/ha) (Fig. 3); at the Hanoi location (5.96 t/ha on average) and at the Dak Lak site (average of 5.6 t/ha).

FIG. 3. Field trial of promising mutant lines in Lao Cai province.

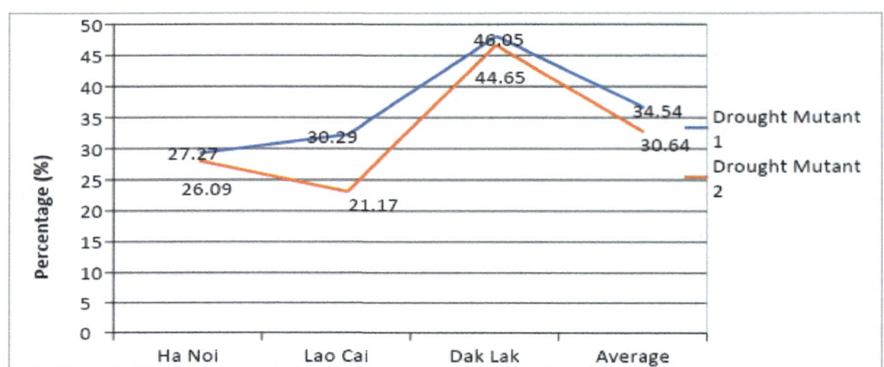

FIG. 4. Ratio of improving yields of two promising lines compared with control.

The yields of both promising mutant lines were improved by more than 30% compared to the yield of the original variety (Fig. 4). The average improved ratio of the yield of Drought Mutant 1 was 34.54% and of Drought Mutant 2 was 30.64%. Drought Mutant 1 had a higher yield than Drought Mutant 2 at all three testing sites. These promising lines will be tested on a larger scale and in more ecosystems for the final goal of releasing to farmers.

5. PRE-FIELD SCREENING PROTOCOLS FOR DROUGHT RESILIENCE IN RICE MUTANTS AT THE SEEDLING STAGE

Drought tolerance is a complex trait, regulated by multiple genes and highly sensitive to many factors. Selecting drought tolerant mutant rice varieties requires screening of large populations to have a chance of finding the target mutation. Thus, each screening technique helps to downsize the selected population in both the early stage and the early generation which will save cost and improve the efficiency of selection.

The hydroponic method is useful to grow many plants on a large scale and also applied for many themes. It is a convenient approach because it is conducted in controlled conditions, input factors are calculated exactly, side effects will be bound and easier to control. The root system has a major role in regulating responses to drought stress. Screening for salt tolerance in rice is conducted in the seedling stage very successfully, with the entire protocol using the hydroponic approach. Testing seeds are grown in the nutrient solution Kimura B and then NaCl is added to induce the stress for selection. To screen the response of rice plants for drought stress, polyethylene glycol (PEG) is added and at the target concentration based on the level of tolerance for a particular variety. Pores of walls of hair cells were determined to be 35–40 angstrom (Å). Thus, each molecule with a diameter larger than this size [2], for example PEG, could not penetrate into the cell. It is reported that PEG induces significant water stress in plants, although on the other hand it helps to protect cells from the effects of osmotic shock through interaction and to stabilize membrane components [3].

In this study, we screened the potential tolerance of mutant rice lines to artificial water deficit through the hydroponic method. For this, plants of 106 mutant lines were grown in hydroponic conditions for 14 days and then PEG 6000 was added for treatment in the next 14 days before evaluation.

5.1. Methodology

A total of 106 M_3 mutant lines derived from Seng Cu, the main material, their wild types and controls were screened in this study (Table 4). The wild type, Seng Cu, is a local variety, with good quality, low seed density, few tillers and narrow adaptation. Controls included improved varieties: susceptible control (DT18) and tolerant control (CH345). CH345 was obtained from crossing between a commercial variety (as recipient) and a local variety (as donor with mapped QTLs) and was confirmed as carrying QTLs of drought. DT18 was the improved variety with a quite high yield and good quality but sensitive to bacterial leaf blight.

TABLE 4. RICE LINES TESTED IN THIS STUDY

No.	Name of varieties	Description	Reference
1	CH345	Improved variety, drought tolerance	AGI, Viet Nam
2	DT18	Improved variety, quite high yield, good quality	AGI, Viet Nam
3	Seng Cu	Local variety, high quality, low yield, awn seed	Lao Cai, Viet Nam
4	Mutant lines	106 M_3 mutant lines from Seng Cu	AGI, Viet Nam

To carry out the hydroponic screening, plantlets were grown in artificial nutrient medium [4]. Seeds were selected through main characteristic of being healthy. To screen drought tolerance by PEG hydroponics, germinated seeds were placed in the holes of fabricating platforms with a nylon mesh underneath. These floats with seeds were kept in plastic trays containing Yoshida nutrient medium with the roots of plants sunk in the solution. The nutrient solution was renewed every week. With treatment, when the plants grew three–four leaves (often after two weeks), PEG 6000, a high molecular weight, non-penetrating inert osmotic substance was added with concentration 20% equal at OP of −4.05 to −4.95 MPa in the hydroponic solution and sustained for 14 days (two weeks). The non-stressed plants (control) were grown in a normal strength of the hydroponic nutrient solution. The pH of both control and treatment were adjusted at 5.8

every day. The drought tolerance was evaluated (Table 5) and scored after two weeks of artificial water-deficit stress as per the Standard Evaluation System (SES) of the International Rice Research Institute (IRRI) [5].

TABLE 5. GRADING OF DROUGHT TOLERANCE IN HYDROPONICS AS PER SES (IRRI) [5]

Grade	Descriptions
0	Leaves normal and healthy
1	Leaf starts to fold and tips of leaf are dead
3	1/4 of most leaves are dead
5	1/4 to 1/2 of leaves are completely dead
7	More than 2/3 leaves of plant are completely dead
9	Almost all parts of the plant are dry and dead

5.2. Results

A total of 106 M_3 mutant lines and controls were screened for drought tolerance at seedling stage by the hydroponic method with PEG 6000 addition. After 14 days treatment, symptoms appeared on plants and SES grading (Table 5) was made on control (parent) and mutant lines and the results are presented in Table 6 and Fig. 5.

TABLE 6. ASSESSMENT OF DROUGHT TOLERANCE AT THE SEEDLING STAGE IN M_3 MUTANT LINES

Variety	Number of lines			
	Total	Tolerant (grade 1–3)	Mediate (grade 5)	Susceptible (grade 7–9)
CH345 (positive control	1	1	-	-
DT18 (negative control)	1	-	-	1
Seng Cu (wild type)	1	-	1	-
Mutant lines	106 (100%)	24 (22.64%)	33 (31.13%)	49 (46.23%)

FIG. 5. Symptoms on rice plants screened by hydroponics with the addition of PEG 6000 after 14 days treatment.

The tolerant grade included lines with slight symptoms (no infection or less than ¼ tip of leaf dead) and almost parts of plant being fresh as normal (score 1–3), whereas lines presenting the symptoms with ¼ to ½ length of leaves dead are in the medium grade (score 5), and lines showing the serious infection with large area of the leaves showing death and almost parts of plant dried are in the susceptible grade (score 7–9). Based on these, our results were: the symptom of CH345 variety (positive control) was slight, with almost plants staying green and just minor tips of leaves being dried. Conversely, the symptom for DT18 was serious, with all plants being dried, collapsed and completely non-green remained. The wild type (Seng Cu variety) had medium symptoms, with ½ length of leaves dried and almost all primary leaves folded down. The symptoms of the mutant lines were different. They were classified into three grades: 24 lines in the tolerant grade (score 1–3), 33 lines in the mediate grade (score 5) and 49 lines in the susceptible grade (score 7–9).

Simulation of drought stress, by PEG inducing drought stress on the plants, was applied widely. PEG6000 was also used successfully to determine the drought stress of bean plants [6], wheat plants [7] and rice plants [8, 9]. In our study, completely different symptoms of the positive control (CH345 line), the negative control (DT18 variety) and the clear separation of symptoms of mutant lines were strong evidence and basis to define the efficiency of this hydroponics protocol in screening drought stress in mutant rice lines.

6. CORRELATION BETWEEN DROUGHT TOLERANCE AND ROOT TRAITS

The hydroponic method of screening for drought tolerance of mutant lines with two sets of tolerant and sensitive lines was done to conduct the experiment of root system evaluation (Fig. 6) with the aim to find out the correlation between drought tolerance and root traits at an early stage. Characteristics were evaluated, including: length of main root; number of lateral root; length of lateral root; density of hair root; fresh weight of root (FWR); dried weight of root (DWR); fresh weight of shoot (FWS); dried weight of shoot (DWS). Results are shown in Table 7.

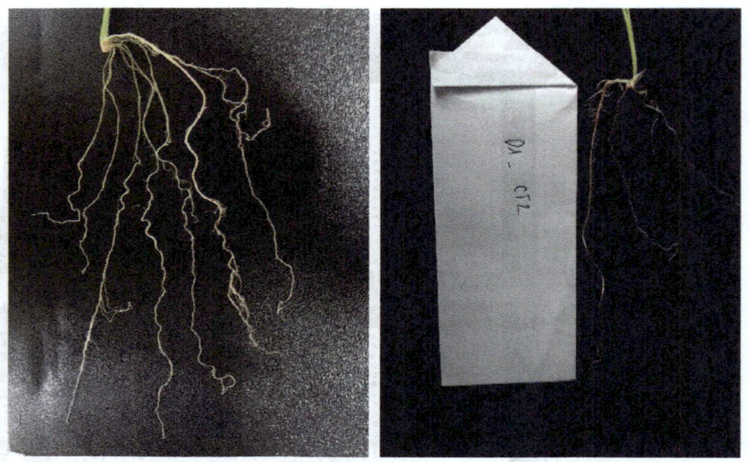

FIG. 6. Root system of the (left) tolerant and of the (right)sensitive lines.

The data showed that drought tolerance had both positive and negative correlations with scored traits. Drought tolerance was assessed as score 1, 3, 5, 7, 9 with decreasing tolerance. Thus, negative correlation indicated that germplasm had higher values at those criteria demonstrating better tolerance. Conversely, positive correlation was noted when germplasm showed higher values for traits for lower tolerance to drought stress. This meant that drought tolerance showed the negative correlation significantly with number of lateral roots (–0.402), density of root hairs (–0.211), root dry weight (–0.237), ratio of DWR/DWS (–0.370) and ratio of DWR/FWR (– 0.514). The criteria for positive correlation significantly with lower drought tolerance included length of lateral root (0.575), root fresh weight (0.409) and ratio of FWR/FWS (0.35).

TABLE 7. CORRELATION BETWEEN DROUGHT TOLERANCE AND ROOT TRAITS

	Drought tolerance	Length of main root	Number of lateral root	Length of lateral root	Density of hair root	Fresh weight of root (FWR)	Dried weight of root (DWR)	Fresh weight of shoot (FWS)	Dried weight of shoot (DWS)	Ratio of FWR/FWS	Ratio of DWR/DWS	Ratio of DWR/FWR	Ratio of DWS/FWS
Drought tolerance	1												
Length of main root	0.040	1											
Number of lateral root	**-0.402**	-0,174	1										
Length of lateral root	**0.575**	0,452	-0,226	1									
Density of hair root	-0.211	-0,193	-0,061	-0.484	1								
Fresh weight of root	**0.409**	0,230	0,035	0.267	-0.033	1							
Dried weight of root	-0.237	0,453	0,198	0.044	0.036	0.484	1						
Fresh weight of shoot	0.158	0.242	0.342	0.083	0.255	0.579	0.292	1					
Dried weight of shoot	0.072	0.209	0.507	0.192	0.048	0.277	0.311	0.713	1				
Ratio of FWR/FWS	0.350	0.033	-0.213	0.239	-0.253	0.720	0.352	-0.138	-0.266	1			
Ratio of DWR/DWS	-0.370	-0.091	0.024	-0.242	-0.171	-0.121	0.157	-0.387	-0.546	0.244	1		
Ratio of DWR/FWR	**-0.514**	0.376	0.293	-0.111	0.084	0.076	0.898	0.107	0.276	0.018	0.242	1	
Ratio of DWS/FWS	0.075	0.166	0.487	0.210	-0.059	0.137	0.283	0.506	0.957	-0.268	-0.544	0.307	1

6.1. Conclusions

Simulation of drought stress by PEG (20% PEG 6000) using a hydroponics set up at the seedling stage for 14 days led to the screening of 24 tolerant lines, 33 medium tolerant lines and 49 susceptible lines. The root system played a key role in responding to drought stress. Since field based root evaluation is often time-consuming, laborious and requires more time and associated costs, the characteristics of the root hairs, root angle and root architecture can be considered for laboratory level hydroponics screening. Several methods are available for accurately assessing the root parameters such as the basket method, which is especially useful for the root growth angle [10] and, the image based method for high throughput screening [11]. Early stage evaluation of root traits in gel tubes or hydroponics for drought screening [12] indicated that it is a rapid screening strategy for screening for drought tolerance. The technique is simple and low cost to evaluate root criteria, which are highly difficult to score, such as root hair and root angle. The results also suggest that the root system with larger dried weight, higher density of hair root, lower water content in root, shorter lateral root but larger quantity will demonstrate a better responsiveness to water deficit. The strategy of a root system with short lateral roots may help to improve the uptake efficiency of water in normal condition while saving energy for the elongation of main root in case of drought stress.

7. FIELD SCREENING PROTOCOLS FOR DROUGHT TOLERANCE IN LOWLAND RICE

Rice consumes almost 80% of the total irrigation freshwater resources during its growth duration [13]. Hence, water stress will affect rice at morphological (reduced germination, plant height, plant biomass, number of tillers, various root and leaf traits), physiological (reduced photosynthesis, transpiration, stomatal conductance, water use efficiency, relative water content, chlorophyll content, photosystem II activity, membrane stability, carbon isotope discrimination and abscisic acid content), biochemical (accumulation of osmoprotectants like proline, sugars, polyamines and antioxidants) and molecular (altered expression of genes which encode transcription factors and defence related proteins) levels and finally affect yield [14]. There are different conclusions about the highest sensitive stage of rice plants to drought conditions. However many authors agreed that the reproductive stage is the most sensitive stage for water deficit [15– 18]. In this study, screening for drought stress tolerance was conducted at the flowering stage to isolate tolerant mutant line(s) at the field level and to assess the impact of drought on potential yield and morphological traits.

7.1. Methodology

The positive check variety, CH345, the negative check variety, DT18, and 459 mutant progenies (derived from Seng Cu, SC) were planted as rows in augmented block design with control plants in each block. At the reproductive stage, rice is highly sensitive between 16 days before flowering and 10 days after flowering. In the lowland field, depending upon the moisture holding capacity of the soil, drought appears in severity from 15 to 20 days. Hence, the field was drained out of water 25 days after transplanting. Both control and drought trials received the same dose of fertilizers (100 kg N: 60 kg; P_2O_5: 90 kg K_2O) per ha in four supplied times. The first time was before transplanting with (15 kg N + total P_2O_5). The second time was at 10 days after transplanting with (28 kg N + 15 kg K_2O). The third time was at 20 days after transplanting with (42 kg N + 30 kg

K_2O); The fourth time was with (15 kg N + 45 kg K_2O) at the initial flowering stage with control and after rewatering with drought trials (Table 8). Temperature and soil moisture were recorded every day in experiment time. Soil moisture was measured in 30 cm soil depth after stopping watering at 0900–10:00 am for each 10 m^2 plot. After 14 days, the leaf rolling criteria were evaluated. After 21 days, the leaf death criteria were evaluated. Water was flood supplied to assess the plant recovery after one week. Data on leaf rolling, leaf death and plant recovery were recorded based on the IRRI SES [5].

TABLE 8. THREE CRITERIA FOR EVALUATION OF THE RESPONSE OF RICE PLANTS TO DROUGHT

Grade	Leaf rolling	Leaf death	Recovery (%)
0	Leaves normal, healthy	No sign of death	-
1	Leaf starts to fold	Tips of leaf are dead	90–100
3	V shaped leaves	1/4 the length of most leaves is dead	70–89
5	U shaped leaves	1/4 to 1/2 the length of leaves is completely dead	40–69
7	Curled leaves O	More than 2/3 of the length of leaves is completely dead	20–39
9	Leaf roll tight	Almost all leaves are dead	0–19

Source: IRRI SES [5].

Field experiments were conducted in the 2020 dry season at Lien Ket II village, Buon Tria commune, Lak province, Dak Lak. Lak Province, located in the east of the Truong Son mountain range between the Buon Ma Thuot highland and the Chu Jang Sin mountain region, is influenced by the southwest monsoon with characteristics of humid and tropical climate. There are two distinct seasons each year: The rainy season starts from May to the end of October, with over 94% of the annual rainfall. The dry season is from November to May with negligible rainfall, with hardly any rain in February. Data were collected on five random plants for each check and M_3 mutant lines of each treatment. Duration was recorded as the number of days from seedling to harvesting. Effective panicles were recorded from every culm. Each panicle was hand-threshed and the unfilled spikelets were separated from filled spikelets through a blower. A grain moisture content of 14% was applied to measure 1000 grain weight and plant yield. Data were analysed by analysis of variance (ANOVA) using Excel 2010 software and means were compared based on the least significant difference test at the 5% probability level. The correlation among drought grade and growth characteristics of these testing rice lines was measured by using the CORR model.

7.2. Results

The experiment in this study was conducted in Lien Ket II village, Buon Tria commune, Lak province, Dak Lak during March 2020. Temperature and soil moisture were recorded daily throughout the treatment duration. Soil moisture was measured at a 30 cm soil depth after stopping watering at 09:00–10:00 am for each 10 m^2 plot. The daily average data showed that temperature was typically stable around 30°C but soil moisture decreased or increased due to the schedule of water supply.

Soil moisture decreased 10% (from 60% to 50%) in 10 days after draining but it declined rapidly from 50% to 30% in the next 10 days. The significant water shortage duration, with less than 40% of soil moisture, was between day 17 and day 22 and it was the same time as the flowering stage of rice plants. After watering again on day 23, soil moisture increased rapidly to 60% in one day and gradually reached the maximum capacity of soil surrounding 70% in the next two days. This period with plentiful water helped plants to recover after stress. A total of 459 mutant lines and checks were scored for three criteria: leaf rolling; leaf death; and plant recovery, respectively corresponding to three point times of two weeks, three weeks and four weeks after draining. The results of this screening are shown in Figs 7(a), 7(b) and Table 9.

(a)   (b)

FIG. 7. Effect of drought stress at the reproductive stage in lowland field of mutant rice lines; (a) rice plants flower in drought stress; (b) the dry cracked ground and stay-green plant.

TABLE 9. RESULTS OF DROUGHT SCREENING OF MUTANTS AT THE FLOWERING STAGE

Name	Grade	Number of lines	Percentage (%)
Tolerant check (CH345)	Tolerant	-	
Sensitive check (DT18)	Sensitive	-	
Wild variety (Seng Cu)	Mediate	-	
Tested mutant lines	Highly tolerant	6	1.3%
	Tolerant	58	12.6%
	Mediate	262	57.1%
	Sensitive	116	25.3%
	Highly sensitive	17	3.7%

The results of this experiment showed that there were 262 lines at medium level (57.1%), similar to wild type; 116 sensitive lines with score 7 (25,3%); 58 tolerant lines with score 3 (12.6%); 17 highly sensitive lines with score 9 (37%); and 6 'good tolerant' lines with score 1 (13%). Six good tolerant mutant lines were: 93(2)-1; 93(2)-2; 93(2)-3; 68-2; 68-3; and 74-1. These lines are very valuable for further breeding to establish a new variety with good resilience to drought stress.

8. EFFECT OF DROUGHT STRESS AT REPRODUCTIVE STAGE ON MAIN AGRO-MORPHOLOGICAL TRAITS

Data on checks CH345 and DT18, the parent variety Seng Cu (SC) and six mutant lines with high drought tolerance, 68-2; 68-3; 74-1; 93(2)-1; 93(2)-2; 93(2)-3; 68-2; 68-3; 74-1, were collected in both control and drought plots to assess the effect of water stress. Results indicated that drought stress impacted more seriously on traits like grain number, full grain number, 1000 grain weight and plant yield (Fig. 8). Drought stress caused the most serious negative effect on the sensitive check variety DT18: decreasing 12.9 % in 1000 grain weight; 23.1% in full grain number and

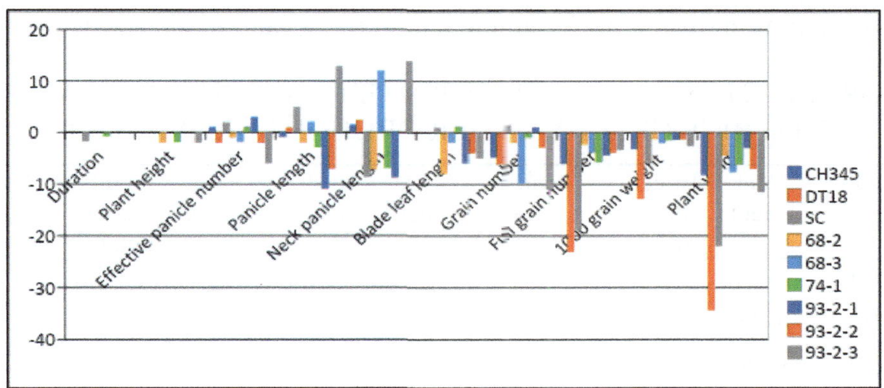

FIG. 8. Percentages of changes of agro-morphological traits of rice lines under drought treatment compared with control conditions.

34.4% in yield. The decrease of those traits on the wild type variety Seng Cu and on the tolerant check variety CH345 were lower. It was found that the effect of drought was variable among rice lines and among traits. However, to know if these differences between control and treatment for traits are due to stress or no stress, data were analysed using two-factor analysis of variance (ANOVA) without replication with level of α = 0.05 (Table 10). Data showed that differences of traits: grain/panicle, of full grain/panicle, of 1000 grain weight and of plant yield between the control plot and drought plot were statistically significant at level α = 0.05. Whereas differences of traits: duration, plant height, effective panicle/culm panicle length, blade leaf length between control plot and drought plot, were not statistically significant at level α = 0.05.

TABLE 10. ANOVA ANALYSIS FOR AGRO-MORPHOLOGICAL TRAITS OF RICE CHECKS AND MUTANT LINES UNDER CONTROL AND DROUGHT TREATMENT (A = 0.05)

Mean	Duration (days)	Plant height (cm)	Effective panicle/ culm	Panicle length (cm)	Neck panicle length (cm)	Blade leaf length (cm)	Grain/ panicle	Full grain/ panicle	1000 grain weight (g)	Plant yield (g)
Control	110.67 ns	108.00 ns	7578 ns	28.12 ns	2.80 ns	34.32 ns	197.19*	149.87*	26.26*	29.65*
Drought	110.33 ns	107.29 ns	7500 ns	28.02 ns	2.82 ns	33.43 ns	188.83*	138.45*	25.35*	26.79*
P-value	0.195	0.094	0.385	0.885	0.838	0.053	0.029	0.012	0.013	0.001

ns: No statistically significant difference; *: statistically significant difference.

Correlations for measured traits in this study showed that there were typically highly significant correlations among agro-morphological traits (Table 11) with absolute values more than 0.9 such as: the correlation of plant yield and effective panicle (0.96); and of full grain number and grain number (0.95). Correlations of drought grade with other traits were marked with a red rectangle. Herein, the result showed the high negative correlations between drought score with plant yield (– 0.78); effective panicle number (–0.66) and full grain number (–0.59). It means that under water shortage, rice lines carrying the better drought tolerance, will establish better values of full grain number, effective panicle number and plant yield. Other negative correlations between drought score with grain number (–0.35), with 1000 grain weight (–0.3) and at lower significant level with plant height (–0.14), panicle length (–0.10) were lower. Three positive correlations were between drought score with duration (0.25); neck panicle length (0.24) and blade leaf length (0.23).

TABLE 11. CORRELATION ANALYSIS FOR AGRO-MORPHOLOGICAL TRAITS AND DROUGHT SCORE OF RICE CHECKS UNDER DROUGHT TREATMENT

	Duration	Plant height	Effective panicle number	Panicle length	Neck panicle length	Blade leaf length	Grain number	Full grain number	1000 grain weight	Plant yield	Drought grade
Duration	1,00										
Plant height	0,69	1.00									
Effective panicle number	-0.13	0.29	1.00								
Panicle length	0.46	0.23	0.25	1.00							
Neck panicle length	-0.40	-0.69	-0.27	-0.15	1.00						
Blade leaf length	0.46	0.44	-0.17	-0.01	-0.24	1.00					
Grain number	-0.69	-0.22	0.48	-0.52	0.37	-0.04	1.00				
Full grain number	-0.75	-0.23	0.57	-0.38	0.29	-0.21	**0.95**	1.00			
1000 grain weight	0.40	0.20	0.08	0.39	-0.57	-0.27	-0.67	-0.48	1.00		
Plant yield	-0.27	0.16	**0.96**	0.14	-0.26	-0.34	0.48	0.63	0.20	1.00	
Drought grade	0.25	-0.14	**-0.66**	-0.10	0.24	0.23	-0.35	**-0.59**	-0.34	**-0.78**	1.00

Rice grain yield is a complex trait that is associated with many component traits, in that it comprises four main indices: panicle number (panicle per culm), grain number (grain per panicle), grain weight (1000 grain weight) and percentage of full grain. Normally, each panicle of each variety includes a definite range of grain number. Thus, any minor change of panicle number will result in a major change of grain number. The ideal rice varieties have five–six effective panicles per culm, around 200 grains per panicle, a percentage of full grain of about 80% and a 1000 grain weight around 25 g, and when planting them at a density 40–45 culm per metre, they will establish a high potential yield. Herein, the difference of effective panicle number (or effective tiller number) between non-drought and drought treatments was not statistically significant, although previous articles stated the opposite [19–22]. It may be suggested that due to the less drought at the stage of panicle production, there could be a non-significant impact and hence was able to grow normally.

In this study, differences between control and drought treatments of grain number, filled grain number, 1000 grain weight and plant yield were statistically significant. It shows the lack of water supply in the reproductive phase will have a critically negative impact on yield, as previously stated [16–18]. The first sign that the rice plant is getting ready to enter its reproductive phase is a bulging of the leaf stem that conceals the developing panicle, called the 'booting' stage. This stage requires plenty of water and the period in which plants form characteristics of panicles (size panicle, number of spikelets per panicle and number of grain per panicle). Herein, the booting phase of almost all test lines in this study occurred in the first duration of drought treatment, hence traits of grain number were reduced significantly but were not the worst. The flowering stage of test lines begins at 80 days after seedling and can continue for about 7 days. Thus, it is indicative of severe drought. As the flowers open they shed their pollen on each other so that pollination can occur to achieve fully fill grain, hence this severe drought has impacted critically on fill grain number and 1000 grain weight.

9. CONCLUSIONS

The methodology of drought screening at the reproductive stage was applied successfully on the mutant population in this study. The severe drought happening at this stage impacts critically on associations of yield such as: grain number, fill grain number, and grain weight for sensitive variety, but there is a less significant effect for the tolerant variety. The hydroponic method with PEG 6000 was applied at an early stage to screen mutant rice lines and it showed the clear separation of drought symptoms among mutant lines and enhanced the efficiency of drought screening and contributed to the establishment of two mutant promising rice lines that had good resilience against water stress in rice.

REFERENCES TO CHAPTER 2–8

[1]   ROJAS, O., Agricultural extreme drought assessment at global level using the FAO-Agricultural Stress Index System (ASIS), Weather Climate Extremes **27** (2020) 100184; https://doi.org/10.1016/j.wace.2018.09.001.

[2]   CARPITAL, N., SABULARSE, D., MONTEZINOS, D., DELMER, D.P., Determination of the pore size of cell walls of living plant cells, **52** 5 (1973) 1154-64.

[3]   RUBINSTEIN, B., TURNER, N.C., Regulation of H+ excreation. Effects of osmotic shock, Plant Physiol. **69** (1982) 939–944.

[4]   YOSHIDA, S., FORNO, D.A., COCK, J.H., Laboratory Manual for Physiological Studies of Rice, International Rice Research Institute, Manila (1971).

[5]   INTERNATIONAL RICE RESEARCH INSTITUTE, Standard Evaluation System of Rice, IRRI, Manila (2002).

[6]   TURKAN, I., BOR, M., OZDEMIR, F., KOCA, H., Differential responses of lipid peroxidation and antioxidants in the leaves of drought-tolerant *P. acutifolius* Gray and drought-sensitive *P. vulgaris* L. subjected to polyethylene glycol mediated water stress, Plant Sci. **168** (2005) 223–231.

[7]   LANDJEVA, S., NEUMANN, K., LOHWASSER, U., BORNER, M., Molecular mapping of genomic regions associated with wheat seedling growth under osmotic stress, Biol. Plan. **52** (2008) 259–266.

[8]   KADHIMI AHSAN A., et al., Effect of irradiation and polyethylene glycol on drought tolerance of MR269 genotype rice (*Oryza sativa* L.), Asian J. Crop Sci. **8** 2 (2016) 52–59.

[9]   NAHAR, S., SAHOO, L., TANTI, B., Screening of drought tolerant rice through morpho-physiological and biochemical approaches, Biocatalysis Agric. Biotechnol. **15** (2018) 150– 159.

[10]   UGA, Y., KITOMI, Y., ISHIKAWA, S., YANO, M., Genetic improvement for root growth angle to enhance crop production, Breed. Sci., **65** (2015) 111–119.

[11]   DAS, A., et al., Digital imaging of root traits (DIRT): A high-throughput computing and collaboration platform for field-based root phenomics, Plant Methods **11** (2015) 51.

[12]   BENGOUGH, A.G., et al., Gel observation chamber for rapid screening of root traits in cereal seedlings, Plant Soil **262** (2004) 63–70.

[13]   BOUMAN, B.A.M., TOUNG, T.P., Field water management to save water and increase its productivity in irrigated lowland rice, Agric. Water Manage. **49** (2001) 11–30.

[14]   PANDEY, V., SHUKLA, A., Acclimation and tolerance strategies of rice under drought stress, Rice Sci. **22** 4 (2015) 147–161.

[15]   PALEG, L.G., ASPINALL, D., The Physiology and Biochemistry of Drought Resistance in Plants, Academic Press, New York (1981).

[16]   LIU, J.X., et al., Genetic variation in the sensitivity of anther dehiscence to drought stress in rice, Field Crops Res. **97** (2006) 87–100.

[17]   ZINOLABEDIN, T.S., HEMMATOLLAH, P., SEYED, A.M.M.S., HAMIDREZA, B., Study of water stress effects in different growth stages on yield and yield components of different rice (*Oryza sativa* L.) cultivars, Pak. J. Biol. Sci. **11** (2008) 1303–1309.

[18]   YANG, X., WANG, B., CHEN, L., LI, P., CAO, C., The different influences of drought stress at the flowering stage on rice physiological traits, grain yield, and quality, Sci. Rep. **9** (2019) 3742; https://doi.org/10.1038/s41598-019-40161-0.

[19]   FUKAI, S., PANTUWAN, G., JONGDEE, B., COOPER, M., Screening for drought resistance in rainfed lowland rice, Field Crops Res. **64** (1999) 61–74.

[20]   INTHAVONG, T., TSUBO, M., FUKAI, S., A water balance model for characterization of length of growing period and water stress development for rainfed lowland rice. *Field Crops Res.* 121, (2011)291–301;

[21]   JI, K., et al., Drought-responsive mechanisms in rice genotypes with contrasting drought tolerance during reproductive stage, Field Crops Res. **169** (2012) 336–344.

[22]   TRIJATMIKO, K.R., Meta-analysis of quantitative trait loci for grain yield and component traits under reproductive-stage drought stress in an upland rice population, Mol. Breed. **34** (2014) 283–295.

2–9. PRE-FIELD SCREENING FOR SEEDLING STAGE DROUGHT RESILIENCE IN SORGHUM MUTANTS

G. LI[1], L. ZHU[2], H. GUO[1], L. LIU[1]

[1]Engineering Laboratory for Crop Molecular Breeding,
National Key Facility for Crop Gene Resources and Genetic Improvement,
Institute of Crop Science, Chinese Academy of Agricultural Sciences,
Beijing, China

[2]Biotechnology Research Institute,
Chinese Academy of Agricultural Sciences,
Beijing, China

Abstract

Drought stress seriously restricts crop yield and quality. As the fifth largest cereal crop in the world, sorghum has strong drought tolerance, but there are great differences in drought tolerance among different genotypes. Thus it is of great significance to establish an efficient and economical method for screening and identifying drought resistance. In the study, the sorghum mutants obtained by EMS and radiation mutagenesis were first evaluated for drought resistance at the germination and seedling stages under PEG-6000 simulated drought stress. The results showed that using 15% PEG to simulate the effects of drought stress on the phenotypes of mutants at the germination stage and seedling stage, was an effective method for rapid and large scale screening of mutants with different drought resistance. These results can provide reference for the subsequent screening and identification of drought resistance of sorghum mutants in the field.

Key words. Sorghum bicolour, PEG drought simulation, drought resilience, mutation breeding, EMS mutant, seedling stage.

1. INTRODUCTION

Unlike hybridization and selection, mutation breeding has the advantage of improving the defects in an elite cultivar, without losing its agronomic and quality characteristics. Since the first release of mutant cultivars that resulted from basic mutation research, mutation breeding has found a niche in plant breeding due to its advantages. In mutation breeding, most of the selected mutants are resulted from easily recognizable characters, such as early flowering, early maturity, changed plant architecture and height, fruit and seed characteristics or resistance to stresses (including biological and abiotic stresses) that can be screened easily. This means that mutants can be easily obtained by simple and/or high-throughput screening for the target traits with only appropriate selection or screening methods.

Sorghum is one of the most important drought resistant crops, cultivated mainly in arid and semi-arid areas. Research has found there is a great variation among sorghum genotypes, although sorghum itself has higher drought resistance than other crops and there still is a great potential for improvement in drought tolerance of sorghum to obtain a higher yield. According to the International Crops Research Institute for the Semi-Arid Tropics (ICRISAT) and the US Sorghum Breeding Program, the four growth stages of sorghum are recognized as the most vulnerable to moisture stresses: (1) germination and seedling emergence; (2) early seedling stage (from seedling emergence to panicle initiation); (3) mid-season (from panicle differentiation to flowering); and

(4) post-flowering to grain filling stage [1].There are a number of methods for evaluation of drought tolerance in crops based on the performance of numerous seeds or seedlings from the same accession. It is not appropriate for early screening for drought tolerant mutants in mutation breeding when every single progeny is different in genetic base and a large population needs to be screened. In this paper we describe a pre-field screening method for drought tolerance at germination and early seedling stage in sorghum.

Simulating drought using polyethylene glycol (PEG) solutions has been widely used in drought resistance screening of crops since the 1960s [2–5], and has been adopted here again with minor modification. It has been reported that PEG has a significant water stress induction effect and has no toxic effects on plants [6]. It is known that PEG molecules with a molecular weight greater than 3000 apparently cannot be absorbed into the cell wall. In this protocol, PEG-6000 was used to simulate drought stress. PEG treatment induces a plant to produce a response similar to that induced by natural drought, for example, causing a depression in seed germination, seedling vigour, and root and shoot growth [7–9]. According to the related research, around 20% PEG-6000 solution was optimal for screening for drought tolerance at 25^0C. Mustamu et al. [10] tested five concentrations (10, 20, 30, 40, and 50%) by using water as the control, and found that 50% could discriminate well for drought tolerant and susceptible species. Five osmotic potentials (0.0, –1.8, –3.6, –7.2 and –10.8 bar) generated by PEG 6000 were tested by AVCI et al. [11] and the results showed that –3.6 bar seemed to be suitable for evaluation of drought tolerance. Hasanuzzaman et al carried out an experiment with five different concentrations of 10%, 15%, 18%, 20% and 25% (wt/vol.) PEG 6000 on two genotypes, respectively, and found that 18% PEG concentration was the optimal one [12]. For more optimal concentration, 16%, 17%, 18%, 19% and 20% are recommended to be tested with water as the control before determination of the concentration to be used for screening for drought tolerance.

A convenient, reproducible, reliable and rapid method for screening drought tolerant mutants from a large number of mutated progenies is required for plant breeders. Several methods have been used to evaluate drought tolerance and water use efficiency that involve measurement of water potential, relative turgidity and diffusion pressure deficit, chlorophyll stability index and carbon isotope discrimination, etc. However, most of these methods are expensive and time consuming and, therefore, are not very efficient for screening large numbers of plants in segregating mutation populations. Also, field screening is difficult due to uncertain environmental conditions, including rainfall and different photoperiod and temperatures in the dry season. Singh et al. developed a simple box screening method of wooden boxes for drought tolerance in cowpea [13]. In the present study, we have used this method in sorghum.

2.   METHODOLOGY

In this study, the seeds of sorghum variety Jiutian 1 were treated with 0.2% EMS (ethyl methyl sulfonate) for 20 h, and then sown in the field after treatment with EMS mutagens. The M_0 generation was self-crossbred and harvested individually. The obtained M_1 generation seeds were planted in the field, and the phenotypes of M_1 were investigated and self-crossed individual plants were harvested. Through continuous multi-generation self-crossing, M_4 generation lines with stable inheritance phenotypic traits were selected for drought resistance screening and identification. M_1–M_4 of gamma irradiated sorghum restorer line Jinliang 5 and LR9198. Variety Z1H08894 was used as the tolerant check, and HW5s as the susceptible check.

2.1. Pre-field screening for mutants for drought tolerance at the germination stage

Four sorghum varieties with different drought tolerance were selected in this experiment. Eighty disease-free sorghum seeds with uniform size and fullnesswere selected (20 seeds for each treatment). The seeds were thoroughly disinfected with 2% NaClO for 10 min and then washed repeatedly with distilled water four times. The sterilized sorghum seeds were placed in petri dishes with different PEG-6000 concentrations (0%, 10%, 15% and 18%) to screen the optimal PEG-6000 concentration for the germination test. The seeds were first incubated in darkness for 24 h, and then incubated at 25^0C for a 12 h/12 h photoperiod for 10 days from the second day. The number of germinated seeds was recorded on the second, fourth, sixth, eighth and tenth day from the date of germination. Seeds are considered germinated when the length of the emergent radicle reaches 2 mm.

The experiment was repeated three times, and the solution and filter paper were replaced every two days to avoid inaccuracy in the experiment due to the increase of PEG concentration caused by water evaporation. The most discriminative PEG stress concentration was selected by comparing the germination rate of different sorghum seeds under different PEG concentration at germination stage.

2.2. Screening and identification of drought resistance of sorghum during germination stage

Forty disease-free seeds with uniform size and fullnesswere selected from each sorghum M_4 mutant line (20 seeds per treatment). The selected sorghum seeds were thoroughly disinfected with 2% NaClO for 10 min, and then repeatedly washed with distilled water four times. The sterilized seeds were then placed in petri dishes with 10 mL distilled water (CK) and the optimal concentration of 15% PEG, respectively. The seeds are first incubated in the darkness for 24 h, and then incubated at 25^0C for a 12 h/12 h photoperiod for 10 days from the second day, and the solution and filter paper were replaced every two days. The germination rate of different sorghum mutants at the 2nd, 4th, 6th, 8th and 10th day were recorded, respectively. The experiment was repeated three times. The drought resistance of different sorghum mutants was evaluated by calculating the relative germination potential (RGP), relative germination rate (RGA) and germination drought resistance index (GDRI) (Fig. 1).

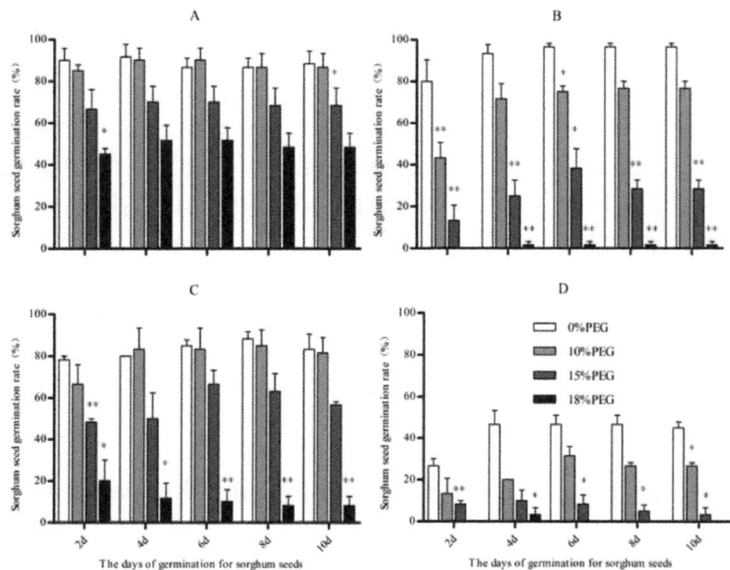

FIG. 1. Effect of different PEG concentrations on the germination of sorghum seeds.
*Note: A, B, C, and D represent the germination rate of different sorghum varieties at different PEG concentrations. 2 d: second day; 4 d: fourth day; 6 d: sixth day; 8 d: eight day; 10 d: tenth day. *: difference is significant at P<0.05 level; **: difference is significant at the P<0.01 level.*

The equations calculating RGP, RGA and GDRI are as follows:

— Germination potential (GP (%)) = (No. of germinated seeds on 4th day/No. of tested seeds) × 100;
— Relative germination potential (RGP (%)) = (germination potential of treatment/germination potential of control) × 100;
— Germination rate (GR (%)) = (No. of germinated seeds on 10th day/ No. of tested seeds) × 100;
— Relative germination rate (RGR (%)) = (germination rate of treatment/germination rate of control) × 100;
— Germination index (GI) = 1.00*ND2 + 0.75*ND4 + 0.50*ND6 + 0.25 *ND8.
— Promptness index (PI) = seed germination index of drought treatment/seed germination index of control.

where ND2, ND4, ND6 and ND8 represent the seed germination rate at the 2nd, 4th, 6th and 8th day, respectively.

If the detected indices are positively correlated with drought resistance, the formula is used as a subordinate function value (Fij) = (Xij – Xjmin)/(Xjmax – Xjmin). If the detected indices are negatively correlated with drought resistance, the formula is used as a subordinate function value (Fij) = 1 – (Xij – Xjmin)/(Xjmax – Xjmin):

Average subordinative function value (ASFV) = (1/n) $\sum$Fij

where Fij represents the subordinate function value of the jth index of the ith germplasm, X_{ij} is the measured value of an index of a certain germplasm, X_{jmax} is the maximum value of this index, X_{jmin} is the minimum value of this index and n is the number of measured indexes, respectively.

2.3. Screening and identification of drought resistant sorghum mutants during seedling stage (PEG method)

The treatment methods of sorghum seed germination were the same as described above. After seed germination, the sorghum three-leaf seedlings with the same growth stage were selected and treated with control (distilled water) and 15% PEG simulated drought, respectively, with ten seedlings in each treatment and three replicates. The drought resistance of different sorghum mutants were evaluated by a wilting index at 7 days after PEG treatment (Fig. 2). The calculation method of wilting index is as follows: Wilting index (WI) refers to the wilting degree of plant stem and leaf under drought condition at seedling stage. The wilting index is generally divided into five levels. The lower the wilting index, the higher the drought resistance level [14].

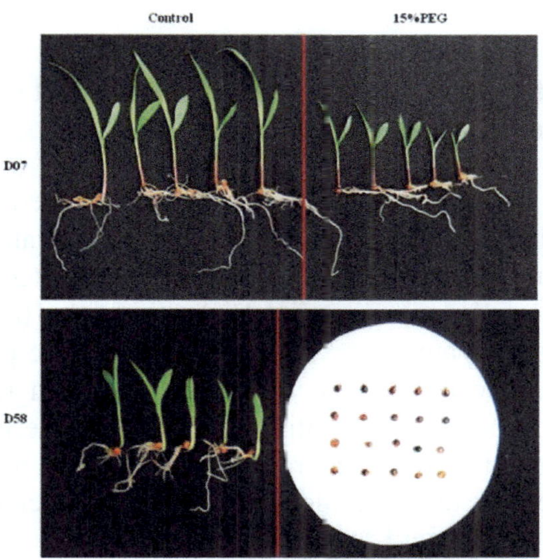

FIG. 2. Comparison of phenotypes of different sorghum mutants at seed germination stage under 15% PEG treatments. D07 and D58 represent different drought resistant mutants, respectively.

2.4. Screening and identification of drought resistance at germination and seedling stage in sorghum

The drought resistance of different sorghum mutants at germination stage is evaluated according to the relative germination potential, relative germination rate and germination drought resistance index by PEG simulation. Combined with the wilting index at seedling stage, the drought resistance of different sorghum mutants can be systematically evaluated by using the subordinate function method (Fig. 3). This comprehensive drought resistance screening evaluation method has good reliability, especially for large screening population, is relatively fast, simple and easy to operate.

FIG. 3. Comparison of phenotypes of different drought resistant sorghum mutants at three-leaf seedlings stage under 15% PEG drought stress. Drought resistant mutant (left); drought sensitive mutant (right).

2.5. Pre-field screening for mutants for drought tolerance at seedling stage (box screening method)

The selected sorghum seeds from each M_2 or M_3 line were sterilized in sodium hypochlorite (1%) solution and then washed three times with deionized distilled water. Three controls, including drought resistant (CKr), drought susceptible (CKs) and non-irradiated initial seeds (CKo) were set, respectively. The sterilized seeds were sown in the boxes with 10 cm between rows and 5 cm between plants within the row. Water was sprayed daily until the partial emergence of the first trifoliate, then was stopped. The permanent wilting rate of each line was recorded at various intervals until all the plants of the most susceptible lines died. Watering was then resumed and the per cent plant recovery for each line were recorded. Drought stress may be continued accordingly, until most CKo seedlings die. The drought tolerant mutants were selected from them. All survived seedlings were transplanted in the field or remained in the boxes to grow mature under regular management.

3. RESULTS

3.1. Effect of different PEG concentrations on the germination of sorghum seeds

According to the results of drought treatment with different concentrations (0, 10%, 15% and 18%) of PEG, it can be seen that the germination of four sorghum variety seeds was inhibited more severely with the increase of drought intensity, in which sorghum variety A was less affected by drought and nearly half of its seeds germinated under 18% PEG treatment. However, variety B and D were significantly affected by drought. With the increase of PEG concentration, the germination rate of seeds on different days decreased significantly compared with the control. Under the treatment of 18% PEG, almost all seeds were inhibited, and the germination rate was almost zero. In addition, the seed germination rate of variety C under the treatment of 18% PEG was 1/4, which indicated that 18% PEG was extremely unfavourable to the germination of sorghum seeds, and it was not suitable for the screening of drought resistance in large quantities during germination. Moreover, the seed germination rate of varieties A and C under 10% PEG stress was similar to or even slightly higher than that of under the control condition, which indicated that a certain concentration of PEG can promote the growth of seeds to some extent, which was consistent with previous studies [15, 16]. The result indicated that 10% PEG was not suitable for mass screening of drought resistance during germination. While under 15% PEG treatment, the seed germination rate was basically about 50%, which had better differentiation than that under the control, indicating that 15% PEG concentration was more suitable for batch screening of drought resistance of sorghum during germination stage (Fig. 4).

3.2. Screening and identification of drought resistance of different sorghum mutants during germination stage

The drought resistances of different M_4 sorghum mutants during germination stage were identified by 15% PEG treatment. The results showed that the responses of different mutants to PEG treatment were diverse. For example, the growth of D07 mutant was inhibited to some extent under 15% PEG treatment, but did not affect its survival, indicating that D07 was relatively drought resistant. While the seeds of D58 almost failed to germinate under 15% PEG treatment, indicating that 15% PEG completely inhibited the seed germination of D58, and showing that D58 mutant was more sensitive to drought stress (Fig. 2).

3.3. Screening and identification of drought resistance of different sorghum mutants

Using 15% PEG to simulate drought treatment of sorghum seedlings at three leaf stage, the results showed that the effects of 15% PEG drought stress on the drought resistance of sorghum mutants were different. The growth of drought resistant mutants was not significantly inhibited, while the leaves of drought sensitive mutants showed obvious wilting and chlorosis, especially root development, which was significantly inhibited under 15% PEG treatment (Fig. 3).

3.4. Comprehensive evaluation of drought resistance of different sorghum mutants at germination and seedling stages

Under 15% PEG simulated drought stress, the seed germination of different sorghum mutants was affected to different degrees. Based on the subordinate function method, the relative germination potential, relative germination rate and promptness index(PI) of different sorghum mutants were analysed, and the average subordinative function value (ASFV) could be used to evaluate the drought resistance of different sorghum mutants. The larger the ASFV, the stronger the drought resistance. Moreover, the ASFV of D07 and D38 were larger, indicating that these mutants were relatively drought resistant, while the ASFV of D47 and later sorghum mutants were less than 0.07, indicating that they were relatively sensitive to drought stress (Fig. 4). At the same time, the effects of 15% PEG drought stress on the growth of different sorghum mutants at seedling stage were analysed based on the wilting index (WI).There were ten mutants identified with better drought resistance at seedling stage, including D05, D10, D11, D16, D18, D24, D43, D60, D62 and D65 (Fig. 4).

According to the preliminary screening results of drought resistance of different sorghum mutants at germination and seedling stages, there were five drought resistant mutants identified at both germination and seedling stages (D04, D14, D19, D22 and D37), and there were six sensitive materials identified at both germination stage and seedling stage (D40, D41, D42, D49, D51 and D53), respectively.

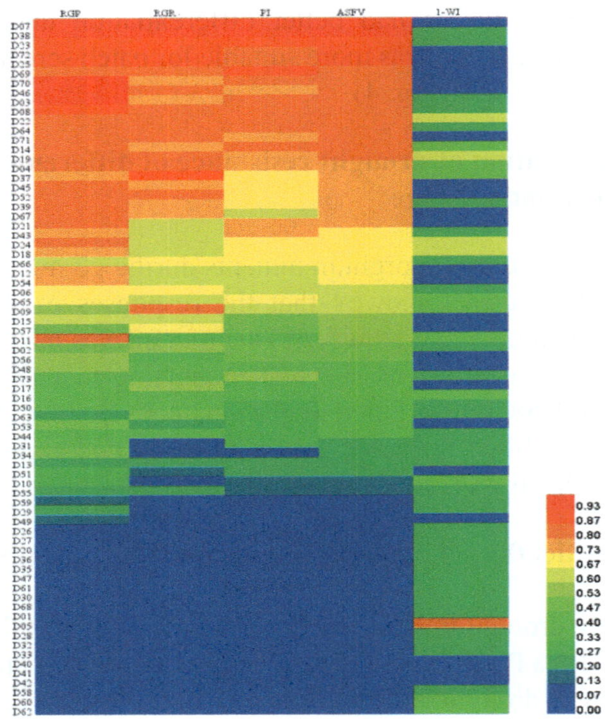

FIG. 4. Subordinate function value of drought resistance index at the germination stage and the seedling stage WI level of different sorghum mutants. Note: RGP: relative germination potential; RGR: relative germination rate; PI: promptness index; ASFV: average subordinate function value; WI: wilting index level assignment at seedling stage.

3.4.1. Analysis of results and choice of the mutants

The results are expressed in terms of a promptness index (P.I.), as described by Bouslama and Schapaugh [3]. The P.I. was calculated as follows:

$$P.I. = nd_2 (1.00) + nd_4 (0.75) + nd_6 (0.50) + nd_8 (0.25)$$

where: P.I. = promptness index, nd2, nd4, nd6, and nd8 = per cent of seeds observed to germinate on the 2nd, 4th, 6th, and 8th day of observation, respectively. A germination stress index (GSI) was expressed in per cent as follows:

Promptness index of stressed seeds (PIS) / Promptness index of control seeds (PIC) × 100.

Considering the difference of evaluation of pure line or hybrid for drought tolerance and that of induced population M_2 or M_3, headlines with drought tolerance will be screened out, single seedling with better performance is selected for transplanting in the field or pot to grow to mature. A quantitative trait mutant cannot be easily detected with a high level of confidence owing to its interaction with environmental factors, therefore, a different procedure is recommended for selecting such mutants. Selection of mutants on the basis of the mean value of M_2 plant derived progeny in the M_3 instead of the value of the M_2 plant is a possible solution [17].

3.5. Result of pre-field screening for mutants with drought tolerance at seedling stage (box screening method)

The M_2 generation was screened for drought tolerance at seedling stage by sowing seeds in plastic boxes. Albino seedlings were easily found, in 3 normal:1 albino ratio, which suggests a single gene mutation. A total of 15 272 M_2 seedlings of Jinliang 5 irradiated by three doses was screened for drought tolerance, and 1239 seedlings survived (Table 1); the survival rate was 8.11%.

TABLE 1. SCREENING OF M_2 SEEDLINGS FOR DROUGHT TOLERANCE (CV, JINLING 5)

Dose (Gy)	M_1 head seed setting	Seedlings	Survival seedlings	Seedling survival rate after drought stress	Plants set seed	Rate of plants harvested
300	≥20%	3928	122	3.11	51	1.30
	<20%	4311	623	14.45	266	6.17
400	≥20%	3040	147	4.84	41	1.35
	<20%	3387	271	8.00	84	2.48
500	<20%	606	76	12.54	33	5.45
	Total	15272	1239	8.11	475	3.11

4. CONCLUSIONS

Drought is one of the important abiotic stresses that restrict agricultural production. Therefore, research on drought resistance of crops is of significance in both theoretical and practical aspects. As the fifth largest cereal crop in the world, sorghum has strong drought tolerance, but there are great differences in drought tolerance among different genotypes, so it is very important to establish an efficient and economic drought tolerance screening and identification method. In this study, the sorghum mutants obtained by EMS and radiation mutagenesis were first evaluated for drought resistance at the germination and seedling stages under simulated drought environment with PEG-6000. The results showed that using 15% PEG to simulate the effect of drought stress on the phenotype of mutants at germination stage and seedling stage was an effective and rapid method to pre-field screen a large number of mutants with different levels of drought resistance. These results could lay a foundation for the subsequent screening and identification of drought resistance of sorghum mutants in the field.

REFERENCES TO CHAPTER 2–9

[1] REDDY, B.V.S., RAMESH, S., REDDY, P.S., RAMAIAH, B., Crop Production in Stress Environments — Genetic and Management Option. Genetic Options for Drought Management in Sorghum, Agrobio Publishers, Jodhpur (2007) 61–84.

[2] PARMAR, M.T., MOORE, R.P., Effects of simulated drought by polyethylene glycol solutions on corn (*Zea mays* L.) germination and seedling development, Agron. J. **58** (1966) 391–392.

[3] BOUSLAMA, M., SCHAPAUGH, W.T., Jr., Stress tolerance in soybeans. I. Evaluation of three screening techniques for heat and drought tolerance, Crop Sci. **24** (1984) 933–937.

[4] GOVINDARAJ, M.P., SHANMUGASUNDARAM, P., SUMATHI, MUTHIAH, A.R., Simple, rapid and cost effective screening method for drought resistant breeding in pearl millet, Electr. J. Plant Breed. **1** (2010) 590–599.

[5] SHAH, T.M., et al., Selection and screening of drought tolerant high yielding chickpea genotypes based on physio-biochemical indices and multi-environmental yield trials, BMC Plant Biol. **20** 1 (2020) 171; https://doi.org/10.1186/s12870-020-02381-9.

[6] EMMERICH, W.E., HARDEGREES, P. (1990) Polyethylene glycol solution contact effects on seed germination. Agron. J., 82:1103-1107

[7] BLUM, A., SINMENA, B., ZIV, O., An evaluation of seed and seedling drought tolerance screening tests in wheat, Euphytica **29** (1980) 727–736; https://doi.org/10.1007/BF00023219,

[8] DHANDA, S.S., SETHI, G.S., Tolerance to drought stress among selected Indian wheat cultivars, J. Agric. Sci. **139** 3 (2002) 319–326.

[9] MUJTABA, S.M., FAISAL, S., KHAN, M.A., MUMTAZ, S., KHANZADA, B., Physiological studies on six wheat (*Triticum aestivum* L.) genotypes for drought stress tolerance at seedling stage, Agric. Res. Technol. **1** 2 (2016) 001–005.

[10] MUSTAMU, E.N., et al., Drought stress induced by polyethylene glycol (PEG) in local maize at the early seedling stage, Heliyon **9** 9 (2023) e20209. https://doi.org/10.1016/j.heliyon.2023.e20209

[11] AVCI, S., İLERİ, O., DEMIR KAYA, M., Determination of genotypic variation among sorghum cultivars for seed vigor, salt and drought stresses, J. Agric. Sci. **23** (2017) 335–343.

[12] HASANUZZAMAN, M., SHABALA, L., BRODRIBB, T.J., ZHOU, M., SHABALA, S., Assessing the suitability of various screening methods as a proxy for drought tolerance in barley, Funct. Plant Biol. **44** (2017) 253–266.

[13] SINGH, B.B., MAI-KODOMI, Y., TERAO, T., A simple screening method for drought tolerance in cowpea, Ind. J. Genetics Plant Breed. **59** 2 (1999) 211–220.

[14] WANG, L., WU, J., JING, R.L., CHENG, X.Z., WANG, S.M., Drought resistance identification of mungbean germplasm resources at seedlings stage, Acta Agron. Sin. **41** 1 (2015) 145–153.

[15] ZHANG, Y.J., ZHAO, L.L., WANG, P.C., CHEN, C., KANG, F.R., Construction and comprehensive evaluation of drought resistance index system in oatmeal during germination, J. Nucl. Agric. Sci. **31** 11 (2017) 2236–2242.

[16] ZHAO, F.Y., TIAN, X.H., DU, W.H., Screening on drought simulation conditions and drought-resistant parameters at germination stage of triticale, Agric. Res. Arid Area **35** 2 (2017) 96–102.

[17] HALLAJIAN, M.T., "Mutation breeding and drought stress tolerance in plants", Chapter 13. Drought Stress Tolerance in Plants (HOSSAIN, M., WANI, S., BHATTACHARJEE, S., BURRITT, D., TRAN, L.S., Eds), Vol. 2, Springer, Cham (2016) 359–382; https://doi.org/10.1007/978-3-319-32423-4_13

2–10. SCREENING FOR DROUGHT TOLERANCE IN MUTAGENIZED POPULATIONS OF POST-RAINY SORGHUM

A. BADIGANNAVAR[1*], G. GIRISH[2], T.R. GANAPATHI[1], F. SARSU[3]

[1]Nuclear Agriculture and Biotechnology Division,
Bhabha Atomic Research Centre, Mumbai, India

[2]Agriculture Research Station, University of Agriculture Sciences,
Hagari, India

[3]Plant Breeding and Genetics Section,
Joint FAO/IAEA Centre of Nuclear Techniques in Food and Agriculture,
International Atomic Energy Agency, Vienna

Abstract

Climate change has a significant impact on crop production globally. Drought is the most extreme environmental stress during crop growth, severely affecting grain and fodder yields. Numerous studies in the past have revealed the potential contribution of drought responsive traits in mitigating pre-/post-flowering drought stress. Among the cereals, sorghum is cultivated worldwide for its grain, fodder, biomass and bio-ethanol production. Although it is a drought tolerant C_4 crop, most of its cultivated species suffer from terminal drought stress. The use of radiation and chemical mutagens have major impact on creating variability and modifying drought responsive traits such as leaf/stem waxiness, leaf rolling, stay-green and earliness. Screening a large mutagenized population of M-35-1, a popular farmer variety over M_3, M_4 and M_5 generations for pre-/post-flowering drought stress has resulted in yield penalty of 25–30% during critical growth stages. But promising mutants showed earliness and stay-green, with retention of at least 60% of green leaf areas at maturity, which had potential in recording moderate grain yield under stress. Such mutants, where drought escape and avoidance mechanisms exist, may translate into better yield levels for marginal farmers in the rainfed system.

Key words: Cereals, drought stress, chemical mutagen, physical mutagen, sorghum.

1. INTRODUCTION

Climate change has triggered a series of consequences, such as a rise in the average temperature, frequent floods, drought and mean sea level rise affecting the agricultural production of various food crops. Based on a report of the Intergovernmental Panel on Climate Change (IPCC), an increase in the global temperature by 1.5°C between 2030 and 2052 would severely affect growth and productivity of food and fodder crops [1]. Therefore, in order to maintain sustainable food production, there is a need to develop abiotic stress tolerance in crops, especially drought, heat and salinity, by exploring various breeding, biotechnological and molecular tools. Among all the abiotic stresses, crops are highly vulnerable to moisture stress conditions during critical growth stages, directly impacting food and fodder production [2]. Among cereal crops, sorghum is the fifth most important, feeding millions of people in African and Asian countries. It is mainly grown for bread making, fodder, alcoholic drinks and bio-fuel [3]. India has an area of over 7.06 million ha with a production of 7.45 million t and productivity of 960 kg/ha. Almost 80% of the Indian sorghum production is contributed by Maharashtra, Karnataka and Andhra Pradesh states [4]. It is primarily cultivated in two distinct seasons: June to October (rainy/kharif) and October to February (post-rainy/rabi). The post-rainy season crop is grown mostly on residual soil moisture over five million ha of the Deccan Plateau, serving as a major source of food for consumption.

2.   DROUGHT TOLERANCE IN SORGHUM

Among the cultivated crops, sorghum is one of the drought tolerant C4 grass species that can maintain optimum photosynthesis at low CO_2 concentrations. But post-rainy season sorghum is grown on residual moisture and often experiences severe terminal drought stress, reducing grain yield by 50–60%. Although the effect is dependent on climatic conditions of the geography, terminal drought stress greatly affects grain and fodder yields [5]. In addition, high soil temperatures may trigger the incidence of charcoal rot disease, affecting grain and fodder quality in a rainfed ecosystem [6]. Therefore, in order to sustain post-rainy grain yield, there is a need for genetic improvement of field crops by harnessing natural and induced variability for drought, heat and salinity responsive traits. Various approaches have been used to develop drought and heat tolerant crops, either by hybridization using primary/secondary gene pool species as donor parent for drought traits or by induced mutagenesis. Mutation breeding is a valuable breeding tool to create new variation in self/often cross-pollinated crops to bring heritable changes in qualitative and quantitative traits by employing physical and chemical mutagens [7]. Physical mutagens such as gamma rays, X rays, ion beams and electron beams cause mutations ranging from single nucleotide changes to chromosomal breaks, while chemical mutagens such as ethyl methane sulphonate (EMS) and sodium azide (SA) lead to high density of point/frame shift mutations [8]. Recently, heavy ion beams and fast neutrons have been extensively used to induce high frequency of mutations in field crops due to their high linear energy transfer (LET) and enhanced relative biological effect (RBE). High LET radiations are capable of inducing irreversible dsDNA breaks knocking out major functional genes [9]. Several crops have been subjected to mutation breeding to create genetic variability for quantitative traits, e.g. peanut, sorghum, wheat, rice, barley and soybean [10]. As a result, as many as 3402 varieties in 228 different field crops have been successfully developed and released for farmer cultivation worldwide [11]. The majority of the mutant varieties have been contributed by Asian countries (2087), followed by Europe (960) and North America (211).

In arid and semiarid regions, various factors, e.g. low precipitation, high temperature, light intensity and dry wind, leading to increased evapo-transpiration from soil profiles contribute to drought stress. In such moisture stress conditions, plants respond differentially due to their genetic makeup and express drought responsive traits to cope up with the length, severity and timing of drought stress [12]. In turn, plant species have developed their own mechanisms to withstand drought stress via escape, avoidance and drought tolerance [13]. In cereals, the major drought stress symptoms include loss of leaf turgor, drooping of leaves, yellowing, etiolation, inward curling of leaves and premature leaf senescence [14], while in the later stages of development, stunted growth, poor seed set and unfilled grains, lead to small panicles with marginal grain/fodder yields. In order to develop a drought tolerant variety, several morpho-physiological, anatomical, biochemical mechanisms need to be explored [6]. In cereals, specifically in sorghum, drought stress mainly affects crop growth severely during pre- and post-flowering stages. In order to breed for tolerant genotypes, there is a need to enhance genetic variability for drought responsive traits such as earliness, stem/leaf waxiness, stay-green, deep root system, osmotic adjustments, and quiescence [15], which can take care of pre-/post-flowering drought stress. By exploiting these traits, one can manipulate the genetic architecture of the plant to withstand moisture stress in a rainfed farming system.

Among the several mechanisms available to cope up against drought stress in sorghum, shortening of life cycle, especially early flowering (40–50 days) or early maturity (80–90 days) may help in avoiding the drought stress with some yield penalty. In addition, plants also maintain high photosynthetic efficiency, remobilization of assimilates and high leaf nitrogen levels under stress conditions (Table 1). Stomatal conductance helps in maintaining the water uptake by a well organized root system [16], especially, a deep thick root system with profuse nodular and lateral roots help in better conduct of water requirement by the plant under stress. Furthermore, thick cuticle, waxy layer, leaf rolling, stay-green, high transpiration efficiency and more root volume will help in maintaining high tissue water potential within the plant. Secretion of protective solutes, high proline content, osmotic adjustments and stomatal conductance may also help in tolerating moisture stress [17]. Apart from these mechanisms, stay-green or non-senescence is an important trait associated with drought tolerance in sorghum. Delayed leaf senescence during the milky stage is largely due to an improved balance between the supply and demand of water, as well as the efficiency with which the crop utilizes water to biomass and grain yield [18].

In order to develop drought tolerant mutants, there is a need to develop efficient drought screening techniques, in laboratory/greenhouse and field conditions, which are robust, cost effective and applicable to large plant populations. The putative mutants from M_2 and subsequent generations need to be validated for drought responsive traits, especially morphological, physiological, anatomical and root traits [19]. In addition, biochemical parameters for estimating sugars, proline, methyl glyoxal (MG) and other molecules and presence of quantitative trait loci related to stay-green need to be validated for effective screening for drought tolerance. Transcriptome based analysis of specific genes and proteome of differentially regulated genes under moisture stress (NAC and aquaporins) would further enhance the understanding of gene regulation of stress responsive traits in mutant plants [6]. In this study, we have summarized an efficient drought screening method at laboratory and field level in a sorghum using mutation breeding principles.

2.1. Screening protocols for drought tolerance in sorghum

Induced mutagenesis using physical and chemical mutagens has been useful for generating new alleles for crop improvement at very low frequencies [20]. Therefore, optimization of seed mutagenesis, specifically identifying optimum dose and dose rate, methods of mutant screening and their handling in segregating generations, identification and validation of tolerant mutants play a key role in the success of mutation breeding programme.

2.1.1. Mutation breeding for abiotic stress tolerance

(i)   Selection of plant material

In a traditional plant breeding programme, the ultimate objective is to develop a high yielding variety with durable tolerance to abiotic stresses and improved seed quality parameters. This also applies to mutation breeding programmes wherein historical varieties, elite or advanced breeding lines and even F_1 and F_2 harvested seeds can be subjected to mutagenic treatment to improve drought responsive traits. A generalized protocol for radiation induced breeding of drought tolerance in cereal crops is outlined in (Fig. 1). In brief, based on the seed germination and field emergence rate, 2000 irradiated M_1 plants were maintained in each treatment of either physical/chemical treatment. With the decrease in mutation rate, the M_1 population size needs to be increased. In general, it is recommended to grow more than 1500–2000 M_1 plants in sorghum mutation projects based on the germination percentage and mutagenic dose. Selection for the dominant and recessive mutations would be targeted in the M_1 and M_2 generations, respectively, apart from estimation of lethal dose and mutation frequency. Subsequent to the M_3 generation, drought screening at field level can be initiated (based on seedling screening using PEG/any other osmoticum at laboratory level). Further, validation of the true breeding nature of the putative mutants will be ascertained using systematic pre-/post-flowering drought stress under rain out shelters. In M_5 and subsequent generations, replicated trials would be able to assess these mutants for grain yield and contributing traits. Large scale multi-location yield trials on a farmer's field under natural rainfed conditions would enable us to understand the $G \times E$ interaction and stability of the selected mutants under drought stress. Based on the average performance of the mutants over the seasons and locations, top performing lines can be proposed for commercial release.

(ii)   Selection of mutagen and dose

Although reports reveal the effects of physical mutagens causing deletions of various sizes and chemical mutagens leading to nucleotide substitutions, there has been no conclusive evidence on the advantages and disadvantages of these two types of mutagens. However, physical mutagen (gamma rays) has a higher tendency to produce knock-out and knock-down mutations than chemical mutagens and, Ethyl Methane Sulphonate is the choice if a change-of-function type of mutation is desirable [21]. Once a mutagen has been selected, the use of a proper dose becomes important. The mutagenic dose depends on the dose rate, type of mutagen, dose/concentration, seed moisture content and genetic constitution of the genotype. Several reports in the past have revealed lethal doses at 50% survivability and one can directly consider the optimum dose around the LD_{50} value.

2.1.2.   Laboratory screening for drought stress using polyethylene glycol (PEG)

In the drought stress experiments, initial laboratory screening using osmotic agents will help us predict the level of tolerance for the moisture stress. The genotypes to be tested against moisture stress need to be compared with universally recognized drought tolerant genotypes. Several genotypes have been recognized as drought tolerant due to their inherent tolerance to stress at seedling/vegetative/reproductive stages, such as B-35, BTx642, E-36-1, SC-56 and 00MN7645 [22]. Among these genotypes, B-35, a durra line from Ethiopia derived from BT × 642 has been used as a major source of stay-green genes in breeding stress tolerant genotypes worldwide [23]. PEG-8000 is a polymer used to induce the osmotic potential of a solution, thus modifying plant water deficit in a relatively controlled manner. Being a large particle, it will not penetrate the plant, which makes it an ideal osmoticum for use in the hydroponic root medium. This method gives the baseline tolerance level of genotype for moisture stress and this preliminary information can be used for irrigation schedules in the field level experiments.

2.2.   Field screening for drought stress under rainout shelter :

(i)   Growing and handling M_1 and M_2 populations

Mutagenized seeds can be planted in a field plot of 5 m length with a spacing of 45 ×10 cm in the M_1 generation (Fig. 2). All the agronomic practices need to be followed for successful crop growth in the mutant and control plots. Natural field screening under rainfed conditions for both M_1 and control plants has to be followed. Due to mutagenic effects, plants are expected to have reduced pollen viability and hence the tendency for out-crossing increases significantly. It is important to protect M_1 plants from out-crossing of other varieties by bagging them prior to flowering. Selfed seeds would enable recessive mutant alleles to express and can be easily screened out in the M_2 generation. There should be no selection of 'mutants' in the M_1 populations, except for the dominant mutations and elimination of off-type plants (mixtures, segregates from out-crossed plants). However, a few hundred plants should be assessed, particularly for plant injuries and seed set to ensure that there is significant effect of mutagenic treatment.

The selfed plants from the M_1 generation are harvested individually and advanced to the M_2 generation. In M_2, all the seeds from each panicle need to be planted as progeny rows in a 5 m plot along with the control (unirradiated parent) at every 50th row as per the augmented field plot design. At the seedling stage, it is important to estimate the frequency of chlorophyll deficient mutants, which are often used to calculate mutation frequency. If the M_1 plants have a normal seed set, then their progenies in M_2 generation will have very low frequency of chlorophyll mutations.

(ii)   Screening for drought stress and confirmation of putative mutants

(a)   Screening of mutants for drought stress

In most cases, mutant screening starts in the M_2 generation, assuming a trait of interest is controlled by recessive genes. This is ideal for most of the qualitative traits controlled by polygenes. Putative mutants can also be identified for quantitatively controlled traits, such as plant height, panicle size, grain yield and seed weight. However, as the quantitatively controlled traits could be significantly influenced by environmental effects, screening of such mutants often starts in the M_3 generation using M_{2-3} lines (seeds harvested from M_2 plants and grown in progeny rows). Selection criteria for screening M_3 plants may be leaf rolling coupled with drought responsive traits such as waxy stem/leaves, early flowering and maturity and drought scores based on 1–9 scale (1: no dried leaves and 9: 90% dried leaves). Once the plant is selected for a mutated trait, further analysis should be performed using their progenies (M_3 or M_4 lines). Very often a quantitative trait variant (putative mutant) could not be confirmed in their progeny due to the difference in selected plant/line from the parental line resulting from micro-environmental differences in the field or controlled chambers.

(iii)   Genetic confirmation of induced mutants:

The ideal way to confirm a mutant is to study the genetic basis of the mutated trait. This can be performed by crossing the mutant with its parent and raising the F_1 and F_2 populations. Distinct plants should be recovered in F_2 populations as the mutant shares almost an identical genetic background to its parent line. Therefore, even the quantitative trait in a mutant can be observed on an individual plant basis and they should possess distinct segregation of mutant and wild type plants in F_2 populations (often in traits controlled by single gene). If there is no such clear-cut segregation, the selected putative mutant might have resulted from mixture or out-crossing with other genotypes and hence it is not a true type.

(iv)   Field design and statistical analysis

The detailed field design in different generations with proper statistical techniques is outlined as follows. Irradiated seeds in the M_1 generation can be sown in a replicated design consisting of blocks with control/check varieties to compare any dominant mutations. Whereas the planting progeny rows/spike to row method is usually followed in the M_2 generation. The whole field can be divided into blocks and in each block one can take up planting of $M_{1:2}$ progeny rows with their control parent at every 25/50th row following an augmented block design. Any recessive mutations that do occur in M_2 can be compared with those in parental plants and documented. Since there will be more progeny rows for each treatment with less seeds per plant in the $M_{2:3}$ generation, all the putative mutants can be planted in the augmented block design. In M_4 or later generations, when the homozygosity/familial characters become uniform, one can follow randomized block design (RBD) with commercial/parental checks for an estimation of *per se* performance. If there are more progenies selected for each mutagenic treatment, lattice/alpha designs can also be followed. In the M_5 and subsequent generations, one can follow RBD design for yield evaluation in multi-location trials. In M_6 generation, selected elite mutants (8–10) with checks can be sown in replicated RBD design in two or more locations to analyse the $G \times E$ and stability of the mutants across different environments.

Morphological data consisting of plant habit, days to flower, yield contributing, drought responsive physiological/biochemical traits along with soil moisture and weather data can be recorded throughout the growth stages (Table 1). Such observations on an individual plant basis can be used to calculate mutation frequency, mutagenic effectiveness and efficiency in the M_2 generation. In the advanced generations, one can perform an analysis of variance for each trait and signify the treatment effect. In the M_5 generation onwards, both univariate and multivariate statistical methods can be applied to analyse the significance of treatment or mutant line over check/parental line. In addition, data can also be used to estimate phenotypic (PCV) and genotypic coefficient of variation (GCV), broad sense heritability (defined as the ratio of the genetic variance (σ^2_G) between genotypes to the total phenotypic variance ($\sigma^2_P = \sigma^2_G + \sigma^2_E$)) and genetic gain. The Pearson's correlation coefficients (r) between the traits under irrigated and drought-stress conditions can be estimated. In addition, the stability of the mutants over seasons/locations can be assessed using $G \times E$ interactions, AMMI (additive main effects and multiplicative interactions) and stability analysis following standard procedures.

TABLE 1. LIST OF MORPHOLOGICAL, PHYSIOLOGICAL, BIOCHEMICAL, MOLECULAR AND ROOT RELATED DROUGHT RESPONSIVE TRAITS [6, 24, 25]

Morphological traits	Plant height (cm), days to flower, days to maturity, stem width (cm), No. of leaves, No. of internodes, internodal length (cm), seed weight (g), grain weight under water stress (g), biomass weight under W/stress (g), harvest index (%), panicle length (cm), panicle width (cm), pollen viability
Root related traits	Deeper thicker roots, greater root volume, root fresh weight root dry weight, root pulling resistance, root anatomical features, root anatomical features, root penetration ability
Water status and photosynthetic parameters as markers:	a) **Water potential**: Decrease in water potential (Ψ < -0.8 MPa) is sign of drought stress and loss of leaf turgor
	b) **Relative water content**: Leaf relative water content in stress is determined gravimetrically, which is ratio of dry weight to fresh weight: highly reproducible.
	c) **Measurement of leaf transpiration**, turgor and xylem flow using sensors
	d) **Stomatal conductance (mmol/m^2/s)**: Stomatal closure disrupts CO_2 supply through parenchyma cells affecting Ps.
	e) **Photosynthetic activity**: Determined by quantitative estimation of pigments: Chlorophylls and carotenoids. Decreased chlorophyll level is symptom of oxidative stress and result of photo oxidation and chlorophyll degradation. Higher Chlorophyll content: drought tolerant
	f) **PSII activity**: Severe decrease in PSII activity under stress. Relative chlorophyll content+ PSII efficiency measured by pulse amplitude modulation (PAM) flourometry. Fv/Fm ratio correlated with max. PSII activity.
	g)**Drought induced stomatal closure**- regulated by ABA (increase under drought)
Profiling of metabolites	**Osmotically active and metabolically neutral solutes**: Sugars, amino acids (proline and glycine), betaine polyamines and organic acids
Protective proteins	Under stress, biosynthesis of proteins — Chaperons, LEA proteins, enzymes of antioxidant defence are up regulated
Molecular markers based screening	Stay-green related markers, drought responsive trait specific markers
Oxidative stress associated with drought	a) **Melondialdehyde**: reliable marker of lipid oxidation damage and it increases in stress leaf
	b) **H_2O_2**: In tissues is an indicator of drought stress. It is more stable and estimated better than ROS
	c) **Anti-oxidant enzymes**: central role in detoxification of H_2O_2 under stress. Ascorbate peroxides: neutralize H_2O_2 in plants
Biochemical traits	Proline content, soluble sugars, carbohydrate content, ABA content, protein, methyl glyoxal content
Transcriptome analysis:	Differential expression of specific genes in response to stress: major up regulated genes are *DREB1A, TF-SDIR1, Ubiquitin ligase, P5CS2*
Proteiomics analysis:	a) NAC genes found differentially regulated in stress. Universal stress protein (USP) expressed high in stay-green (B) genotypes
	b) Aquaporins: water and solute transport channels vital for water flux control throughout plant influencing stomatal conductivity, root hydraulic

	capacity and transpiration. *PIP2B* isoforms are related to water transport (*PIP 1 & 2* in Sorghum)

3. DROUGHT SCREENING OF SORGHUM MUTAGENIZED POPULATION: A CASE STUDY

Breeding for hardy, input use-efficient cereal varieties that are drought tolerant with moderately better yields would be the best strategy under a changing climate. In this context, there is a need to develop climate resilient sorghum varieties with improved water use efficiency so as to sustain terminal moisture stress. In the present study, a mutant population of M-35-1 was developed by gamma rays and EMS (300 Gy + 0.1%) and evaluated under irrigated and moisture stress conditions. Laboratory and field level screening of mutagenized populations for pre- and post-flowering drought stress is described in the following sections.

3.1. Laboratory studies

A preliminary study involving M-35-1 (terminal drought susceptible) and a drought tolerant genotype, E-36-1, were used to determine the baseline tolerance level using PEG-8000 as osmoticum. Based on the literature, we considered six treatments, i.e. 0% (0 osmotic pressure), 4% (−0.33 MPa), 8% (−1.01 MPa), 12% (−2.03 MPa), 16% (−0.40 MPa) and 20% (−5.11 MPa). Six day old seedlings were transferred in each treatment and maintained in four replications for comparison. The treatments were imposed in the liquid Murashige and Skoog (MS) media without growth regulators and seedlings were exposed to 16 h of daylight in the plant growth chamber.

Seedling growth and vigour were observed for 15 days after the inoculation of PEG-8000. Based on observations, shoot length, root length, number of leaves and number of root hairs were drastically reduced as the PEG concentration increased in both parents (Fig. 3). M-35- 1 seedlings could tolerate only up to 4% of PEG, while E-36-1 seedlings were healthy even at 8% PEG concentration. After 20 days of transfer, most of the seedlings were dried in the 8–20% PEG treatments. In conclusion, M-35-1 and E-36-1 can tolerate up to 4% and 8% of PEG-8000 treatment, respectively and beyond this osmotic potential, most of the seedlings died. This study also revealed that genotypes can tolerate up to −1.01 osmotic pressure without major growth retardation. Similarly, PEG was found inhibiting coleoptiles length [24] and was positively correlated with the drought tolerance index in rice [26]. It was also found to inhibit germination in sorghum cultivars at different levels [27, 28]. Osmotic potential as induced by the PEG for mimicking drought stress at 0 to −0.8 MPa was the effective level at which germination, its rate index and the amount of water absorbed by the seeds were significantly reduced [29]. Such preliminary laboratory study helps us to decide the tolerance limit of the osmotic pressure by the genotypes used and aid in withholding the irrigation in field level drought stress studies.

3.2. Field screening of M-35-1 based mutants for drought stress

M-35-1 is a popular variety among farmers in rainfed systems and has been their main source of food and fodder. It is widely used for bread making and is very popular among farmers due to its premium seed quality traits. In order to improve its yielding ability under drought stress with all the seed quality traits intact, mutation breeding work was initiated. Initially, selfed seeds of M- 35- 1 were irradiated with gamma rays (100, 200, 300, 400 and 500 Gy at 38 Gy/min dose rate) and EMS (0.1% and 0.2%). Finally, a dose comprising 350 Gy + 0.1% EMS possessing high mutagenic efficiency was explored further. The M_1 and M_2 generations were grown at the Experimental and Gamma Field Facility, Bhabha Atomic Research Centre, Trombay, India, during the post-rainy seasons of 2016–2017 and 2017–2018, respectively. In the M_1 generation, a population consisting of 1550 plants was grown and as many as 527 plants were selected. These plants to row progenies were sown in M_2 and M_3 generations (2017–2019) and data on drought responsive traits were recorded. Elite mutants showing improved drought responsive traits were selected in the M_4 and M_5 generations. These mutants were extensively studied for morpho-yield and root related traits under irrigated and drought stress conditions. Based on the data, few mutants showed drought avoidance/escapism/tolerance mechanisms to adapt to the drought stress conditions. Among them, epicuticular wax on the leaf sheath, peduncle, stem and leaf lamina reduced net radiations by increasing reflectance and cuticular transpiration (Fig. 4). Leaf rolling under terminal moisture conditions is the decisive reference to determining the moisture stress tolerance in the crop plants. M-35-1 based mutants also showed extensive leaf rolling during mid-day, indicating their tolerance to pre-flowering drought stress.

In order to induce drought conditions for efficient screening of mutants, a systematic screening protocol was followed at pre- and post-flowering stages in M_3 and subsequent generations. Two sets of each progeny were planted in a separate block in the M_3, M_4 and M_5 generations. Among the two sets studied, one set was subjected to drought stress during the pre- (45–60 DAS) and post-flowering (75–90 DAS) stages and the other replication was maintained in a well watered condition. The baseline tolerance level for stress is mainly based on the PEG based laboratory study, which helped to decide the irrigation schedule after a brief period of drought stress. Selection criteria may be leaf rolling coupled with drought responsive traits such as waxy stem/leaves, early flowering and maturity and drought scores on 1–9 scale (1: no dried leaves and 9: 90% dried leaves). In order to irrigate the fields after planned drought stress, field level symptoms such as drooping of leaves in the mid-day (check variety), leaf rolling (mutants) and soil tensiometer readings (at least –50 kPa) were confirmed. In the M_4 generation, selected M-35-1 mutants were able to maintain optimal photosynthesis with better expression of drought responsive traits such as leaf rolling and stay-green selected. Seeds were analysed for starch bodies at maturity and found intact with no distinguishable structural changes. Univariate analysis revealed wide variability for all the traits studied under irrigated and moisture stress conditions.

Most of the mutants observed in the irrigated block were tall (263 cm), thick stem (2.24 cm), late flowering (73 days), had a larger panicle area (16.8×5.5 cm^2) and took 113 days to mature and set seeds, as compared to drought stress conditions (Table 2). The grain yield ranged from 63 to 96 g/plant with bold and lustrous seeds (2.5–4.2 g/100 seeds) in well watered conditions as compared to drought stress situations (23–79 g/plant and 2.3–3.4 g/100 seeds, respectively). The number of leaves and total chlorophyll content had no significant difference between these two water regimes, but moisture stress had led to the loss of >60% of green leaf area as compared to irrigated conditions. Some of the high yielding mutants possessing 79 g/plant grains with lustrous bold seeds (3.4 g/100 seeds) were highly efficient in managing the water stress by modifying its flowering time (57 days to flower) and root architecture. Most of these mutants did retain 40% of their green leaf area with profuse lateral and nodal roots to absorb soil moisture for optimum photosynthesis. Based on the correlation coefficients, positive and negative correlations were observed among the quantitative traits (Table 3). Significant positive correlations were observed between dry and fresh root weight (0.73), root volume (0.43) and seminal roots (0.35). FRW was also positively correlated with RL (0.38), RW (0.33) and SR (0.39). Negative correlations were observed between fodder yield and root weight – 0.36; primary roots –0.32 and seed yield –0.33. Such correlations will help breeders during mutant selection and decide the most influential traits for simultaneous trait improvement.

TABLE 2. UNIVARIATE ANALYSES FOR M-35-1 MUTANTS UNDER DROUGHT AND IRRIGATED CONDITIONS IN M_4 GENERATION

Traits	Irrigated/ Drought	Min.	Max.	Mean	SE	Std. dev	Median
Plant height (cm)	I	145.8	317.5	263.4	9.1	56.0	262.1
	D	190.0	240.3	236.4	2.7	15.5	235.0
Days to flower	I	67.0	77.5	72.8	0.5	2.9	73.0
	D	57.0	75.0	64.9	0.6	3.8	65.0
Grain yield (g/plant)	I	63.0	96.0	78.8	1.5	8.7	80.0
	D	23.0	79.0	51.7	2.2	13.8	52.5
Seed weight (g)	I	2.50	4.20	3.7	0.1	0.6	3.7
	D	2.30	3.40	3.1	0.1	0.4	3.8
Chlorophyll content (SPAD)	I	43.2	62.0	51.1	1.0	5.5	51.1
	D	39.9	53.2	44.7	0.8	4.6	55.5
Root length (cm)	I	15.0	59.0	23.1	1.8	9.2	20.5
	D	14.0	30.0	20.5	0.8	4.2	19.0
No. of seminal roots	I	4.00	12.0	8.9	0.6	3.0	9.0
	D	4.00	15.0	9.7	0.4	2.1	8.0
No. of primary roots	I	2.00	12.0	7.3	0.57	0.57	7.0
	D	1.50	9.00	5.7	0.49	2.5	6.0
Root volume (cm)	I	2.0	7.0	5.5	0.4	2.1	6.0
	D	1.6	15.6	9.6	0.5	2.6	5.6
Fresh root weight (g)	I	3.6	10.8	7.8	0.7	3.7	7.0
	D	1.5	9.0	6.8	0.5	2.5	6.0
Dry root weight (g)	I	1.8	6.2	3.3	0.4	2.1	2.9
	D	2.7	8.0	5.0	0.3	1.4	3.2
Disease score[#]	I	2.0	7.0	4.5	0.5	1.6	4.5
	D	4.0	9.0	6.5	0.7	1.8	4.5

*D: Drought; I: Irrigated conditions.
[#] Drought scores based on a 1–9 scale (1: No dried leaves and 9: 90% dried leaves at maturity) ; SE: standard error.

TABLE 3. CORRELATION COEFFICIENTS AMONG THE YIELD CONTRIBUTING TRAITS OF M-35-1 MUTANTS UNDER DROUGHT STRESS CONDITIONS IN M_4 GENERATION

Traits	RL	RW	SR	PR	RV	FRW	DRW	SY	FY
Root length (RL)	1.000								
Root weight (RW)	0.251	1.000							
Seminal root (SR)	0.254	0.199	0.000						
Primary root (PR)	0.090	0.045	-0.125	1.000					
Root volume (RV)	0.432**	0.215	0.368*	0.086	1.000				
Fresh root weight	0.382*	0.332*	0.386*	-0.188	0.371*	1.000			
Dry root weight (DRW)	0.164	0.313*	0.352*	-0.088	0.433**	0.733**	1.000		
Seed yield (SY)	-0.035	0.051	0.013	0.044	-0.005	-0.111	-0.062	1.000	
Fodder yield (FY)	0.042	-0.363*	0.070	0.321*	-0.039	-0.007	-0.015	-0.329*	1.000
Chlorophyll content	-0.008	0.025	0.038	-0.218	-0.268	0.255	-0.027	-0.099	-0.049

*and **: Significant at 0.05 and 0.1%, respectively.

Subsequently, in the M_5 generation, 20 selected mutant progenies were evaluated under irrigated and drought stress conditions (Tables 4 and 5). The analysis of variance (ANOVA) for morpho-yield and drought responsive traits revealed significant differences for the mutants studied, except for the number of leaves and panicle length. The coefficient of variation (CV%) varied from 2.99 to 25.58%. The heritability in a broad sense also varied from 40.82 to 89.84 with seed weight and fodder yield recording the highest heritability values. Specifically, univariate analysis under drought and well watered conditions revealed wide variability for all the traits studied. M-35-1 mutants responded well under pre-flowering drought by early flowering (57–67 DAS) compared to well watered conditions (64–81 DAS). For yield contributing traits, panicle length (17.01 cm), panicle width (4.65 cm) and seed weight (3.11 g/100 seeds) were moderately reduced under moisture stress as compared to well watered conditions. Mutants showed a greener canopy (35–40%) with optimum chlorophyll content (43.9–44.3) under both phases of moisture stresses.

TABLE 4. ANOVA (MEAN SUM OF SQUARE) FOR M-35-1 DERIVED MUTANTS UNDER DROUGHT CONDITIONS IN M_5 GENERATION

Source of variation/traits	Replication	Genotype	Error	CV (%)	h^2bs (%)
df	1	19	19	-	
SPAD values	1191.37**	31.38*	13.93	14.521	55.62
Days to flower	15.62**	4.90**	1.2566	1.757	74.40
Days to maturity	11.025	23.01**	3.6566	2.99	84.11
No. of leaves	5.62*	1.85	1.0987	6.181	40.82
Stem girth (cm)	0.041	0.14**	0.0378	14.484	73.30
Plant height (cm)	1.60	135.73	63.653	3.229	53.11
Panicle length (cm)	3.60	9.60	6.9684	12.639	57.41
Panicle width (cm)	0.256	1.71**	0.4692	19.247	72.68
Seed yield (g/plant)	5.625	626.69**	1.4145	25.585	79.77
Seed weight (g/100 seeds)	0.027	0.33**	0.0064	13.375	88.06
Fodder yield (g/plant)	0.25	4455.05**	7.2105	20.137	89.84

* and ** indicate significance at 5% and 1% level; h^2bs: Broadsense heritability; CV: coefficient of variation; df: degrees of freedom.

TABLE 5. UNIVARIATE ANALYSES FOR M-35-1 MUTANTS UNDER DROUGHT AND IRRIGATED CONDITIONS IN M5 GENERATION

Parameters	Min.	Max	Mean	Min.	Max	Mean
	Drought#			Irrigated		
SPAD values	35.8	53.4	44.35	33.9	48.2	43.95
Days to flower	58	69	62.7	64	81	77.7
Days to maturity	102	118	105.95	101	126	118.15
No. of leaves	10.8	13.4	11.9	10.4	13.6	12.07
Stem girth (cm)	1.33	2.1	1.733	1.25	2.19	1.84
Plant height (cm)	218	253	231.35	233	279	259.55
Panicle length (cm)	13.0	19.4	17.01	14.7	22.8	19.21
Panicle width (cm)	3.0	5.3	4.656	3.4	5.8	4.68

Green canopy (%)	10.0	52.0	28.0	40.0	80.0	58.0
Seed yield (g/plant)	17.6	57.5	31.565	39.2	97.5	66.16
Seed weight (g/100 seeds)	2.16	3.65	3.18	2.82	3.96	3.6
Fodder yield (g/plant)	73	245	118.65	75	260	135.25
Drought score (1–9 scale)	4.0	9.0	5.86	1.0	7.0	4.35

[#] Drought scores based on 1–9 scale (1: No. of dried leaves; 9: 90% dried leaves at maturity).

In the post-flowering drought stress (75–90 DAS), reduced leaf transpiration by excessive leaf rolling, waxy coating, high water use efficiency and efficient stomatal conductance are the mechanisms due to which these mutants could still maintain better seed set and optimum grain yield (58.0 g/mutant against 37.5 g/plant in the parent under drought and 97.5 g/plant in the irrigated condition) (Figs 5 and 6). A rainfed ecosystem is often complex due to certain biotic stresses due to dynamic soil temperatures. Charcoal rot is one of the major diseases prevailing in the post-rainy season under drought stress, when the soil temperature is more than the normal. Therefore, there is need to screen drought tolerant mutants for charcoal rot disease using a standard tooth pick method [30]. In this study, most of the mutants showed thin stems with significantly less charcoal rot disease (reduced *sclerotium* in the stem region without discoloration) and survived even until the maturity stage with less per cent of lodging. Some of the high yielding mutants were carefully uprooted and their root architecture studied (Fig. 7). Mutants possessed a higher number of crown/nodal and seminal roots as against control plants under stress conditions. The root angle was much less and the cone angle was wider as compared to control plants. Steeper root angles usually lead to the formation of deep roots with better uptake of soil resources contributing to drought avoidance mechanisms [31]. Stress induced conditions trigger longer and more profuse seminal roots leading to increased fresh root weight as compared to a well watered condition. This could help in better extraction of water from lateral layers. Longer nodal roots with more root volume contributed significantly in the extraction of soil moisture under stress conditions. Similarly, past studies have revealed the influence of a profuse root system in maintaining optimum stomatal opening at lower leaf moisture content and high osmotic levels [32].

In the climate change scenario, drought and heat stress are the major yield limiting factors for food security. Although moisture stress during critical growth stages results in a yield penalty of 25–50%, supplementing with tolerant genotypes may reduce the yield loss and ensure optimum returns for the rainfed farmer. In the present study, incorporating some of the drought responsive traits in the mutagenized population of M-35-1 resulted in moderate improvement in the grain and fodder yields compared to the parent. Such mutants can be recommended for deep black soils (clay) and can even yield better under residual moisture conditions. Since most of the area under this variety is in rainfed conditions, improved mutant lines can fetch moderate to high yield under multiple drought stress events.

4. CONCLUSIONS AND FUTURE PROSPECTS

The major challenge in present agriculture is to develop heat and moisture stress tolerant varieties that can suit a wide range of agro-climatic conditions. M-35-1 derived mutants from this study have shown earliness as an escape mechanism to combat post-flowering drought stress. Some of the mutants have also shown earliness and stay-green as drought responsive traits (avoidance mechanism), which translated into yield optimum under stress conditions. Cultivation of such improved mutants would fetch moderate to high returns for farmers along with dry fodder for their livestock. Since the drought responsive genes are multi-genic and quantitative in nature, phenotypical expression for stress tolerance is influenced by $G \times E$ effects. Therefore, breeders must ensure that there is a clear and strong association between drought tolerance and stable grain yield in the target stress environments. The expression of tolerance must be readily measurable with adequate replication in time and space, precise phenotyping and appropriate selection methods to ensure stable yield levels under stress conditions.

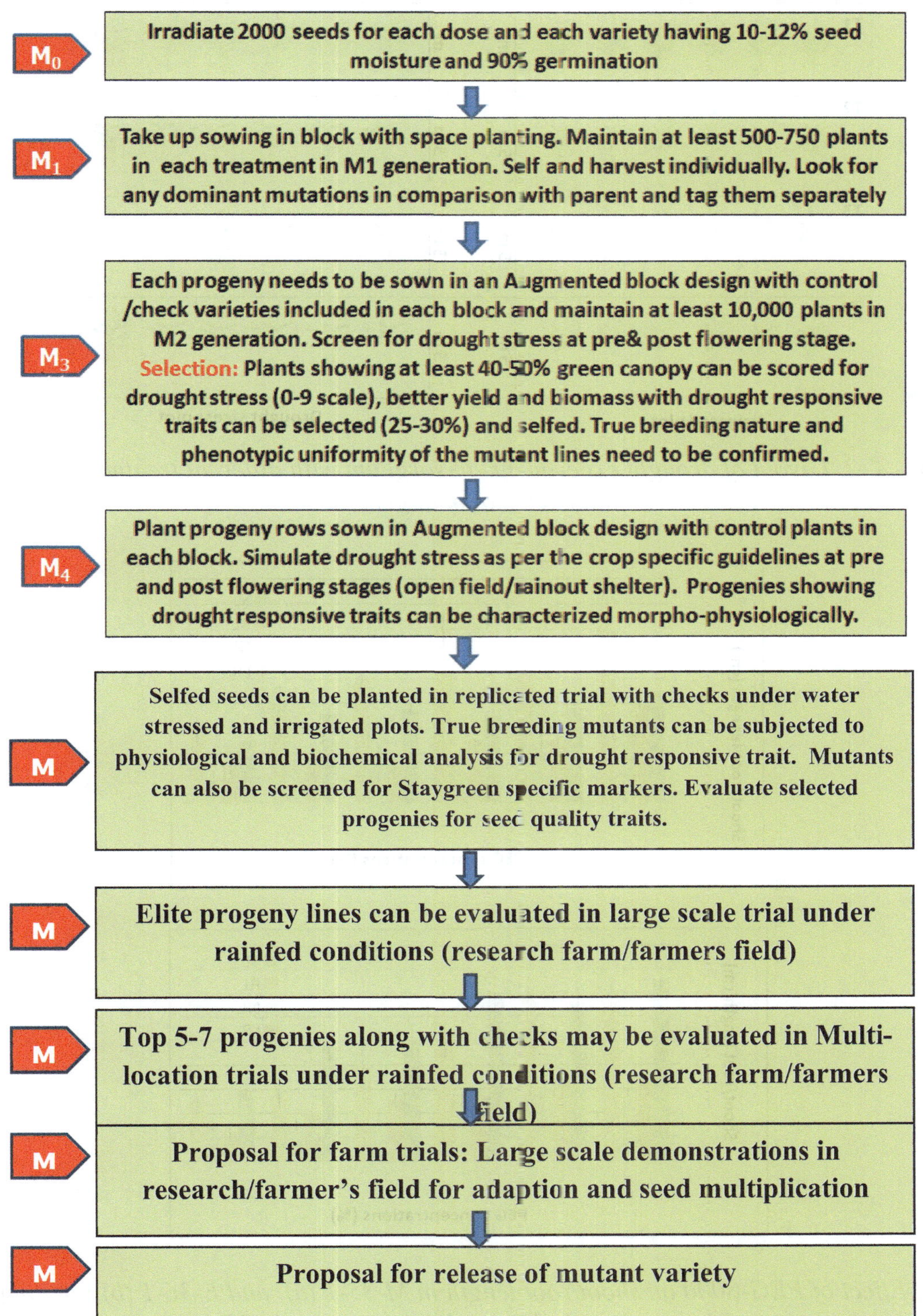

FIG. A–1. Procedure for development, evaluation and release of drought tolerant mutant variety for farmer's cultivation.

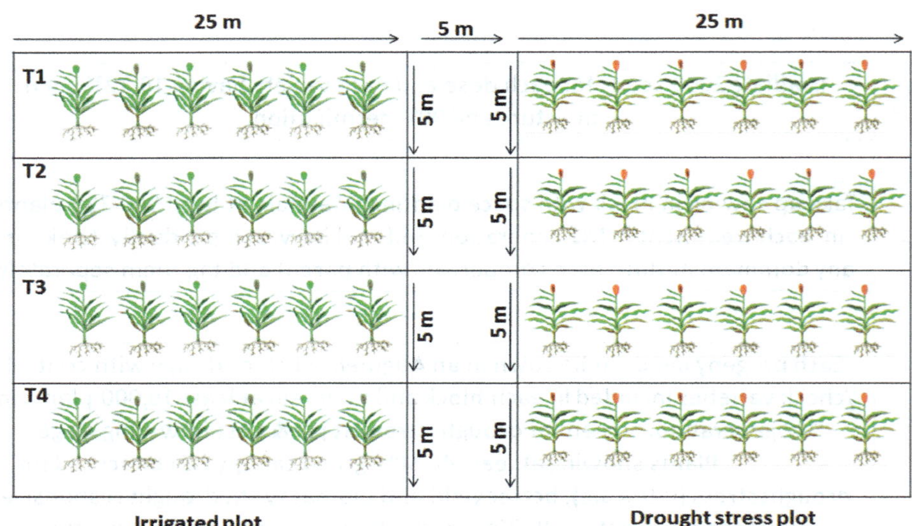

FIG. A–2. Layout of drought experiments in sorghum with different treatments of physical and chemical mutagens.

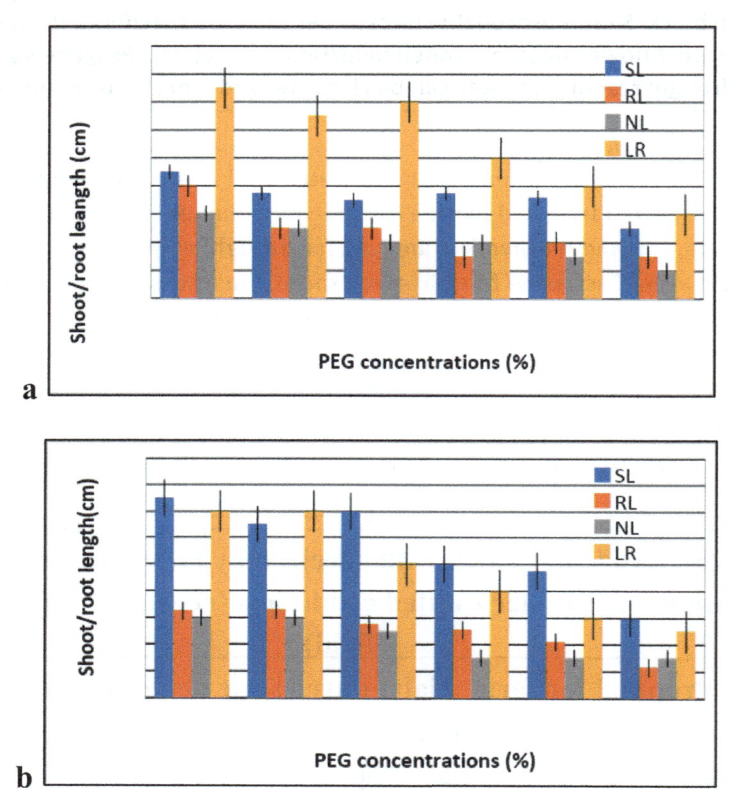

FIG. A–3. Effect of PEG-8000 on shoot/root length in M-35-1 (a), and E-36-1 (b) seedlings (SL: seminal root length (cm); RL: root length; NL: nodular root length; LR: lateral root length (cm).

FIG. A–4. Drought responsive traits observed in the Sorghum mutagenized population.

FIG. A–5. Field view of radiation induced sorghum drought experiment. A; B: Drought induced and irrigated plots; C; D: panicle shape and size of selected mutants under drought stress and irrigated plots.

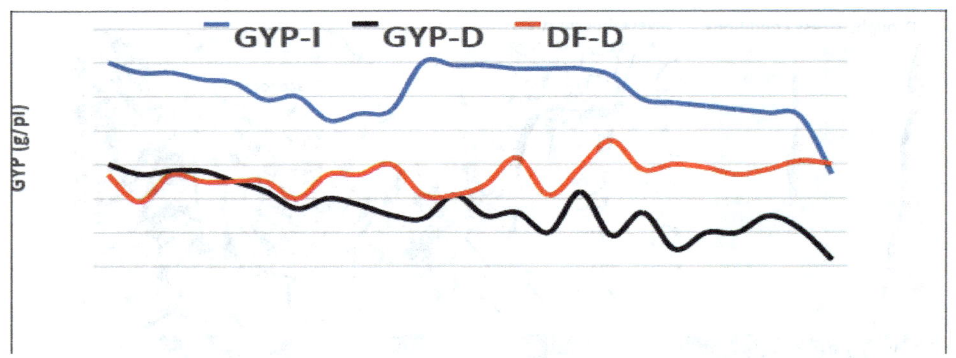

FIG. A–6. Flowering and grain yield of M-35-1 derived mutants screened under irrigated (I) and drought stress (D) conditions. GYP-I and GYP-D: Grain yield per plant in irrigated and drought stress conditions; DF-D: days to flower under drought stress.

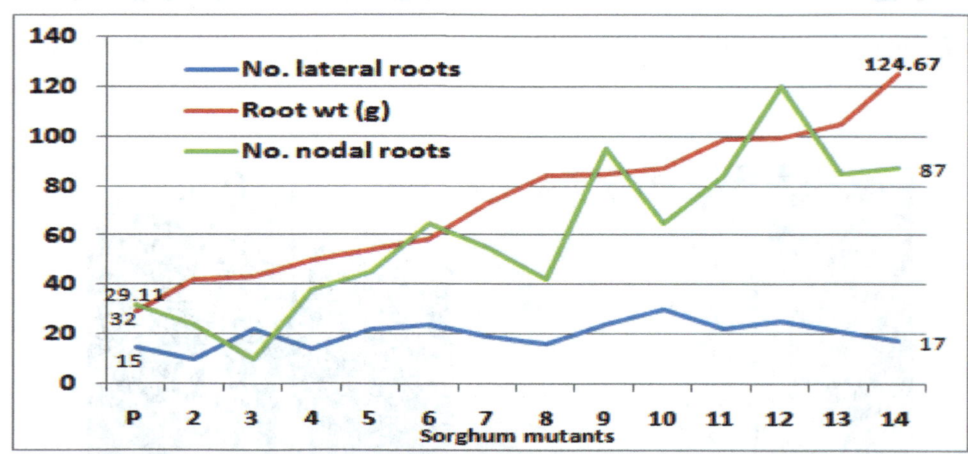

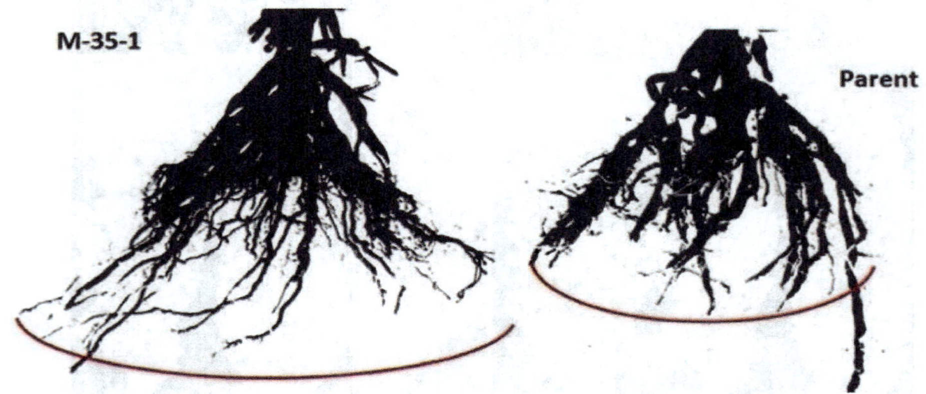

FIG. A–7. Performance of M-35-1 derived mutants for root related traits under drought stress conditions.

REFERENCES TO CHAPTER 2–10

[1]   CHOU, J., et al., Changes in extreme climate events in rice growing regions under different warming scenarios in China. Front. Earth Sci. **9** (2021) 655128. doi=10.3389/feart.2021.655128.

[2]   FRACASSO, A., LUISA, M., TRINDADE, AMADUCCI, S., Drought stress tolerance strategies revealed by RNA-Seq in two sorghum genotypes with contrasting WUE, BMC Plant Biol. **16** (2016) 115–133.

[3]   IMPA, S.M., PERUMAL, R., BEAN, S.R., SUNOJ, V.S.J., JAGADISH, S.V.K., Water deficit and heat stress induced alterations in grain physico-chemical characteristics and micronutrient composition in field grown grain sorghum, J. Cereal Sci. **86** (2019) 124–131; https://doi.org/ 10.1016/j.jcs.2019.01.013.

[4]   INDIAN INSTITUTE OF MILLETS RESEARCH, Annual Report 2019, IIMR, Hyderabad (2019) 112.

[5]   HAMMER, et al., Crop design for specific adaptation in variable dry-land production environments, Crop Pasture Sci. **65** (2014) 614–626.

[6]   BADIGANNAVAR, A., et al., Physiological, genetic and molecular basis of drought resilience in sorghum [*Sorghum bicolor* (L.) Moench], Indian J. Plant Physiol. **23** 4 (2018) 670–688.

[7]   Alam Amirul, Md., et al., Qualitative and quantitative traits associate genetic variability of soybean (*Glycine max*) mutants for expedited varietal improvement program, Legume Res. **46** 9 (2023) 1162–1167; doi: 10.18805/LRF-735.

[8]   THAPA, C.B., Effect of acute exposure of gamma rays on seed germination and seedling growth of *Pinus kesiya* Gord and *P. Wallichiana* A.B. Jacks, Our Nature **2** (2004) 13–17.

[9]   KUMAWAT, S., et al., Fast neutron mutagenesis in plants: Advances, applicability and challenges, Plants **8** (2019) 164; doi: 10.3390/plants8060164.

[10]   KALPANDE, H.V., SURASHE, S.M., BADIGANNAVAR, A., MORE, A., GANAPATHI, T.R., Induced variability and assessment of mutagenic effectiveness and efficiency in sorghum genotypes [*Sorghumbicolor* (L.) Moench], Int. J. Radiat. Biol. (2021); DOI: 10.1080/09553002.2022.2003466.

[11]   FAO/IAEA Mutant Variety Database 2022, IAEA, Vienna (2022) https://nucleus.iaea.org/sites/mvd/SitePages/Home.aspx.

[12]   JOHNSON, S.M., CUMMINS, I., LIM, F.L., SLABAS, A.R., KNIGHT, M.R., Transcriptomic analysis comparing stay-green and senescent *Sorghum bicolor* lines identifies a role for proline biosynthesis in the stay-green trait, J. Exp. Bot. **66** 22 (2015) 7061–7073.

[13]   LUDLOW, M.M., "Strategies of response to water stress", Structural and Functional Responses to Environmental Stresses (KREEB, K.H., RICHTER, H., HINCKLEY, T.M., Eds) SPB Academic, The Hague (1989) 269–281.

[14]   SELEIMAN, M.F., et al., Drought stress impacts on plants and different approaches to alleviate its adverse effects, Plants **10** 2 (2021) 259; https://doi.org/10.3390/plants10020259.

[15] DUGAS, D.V., et al., Functional annotation of the transcriptome of *Sorghumbicolor* in response to osmotic stress and abscisic acid, BMC Genomics **12** (2011) 514.

[16] FAROOQ, M., et al., Plant drought stress: Effects, mechanisms and management, Agron. Sustain. Dev. **29** (2009) 185–212; https://doi.org/10.1051/agro:2008021

[17] HUSEN, A., IQBAL, M., AREF, I.M., Growth, water status, and leaf characteristics of *Brassica carinata* under drought and rehydration conditions, Brazilian J. Bot. **37** 3 (2014) 217– 227; https://doi.org/10.1007/s40415-014-0066-1.

[18] BORRELL A.K., et al., Drought adaptation of stay-green sorghum is associated with canopy development, leaf anatomy, root growth, and water uptake, J. Exp. Bot. **65** 21 (2014) 6251– 6263.

[19] TARDIEU, F., SIMONNEAU, T., MULLER, B., The physiological basis of drought tolerance in crop plants: A scenario-dependent probabilistic approach, Annu. Rev. Plant Biol. **69** 1 (2018) 733–759.

[20] MBA, C., Induced mutations unleash the potentials of plant genetic resources for food and agriculture, Agronomy **3** (2013) 200–231; doi:10.3390/agronomy3010200.

[21] FAO/IAEA, Manual on Mutation Breeding, Third Edition (SPENCER-LOPES, M.M., FORSTER, B.P., JANKULOSKI, L., Eds), Food and Agriculture Organization of the United Nations, Rome (2018) 301.

[22] ABREHA, K.B., et al., Sorghum in dryland: Morphological, physiological, and molecular responses of sorghum under drought stress, Planta **255** (2022) 20; https://doi.org/10.1007/s00425-021-03799-7.

[23] EVANS, J., et al., Extensive variation in the density and distribution of DNA polymorphism in sorghum genomes, PLoS ONE (2013) https://doi.org/10.1371/journal.pone. 0079192.

[24] VERULKAR, S.B., VERMA, S.K., Screening protocols in breeding for drought tolerance in rice, Agric. Res. 3 (2014) 32–40, https://doi.org/10.1007/s40003-014-0094-x.

[25] YANG X, et al., Response mechanism of plants to drought stress, Horticulturae 7 3 (2021) 50, https://doi.org/10.3390/horticulturae7030050.

[26] HU, S.P., YANG, H., ZHOU, G.H., Relationship between coleoptiles length and drought resistance index of rice and their QTLs, Chin. J. Rice Sci. **20** 1 (2006) 19–24.

[27] JAFAR, M.S., NOURMOHAMMADI, G., MALEKI, A., Effect of water deficit on seedling, plantlets and compatible solutes of forage Sorghum cv. Speedfeed, Crop Science (Proc. 4th Int. Congr. Brisbane, 2004); www.cropscience.org.au/icsc2004.

[28] BAYU W, et al., Water stress affects the germination, emergence, and growth of different sorghum cultivars, Ethiop. J. Sci. **28** 2 (2005) 119–128; https://doi.org/10.4314/sinet.v28i2.18248

[29] OLIVEIRA, A.B.D., GOMES-FILHO, E., Germination and vigour of sorghum seeds under water and salt stress, Rev. Brasil. Sementes **31** 3 (2009) 48–56.

[30] BADIGANNAVAR, A., GIRISH, G., GANAPATHI, J., GANAPATHI, T.R., Gamma ray induced genetic improvement of sorghum landraces for grain yield and charcoal rot tolerance, Elect. J. Plant Breed. **9** 3 (2018) 894–898.

[31] UGA, Y., Control of root system architecture by deeper rooting increases rice yield under drought conditions. Nat. Genet. **45** 9 (2013) 1097–102; doi: 10.1038/ng.2725.

[32] YARED, A., STAGGENBOR, S.A., PRASAD, V.P.V., Grain sorghum water requirement and responses to drought stress: A review, Crop Manage. Res. **9** 1 (2010).

PART 3
SECONDARY TRAITS INDICATIVE OF DROUGHT TOLERANCE IN RICE

3–1. SCREENING FOR DROUGHT TOLERANCE IN RICE MUTANTS USING MORPHO-PHYSIOLOGICAL AND MOLECULAR MARKERS

J. BAGRI[1], Y. CHATTERJEE[2], M. MISHRA[2], B. KUMAR[2]
R.N. SAHOO[1], S.L. SINGLA-PAREEK[2], A. PAREEK[1, 3]

[1] Jawaharlal Nehru University, New Delhi, India

[2] International Centre for Genetic Engineering and Biotechnology, New Delhi, India

[3] National Agri-Food Biotechnology Institute, Mohali, India

Abstract

To achieve crop yield stability, it is essential to address the adverse effects of environmental stresses like drought on crop productivity. In this study, a screening of a population of rice mutant lines generated through gamma radiation was conducted, focusing on stress-inducible biochemical and molecular markers. Specifically, the activities of antioxidant enzymes were assessed, such as superoxide dismutase, catalase, ascorbate peroxidase, and lipid peroxidase. Based on the evaluations, it was observed that the M4_1 and M4_2 mutant lines exhibited significantly higher enzyme activities compared to the wild type (WT). Moreover, under drought stressed conditions, these mutant lines showed higher proline content while demonstrating reduced levels of methylglyoxal, lipid peroxidation, and electrolyte leakage in comparison to the WT. Furthermore, the M4_1 and M4_2 mutant lines displayed sustained and increased expression of DREBs, TPS, and GLYs genes, compared to the WT, under drought stress conditions. In conclusion, the study successfully developed a reliable and repeatable drought screening protocol at reproductive stages, employing well-established biochemical and molecular markers. Through this approach, potential drought-tolerant mutant lines (M4_1 and M4_2) with promising characteristics were identified.

Key words : Rice, drought stress, oxidative markers, molecular marker.

1. INTRODUCTION

The mechanisms of drought tolerance of screened lines and varieties are classified in this chapter using: (1) physiological and biochemical indicators that have been linked to plant stress responses to classify the mechanism of drought tolerance of screened lines and varieties; and (2) open rainout-shelter under real stress conditions, which provide conclusive data for screening and selection of promising lines to advance these lines to confirm the methods developed. Data were used to evaluate the rice genotypes that were evaluated, and comparisons to drought tolerant or susceptible standards were made. Drought stress affects leaf chlorophyll content and membrane stability through lipid peroxidation, which results in the generation of peroxide ions and MDA [1]. The electrolyte leakage percentage and MDA content are are good measurable parameters [2] found that under stress, the tolerant genotypes have more antioxidant enzymes and proline and less methylgloxal (MG).

In this study, screening of rice mutant lines generated through gamma radiation was done using physiological and biochemical assessments, including electrolyte leakage, proline content, MG level, antioxidant enzymes (catalase, ascorbate peroxidase, and superoxide dismutase) activities.

2. METHODOLOGY

Rice (*Oryza sativa* L.) seeds were surface sterilized and allowed to germinate in the dark for two–three days and sown to raise a nursery bed for two weeks at 28°/23°C (day/night) with 70%±5 relative humidity and 260–350 μE m^2 s^{-1} light intensity [3, 4]. Two week old seedlings were then transplanted to the field. The experiment comprised a control field and drought stress field, each with six biological replicates of the mutant lines, drought tolerant check (N22), susceptible check (SM) and WT. To analyse the dynamic changes in the physiology of the rice plants under drought stress (DS), three treatments were chosen: (1) control (CO);, (2) DS, when the soil moisture content reached up to severe conditions; and (3) ten days of drought recovery (DR). Before the onset of DS, tissues were also harvested (beginning stress (BS)) to avoid the variation arising due to any other edaphic or topographical factors. The control plots (CO) were puddled and continuously flooded with 1–3 cm of water level for the first 15 days after transplanting. Thereafter, the water level was gradually increased to 5–10 cm for two weeks. NPK (nitrogen–phosphorous–potassium) fertilizer was applied in 10:6:6 ratios to both CO and drought fields. For DS treatment, drought was imposed on plots by withholding water at the pre-flowering stage, i.e. 60 days after transplantation and post-flowering stage, and 70 days (targeting 50% flowering) after transplantation. Soil water potential was monitored regularly using a soil tensiometer (five instruments in each plot or at least one tensiometer/10 m^2 as per the protocol given in the layout) in order to achieve the soil water potential at −35 ±5 kPa level at 30 cm depth (see Chapters 2–3 in Part 2 of this publication).

3. MEASUREMENT OF ELECTROLYTE LEAKAGE:

Electrolyte leakage was carried out as described in the protocol in Ref. [5]. The samples from control and drought-treated from the plant were harvested and cleaned with distilled water to remove any surface adhering ions. The percentage of electrolyte leakage was used to calculate the relative electrical conductivity: percentage of electrolyte leakage = E1/E2*100.

3.1. Oxidative stress marker

Lipid peroxidation was determined by measuring the accumulation of malondialdehyde (MDA) content by thiobarbituric acid reacting substances (TBARS) using the procedure described by Heath and Packer [6].

3.2. Biochemical indicators of drought stress

Proline and methylglyoxal (MG): The methylglyoxal (MG) and proline content were analysed from the leaf. For proline estimation, leaf tissue was ground into fine powder in liquid nitrogen and the sample was then mixed with 5 mL of extraction solvent in the test tube. A further 2 mL of acid ninhydrin (1.25 g ninhydrin in 30 mL of glacial acetic acid and 20 mL of 6M phosphoric acid) and 5 mL of glacial acetic acid were added to each tube and kept in a boiling water bath for 60 min. After that, the test tubes were cooled in an ice bath for 5 min. Further, the content was vigorously mixed with 4 mL of toluene, and the mixture was then warmed up at room temperature. The O.D. of the upper layer was measured at 520 nm using toluene as a blank. The proline concentration (μg/gm of tissue) was determined using a proline standard curve [7].

For methylgoxal estimation, fresh tissue was weighed and crushed in a mortar pestle using liquid nitrogen. Next, 2.5 mL of 0.5M perchloric acid (PCA) was added and mixed well. The mixture was then transferred to a microcentrifuge tube and incubated for 15 min on ice, and extract was centrifuged at 11 000g for 10 min at 4°C. The supernatant was transferred to a fresh tube. If the supernatant was coloured, charcoal (10 mg/mL) was added to decolourize it. Furthermore, the supernatant was mixed well with charcoal, and kept at room temperature for 15 min. The mixture was then centrifuged at 11 000g for 10 min. The clear supernatant was transferred to a fresh tube, and an additional spin was given to remove the residual charcoal from the solution. The solution was neutralized (pH7.0) using 1 M Na_2HPO_4, which was added gradually (initially 20 µL, followed by 2 L at a time). The bubbles of CO_2 gas were allowed to come out. The absorbance was initially measured at 336 nm at 0 hours and set to 'blank' The same reaction mixture was left at room temperature for 3 h, and the absorbance was measured again at 336 nm.

3.3. Profiling of antioxidant enzymes

Soluble proteins were extracted and quantified using the Lowry method, as described in Refs [8, 9]. In a semi-high-throughput set-up, enzyme activities were determined [10, 11]. The activities of ascorbate peroxidase (APX) were determined in extracts obtained from 100 mg of frozen tissue in 1 mL of extraction buffer: 50 mM MES/KOH (pH6.0) containing 0.04M KCl, 2 mM $CaCl_2$, and 1 mM ASC, homogenized by MagNALyser (Roche) The method of Murshed et al. was used to determine APX activities in microplates [9]. The inhibition of nitro-blue tetrazolium (NBT) reduction ($550 = 12.8$ mM^{-1} cm^{-1}) was used to assess superoxide dismutase (SOD) activity [12]. Catalase (CAT) activity was determined by measuring H_2O_2 decomposition at 240 nm ($240 = 39.4$ M1 cm1) [13].

3.4. Results

Our results showed that under DS, grain weight plant height, and panicle number were significantly lower in the drought tolerant mutant lines and drought tolerant check (N22) as compared to the drought susceptible (SM) and WT in both M_3/M_4 generations. The physiological traits in response to DS were drastically affected as compared to plants under non-stress conditions. There was a significant variation among the traits in response to DS.

4. SCREENING OF DROUGHT RESILIENT MUTANTS USING BIOCHEMICAL AND MOLECULAR APPROACHES

Drought stress is one of the most important and complex abiotic stress limiting crop yields. Therefore, it is essential to dissect the complex trait for a better understanding of the mechanisms of drought tolerance by plants grown in the field. In this study, experiments were performed on gamma induced mutant rice to understand their management under water stress conditions. Approximately ten drought tolerant M_4 mutant lines were used to analyse the antioxidant enzyme activities such as SOD, CAT, APX, and lipid peroxidase at pre- and post-flowering stages under both control and drought stress conditions. It is well acknowledged that the expression of antioxidant enzymes encoding genes is induced by abiotic stresses such as drought, salinity, higher temperature, and cold to protect plants from abiotic stress-induced oxidative damage [14–17]. To protect cellular organelles from reactive oxygen species (ROS), plants have evolved a complex enzymatic and non-enzymatic antioxidative machinery to prevent oxidative damage [18]. ROS targets membrane lipid molecules and initiates lipid peroxidation, resulting in cellular damage. Our findings showed that the mutant plants exhibited a significantly lower membrane lipid peroxidation rate, i.e. lower accumulation of malondialdehyde (MDA) content than for sensitive plants as well as tolerant genotypes under DS conditions at both pre-flowering and post-flowering stages, though the pattern of accumulation is distinct for both stages. Evidently, the electrolyte leakage was also significantly lower in mutant lines as compared to WT under DS conditions (Fig. 1B). Interestingly, it has been reported that MDA is the most accurate biomarker for grain yield loss, suggesting that drought-induced lipid peroxidation is the major constraint for grain yield [1].

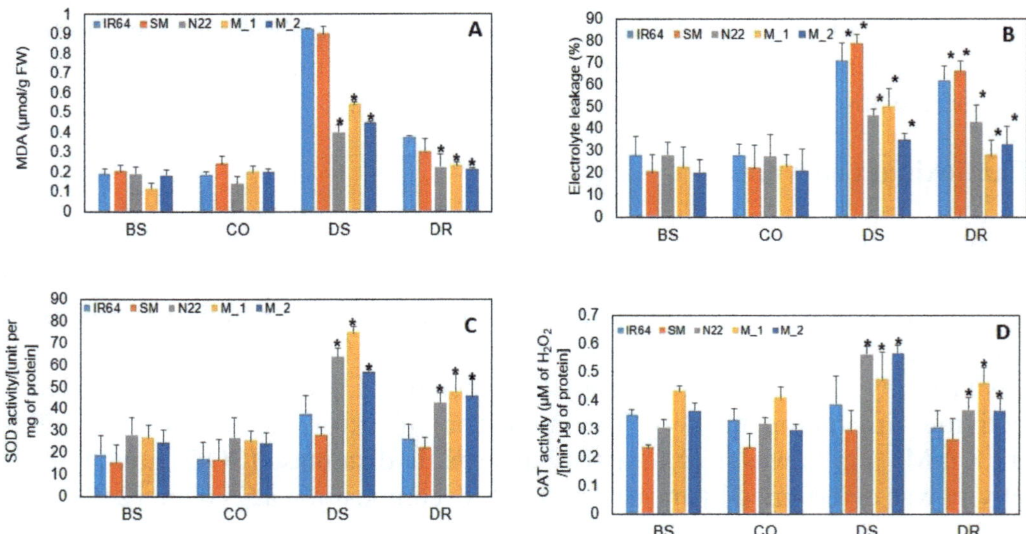

FIG. 1. Assessment of biochemical indices in gamma induced rice (M₄) mutant lines at pre-flowering and post-flowering under control and DS conditions. A: malondialdehyde level (MDA); B: electrolyte leakage (EL); C: super oxidase dismutase activity (SOD); D: catalase activity (CAT). The bar graph represents beginning of stress (BS), control (CON), drought stress (DS) and drought recovery (DR) conditions. Data are means (±SE) of three biological replicates. The asterisk (P<0.05) represents significant differences between the mutant and the WT (IR64).*

On the other hand, antioxidant enzymes, such as SOD, CAT and APX, are also part of defence mechanisms that help the plant combat the adverse effects of abiotic stress. The activities of these antioxidant enzymes SOD, CAT, and APX were found to be highly and constitutively accumulated in the mutant lines as compared to the WT under control conditions (Figs 1B, D and 2A). Under DS, a significant increase in the activity of these antioxidant enzymes was observed as compared to WT at the pre-flowering and post-flowering stages. Nevertheless, the activity of these enzymes was distinct in both stages (Fig. 1A and 1B). Proline and MG were also found to be higher in DS and DR compared to control (Fig. 2 B, C).

226

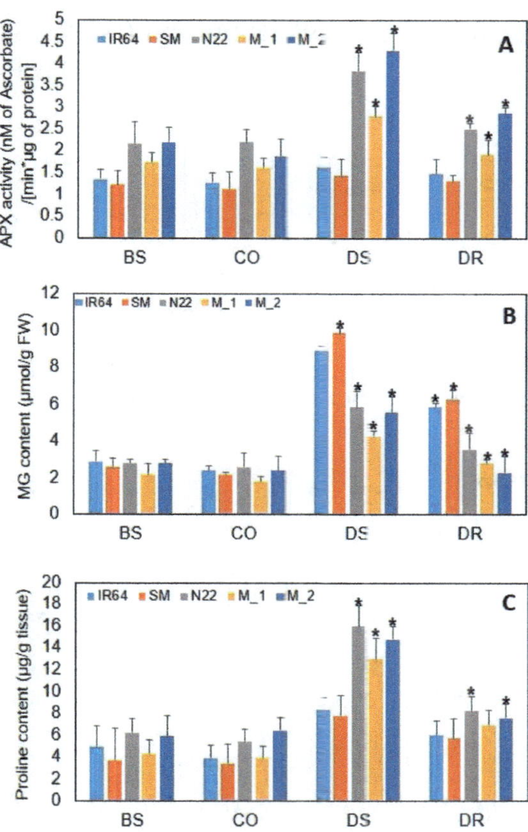

FIG. 2. A–C: Assessment of biochemical indices in gamma induced rice (M₄) mutant lines at pre-flowering and post-flowering under control and drought stress conditions. A-ascorbate peroxidase activity (APX), B-methylglyoxal content (MG), C-proline content. The bar graph represents the beginning of stress (BS), control (CON), drought stress (DS) and drought recovery (DR) conditions. Data are means (±SE) of three biological replicates. The asterisk (P<0.05) represents significant differences between mutant and the WT (IR64).*

Noctor et al. [2] found that the antioxidant enzymes CAT and APX help get rid of hydrogen peroxide (H_2O_2) from organelles to protect them from damage to cells. Both CAT and APX enzyme activity exhibited a comparatively higher population-wide increase under drought stress (Figs 1A and 2A). Conversely, SOD also showed the highest drought induced increment in plants. Results suggest that enhancement in SOD activity is independent of the genotypes, i.e. WT, N22 and/or mutant lines under stress conditions (Fig. 1C). These results suggest that the mutant lines have better detoxifying machinery to scavenge ROS formed during stress. Overall, our results show that there are new breeding targets (possible drought tolerant mutant lines) for improving the stability of rice grain yield under DS both before and after flowering.

4.1. Profiling of DREB, GLY, and TPS genes in the mutant lines

Plants respond and adapt to stresses to survive at physiological and biochemical levels under stress conditions. In plants, abiotic stresses have been shown to regulate the expression of various genes having diverse functions. DREB genes play an important role in the ABA-independent stress-tolerance pathways that induce the expression of stress responsive genes in plants [19, 20, 21].

Taking into account the physiological and biochemical results, we further assessed the expression profiles of DREBs genes (OsDREB1A, OsDREB2A, and OsDREB2B). To assess the transcript levels of all the DREB genes in gamma induced rice mutant lines, WT and N22, a quantitative real-time PCR technique was used (Fig. 3).

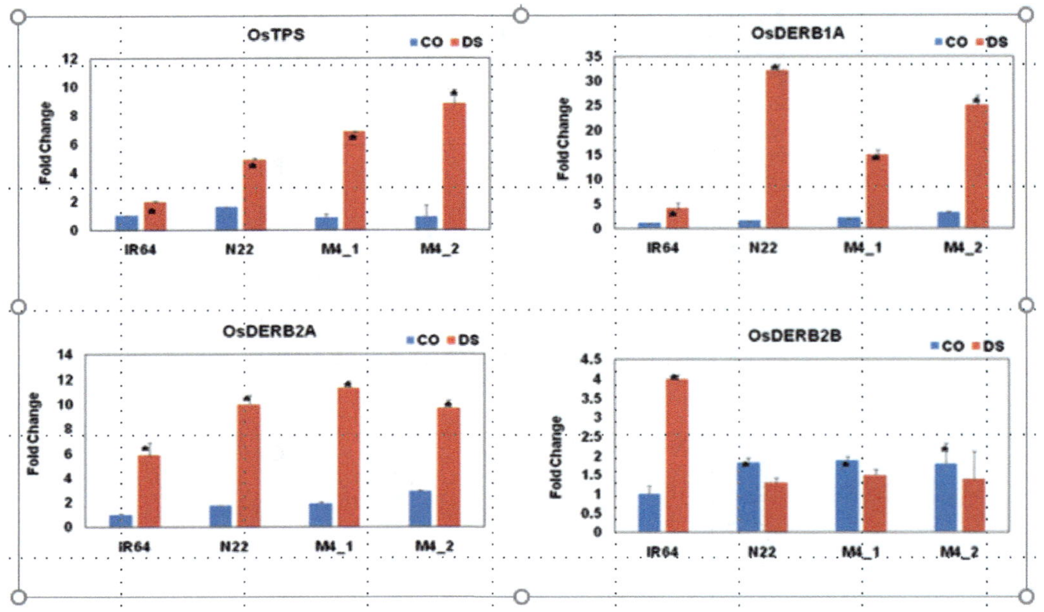

FIG. 3. qRT-PCR shows the fold change in the transcript level of DREB genes in M₄ mutant lines in the pre-flowering stage under control and DS conditions. The bar graph depicts the dynamic change in OsTPS, OsDERB1A, OsDERB2A, and OsDERB2B expression and abundance of transcripts M₄ mutant lines as compared to IR64. Data are means (±SE) of three biological replicates; the asterisk (*) represents significant difference between mutant lines and WT.

Under drought conditions, a significant increase in the expression of OsDREB1A and OsDREB2A in M4_1 and M4_2 mutant lines was found as compared to the WT (Fig. 3). The expression of OsDREB1A was enhanced by ~14- and ~24-fold in the M4_1 and M4_2 mutant line at reproductive stage. Excess MG production and enhanced glyoxalase activity are considered stress indicators in plants [22]. The transcripts of glyoxalase, as well as its activity, have been reported to be upregulated in response to stress [22]. We measured the expression levels of the GLYs genes in mutant lines under control and drought stress at the pre-flowering stage using qRT-PCR (Fig. 4).

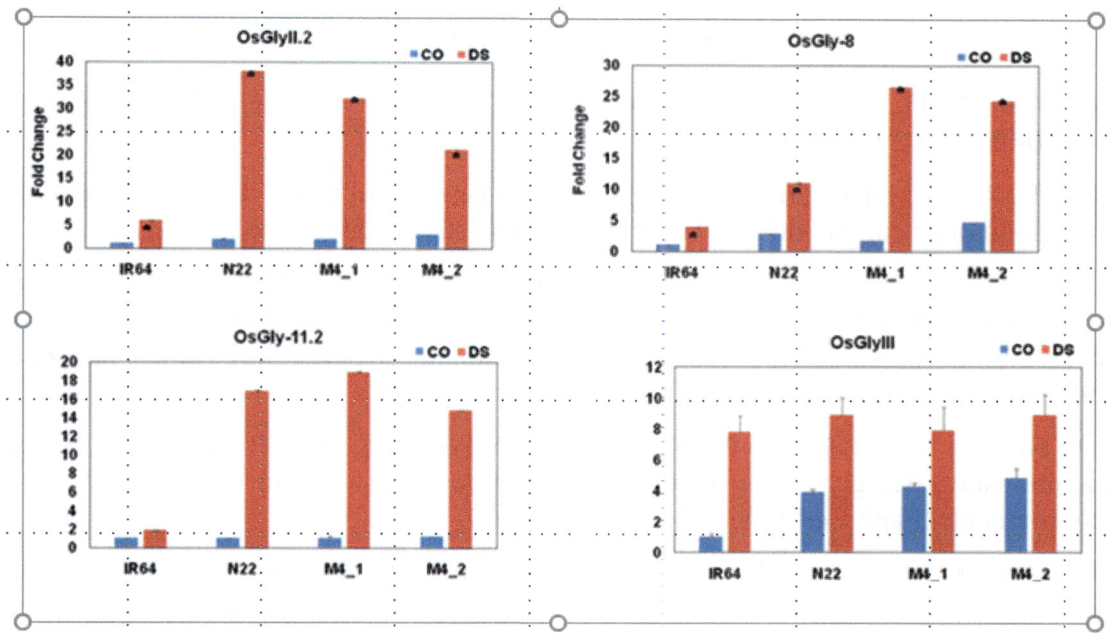

FIG. 4. qRT-PCR shows the fold change in the transcript level of GLY genes in M_4 mutant lines and IR64 in the pre-flowering stage under control and DS conditions. The bar graph depicts the dynamic change in GLY genes (i.e. OsGlyII.2, OsGly-8, OsGly-11.2, OsGlyIII) expression and abundance of transcripts M_4 mutant lines as compared to IR64. The presented data represent the means (±SE) of three biological replicates, and the asterisk (*) indicates significant differences between the mutant lines and WT.

We observed a significant change in the transcript levels of sets of GLY genes among the mutant lines compared to WT under drought stress conditions. Our results show that the OsGLY-8 has higher transcript abundance in M4_1 (26-fold) and M4_2 (24-fold) than the WT. Similarly, in the case of OsGLYI-11.2, OsGLYIII and OsGLYII.2 were significantly higher in the M4_1 and M4_2 lines than WT (Fig. 4). Interestingly, we have also observed significantly lower accumulation of MG levels in these two mutant lines than WT under DS. Under DS, OsTPS transcript abundance is significantly higher in the M4_1 (6.8-fold) and M4_2 lines (~8.8 fold) than the WT. Interestingly, we found varied transcript abundance of OsDREB, OsGLY, and OsTPS genes among M4 mutant lines under both control and drought conditions (Fig. 4).

5. CONCLUSIONS

In this study, morpho-physiology, biochemical, and molecular related traits were studied and it was found that M4_1 and M4_2 are the most tolerant genotypes, even higher than N22 (tolerant) during DS at pre-flowering and post-flowering stage. These rice mutants can be used as potential breeding materials based on their superior traits. Furthermore, the study presents a robust, reliable, and comprehensive methodology for studying plant responses at both pre-flowering and post-flowering stages. This valuable methodology will prove beneficial for breeders and scientists in their pursuit to enhance crop productivity under abiotic stress conditions.

REFERENCES TO CHAPTER 3–1

[1]   MELANDRI, G., et al., Biomarkers for grain yield stability in rice under drought stress, J. Exp. Botany **71** 2 (2020) 669–683.

[2]   NOCTOR, G., MHAMDI, A., FOYER, C.H., The roles of reactive oxygen metabolism in drought: Not so cut and dried, Plant Physiol. **164** (2014) 1636–1648.

[3]   KUMARI, S., SINGH, P., SINGLA-PAREEK, S.L., PAREEK, A., Heterologous expression of a salinity and developmentally regulated rice cyclophilin gene (OsCyp2) in *E. coli* and *S. cerevisiae* confers tolerance towards multiple abiotic stresses, Mol. Biotechnol. **42** (2009) 195–204.

[4]   LAKRA, N., KAUR, C., ANWAR, K., SINGLA-PAREEK, S.L., PAREEK, A., Proteomics of contrasting rice genotypes: identification of potential targets for raising crops for saline environments, Plant Cell Environ. **41** 5 (2018) 947–969.

[5]   BAJJI, M., KINET, J.M., LUTTS, S., The use of the electrolyte leakage method for assessing cell membrane stability as a water stress tolerance test in durum wheat, Plant Growth Reg. **36** 1 (2002) 61–70.

[6]   HEATH, R.L., PACKER, L., Photoperoxidation in isolated chloroplasts. I. Kinetics and stoichiometry of fatty acid peroxidation, Arch. Biochem. Biophys. **125** 1 (1968) 189–198 doi: 10.1016/0003-9861(68)90654-1.

[7]   BATES L.S., ALDREN, R.P., TEARE, I.D., Rapid determination of free proline for water-stress studies, Plant Soil. **39** 1 (1973) 205–207.

[8]   MURSHED, R., LOPEZ-LAURI, F., SALLANON, H., Microplate quantification of enzymes of the plant ascorbate–glutathione cycle, Anal. Biochem. **383** 2 (2008) 320–322.

[9]   LOWRY, O., ROSEBROUGH, N., FARR, A., RANDALL, R., Protein measurement with the Folin phenol reagent, J. Biol. Chem. **193** 1 (1951) 265–275.

[10]   ZINTA, G., et al., Physiological, biochemical, and genome-wide transcriptional analysis reveals that elevated CO_2 mitigates the impact of combined heat wave and drought stress in *Arabidopsis thaliana* at multiple organizational levels, Global Change Biol. **20** 12 (2014) 3670– 3685.

[11]   ABDALLAH, M.S., ABDELGAWAD, Z.A., EL-BASSIOUNY, H.M.S., Alleviation of the adverse effects of salinity stress using trehalose in two rice varieties, South African J. Bot. **103** (2016) 275–282.

[12]   DHINDSA, R.S., PLUMB-DHINDSA, P.L., REID, D.M., Leaf senescence and lipid peroxidation: Effects of some phytohormones, and scavengers of free radicals and singlet oxygen, Physiol. Plant. **56** 4 (1982) 453–457.

[13]   AEBI, H.E., Catalase methods of enzymatic analysis, Crop Sci. **5** 23 (1983) 2789–2794.

[14]   EDWARDS, R., DIXON, D.P., WALBOT, V., Plant glutathione S-transferases: enzymes with multiple functions in sickness and in health, Trends Plant Sci. **5** 5 (2000) 193–198.

[15]   ROXAS, V.P., LODHI, S.A., GARRETT, D.K., MAHAN, J.R., ALLEN, R.D., Stress tolerance in transgenic tobacco seedlings that overexpress glutathione S-transferase/glutathione peroxidase, Plant Cell Physiol. **41** 11 (2000) 1229–1234.

[16] LIANG, D., et al.,. Establishment of a patterned GAL4-VP16 transactivation system for discovering gene function in rice. The Plant Journal, **46** 6 (2006) 1059–1072.

[17] CSISZÁR, J., et al., Different peroxidase activities and expression of abiotic stress-related peroxidases in apical root segments of wheat genotypes with different drought stress tolerance under osmotic stress, Plant Physiol. Biochem. **52** (2012) 119–129.

[18] YOU, J., CHAN, Z., ROS regulation during abiotic stress responses in crop plants, Front. Plant Sci. **6** (2015) 1092.

[19] GILMOUR, S.J., SEBOLT, A.M., SALAZAR, M.P., EVERARD, J.D., THOMASHOW, M.F., Overexpression of the Arabidopsis CBF3 transcriptional activator mimics multiple biochemical changes associated with cold acclimation, Plant Physiol. **124** 4 (2000) 1854– 1865.

[20] QIN, F., et al., Regulation and functional analysis of ZmDREB2A in response to drought and heat stresses in *Zea mays* L, Plant J. **50** 1 (2007) 54-69.

[21] SAKUMA, Y, et al., Functional analysis of an Arabidopsis transcription factor, DREB2A, involved in drought-responsive gene expression, Plant Cell **18** 5 (2006) 1292–1309.

[22] KAUR, C., Glyoxalase and methylglyoxal as biomarkers for plant stress tolerance, Crit. Rev. Plant Sci. **33** 6 (2014) 429–456.

3–2. STOMATA DENSITY AS A DETERMINANT TRAIT OF DROUGHT TOLERANCE AMONG INDUCED MUTANTS OF RICE

N.T. HONG[1], V.T.M. TUYEN[1], N.T.M. NGUYET[1],
L.D. THAO[1], P,X. HOI[1], L.H. HAM[1, 2]

[1] Agricultural Genetics Institute;
Viet Nam National University, Hanoi, Viet Nam

[2] University of Science and Technology,
Viet Nam National University, Hanoi, Viet Nam

Abstract

Secondary traits are plant traits other than crop yield which offer further information on the consequences of drought stress on yield. Such traits can be useful as indirect selection criteria during the screening of mutant populations. There have been interesting observations on the impact of environmental stress on stomata morphology. This study presents results on stomata density as a key parameter in the tolerance of a mutant plant under drought stress. Drought tolerance depends on the plant's ability to reduce transpiration through stomata decrease under drought stress.

Key words: Secondary traits; drought stress; stomata structure; stomata density; mutants.

1. INTRODUCTION

Stomata are microscopic pores on the surfaces of plant leaves. The term 'stoma' means mouth in Greek. Described as a mouth containing an opening surrounded by two lips, a stoma consists of two guard cells that surround a pore [1]. Land plants have stomata on both sides of leaf surfaces. However, the majority of them are located on the underside. Stomata consist of tiny pores that are called stoma and surrounded by a pair of specialized guard cells (Fig. 1). Two guard cells connect each at both ends and look like beans. The cell walls that surround the pore are flexible. The opening and closing of stomata occur according to the expanding or deflating of guard cells. Surrounding the guard cells are subsidiary cells. The subsidiary cells are located between guard cells and epidermal cells. They integrate the efficiency of stomatal functions, additionally they also play the role as a buffer to protect epidermal cells from the pressure of the expanding of guard cells when stomata open [2, 3].

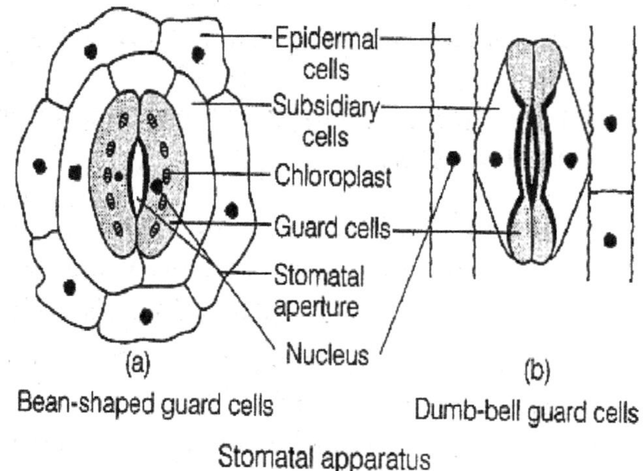

FIG. 1. Simulation structure of stomata (source: https://www.sarthaks.com/190907/stomata-in-grass-leaf-are-1-dumb-bell-shaped).

Stomata play key roles in most life activities of plants such as: gaseous exchange; photosynthesis; and transpiration. As stomata open, gas is exchanged through pores, carbon dioxide is taken in for photosynthesis and oxygen is released out into the environment. They also help to control the cadence of transpiration by opening and closing pores. The transpiration is often intense in hot weather to help cool down the organism. On the other hand, when water evaporates through pores, it creates the negative pressure on the underground system. Hence, water and mineral components from soil will be diffused in plants through roots.

Stebbins and Shah [4] suggested that stomata develop through five stages: (1) formation of the guard mother cell (GMC); (2) GMC mitoses to form subsidiary cells; (3) the GMC and two subsidiaries were matured; (4) GMC was divided to form the guard cells; (5) the specialization was of four cells. In a study on stomatal development of Arabidopsis, Pillitteri and Dong [1] reported that guard cells in Arabidopsis (dicot) were bean shaped and formed randomly on the leaf epidermis [1]. On the other hand, the stomata of most monocot species were arranged with a pre-determined location and typically linear-axial [5]. However, when the processing of stomata development goes to the terminal row, many irregularities like the number of stomata will occur more often. With grass plants, it was shown that stomata were formed in rows along the veins and presented with high density near the main vein.

The cycle of water evaporation and water uptake helps to maintain the water balance in plants. If there is any mismatch in that cycle, such as over transpiration due to the heat or the impossibility of water uptake due to drought soil, this balance will be broken and plants will have problems. This means that if in drought soil conditions, plants with less water evaporation (e.g. closing stomata actively or lower density of stomata) will have the capacity to maintain the balance longer and stay stronger. Thus, lower stomatal density is the target for the goal of reducing water loss and improving drought tolerance of plants. In previous articles, it was stated that the reduction of stomatal density led to increased drought tolerance through the restriction of water loss in both stress and non-stress conditions [6–8]. Although stomata regulate CO_2 access to the photosynthetic tissues of the leaf, the maintenance or improvement of productivity in spite of a reduced rate of photosynthesis in some conditions was established [9, 10]. It was also stated that there were no

negative effects of reducing stomatal density on seed number, seed weight, the harvest index, plant
height, upper ground biomass of barley lines [6].

TABLE 1. STOMATA DENSITY OF SOME MONOCOTYLEDONOUS AND
DICOTYLEDONOUS CROPS

Crops	Total number of stomata/mm^2	
	Upper surface	Lower surface
Wheat	50	40
Barley	70	85
Onion	175	175
Sunflower	120	175
Alfalfa	169	188
Geranium	29	179

Source: https ://byjus.com/biology/stomata/

In addition to the stomata density index approach, the other method to improve drought tolerance
of plants is a speedy index of stomatal responsiveness and movement [11]. The flexibility of
stomata is based on the shape of guard cells and the presence or absence of subsidiary cells [3].
Stomata of any plants with faster stomatal responses, distinct morphological and structural features,
and unique membrane transport and signalling systems were hypothesized to adapt rapidly to
environment changes [12, 13]. In this case, the stomatal opening and closing responses of the grass
family were proved, in several studies, that they were capable of expressing faster and more
efficient stomatal regulation [14–18]. It was explained that grass stomata, with dumbbell-shaped
guard cells and only two flanked subsidiary cells, differ from the bean-shaped type and will transfer
the status of opening/closing more sensitively just by minor changes [5, 19, 20]. However, when
the drought condition is over and water is supplied again, the recovery of plants with leaf water
potential (dicot) was shown to be quicker than that of plants with stomatal closure in early drought
(monocot) [21]. Recent reports suggested that both stomata density and size are correlated to
photosynthetic efficiency and better adaptation to drought [22].

2.   RESULTS

Leaves of drought tolerant mutant line (93-2-3) and control varieties (CH345, DT18) in both
control and drought experiment plants were collected to evaluate stomata at the time of finishing
drought treatment. Surface tissues of testing samples were collected at upper, middle and bottom
positions of leaf and fixed on microscope slides for examination. The numbers of stomata were
examined on three random fields of view (FOV) and the average for the recorded data was
calculated.

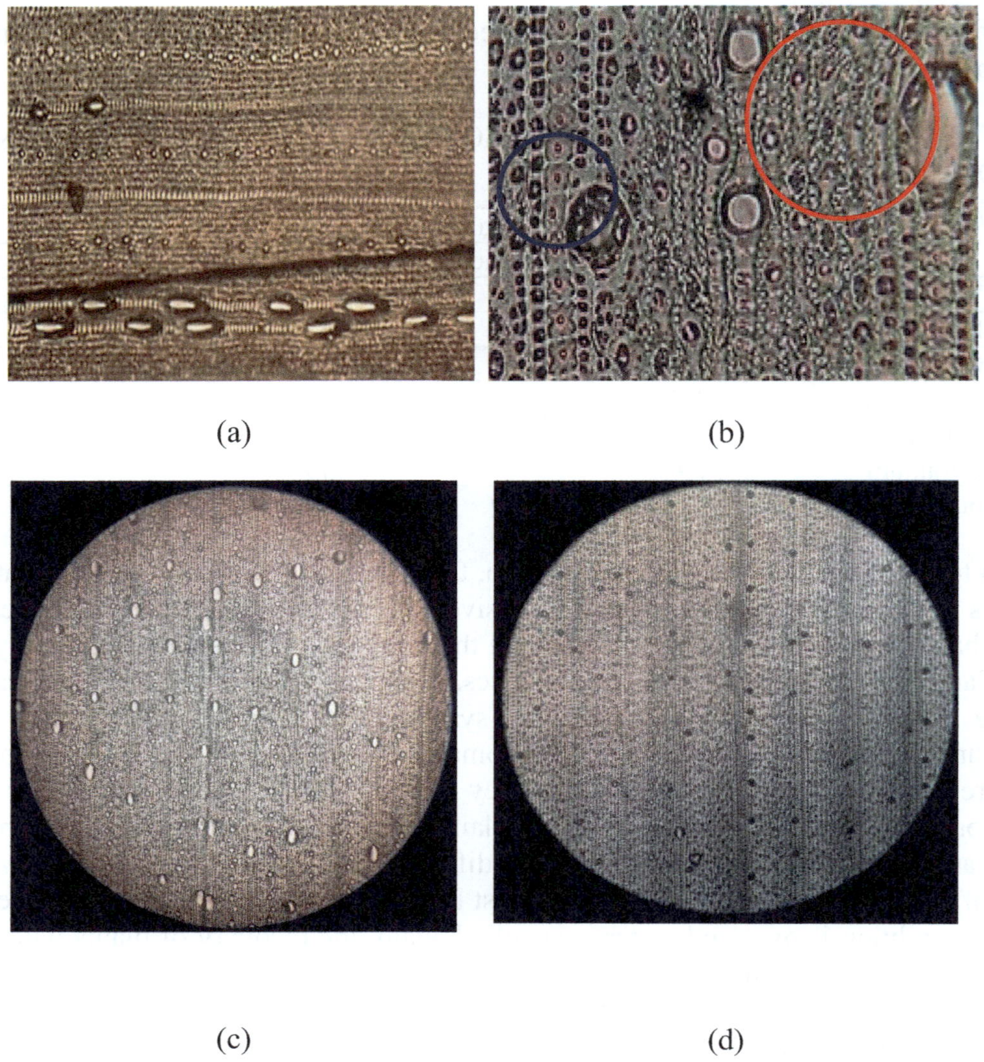

FIG. 2. Microscopic examination of leaf surface tissue (a): Stomata gather in high density near the central vein at 10X magnification; (b): Variations of stomatal shape and size at 40X magnification; (c) stomata of mutant line in control condition; (d) stomata of mutant line in drought condition

Stomata observed near the central vein showed remarkably higher density, larger and longer size (Fig. 2). Stomata were also present in different shapes and sizes in the samples. In Fig. 2(b), the left stomata with a blue outline presents an almost round shape and smaller size . The right one with a red outline is ellipse shaped and larger in size. There are some variations in stomatal present among rice samples (Table 2). In all the samples, total stomata per FOV in the control set were higher than those in the drought set. They range from the lowest 33.02 stomata per FOV of CH345 to the highest 90.90 stomata per FOV of mutant line 68(3) in normal conditions without water stress (Table 2). Conversely, the total stomata per FOV of all rice samples decreased rapidly and significantly in drought treatment, with data ranging from the lowest (2.00) of tolerant mutant line 93(2)-3 and the highest (32.00) for the DT18 variety, the negative control.

TABLE 2. DIFFERENCES IN STOMATAL DENSITY OF RICE LINES BETWEEN NON-STRESS AND DROUGHT STRESS

Rice lines	Total stomata/FOV in control	Total stomata/FOV in drought	Percentage of total stomatal decrease drought (%)
CH345	**33.02**	8.0	75.7
DT18	46.60	**32.00**	**31.33**
SC	41.30	19.30	53.27
68(2)	66.15	9.30	85.94
68(3)	**90.90**	4.00	95.60
74(1)	68.00	13.00	80.88
93(2)-1	51.00	3.30	93.53
93(2)-2	51.33	8.00	84.42
93(2)-3	47.00	**2.00**	**95.74**

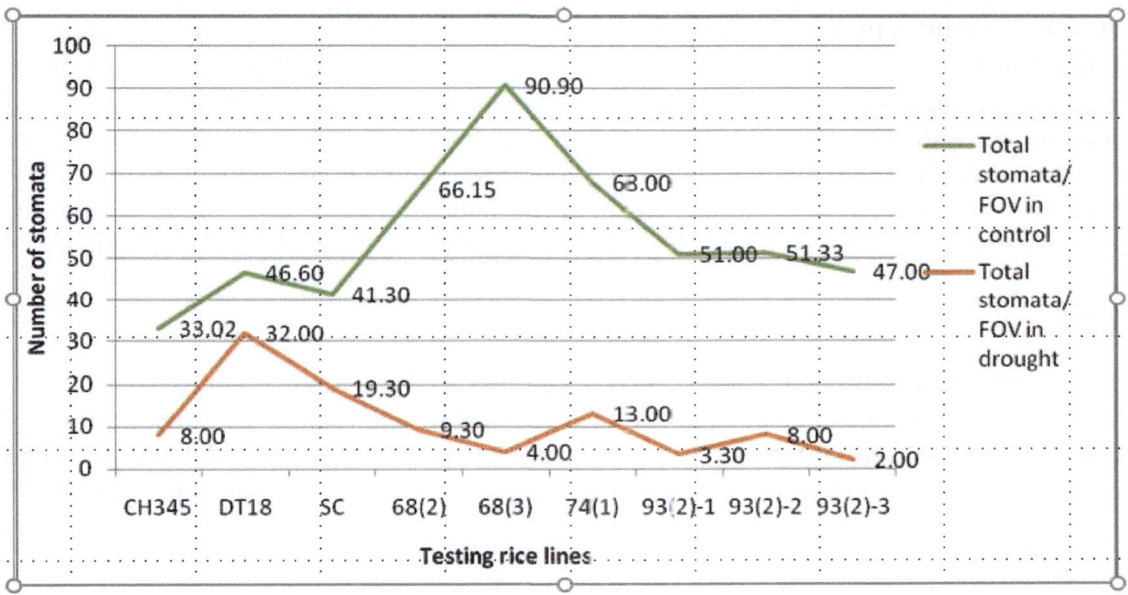

FIG. 3. Decreasing of stomata of testing rice lines in the drought stress compared with control conditions.

All mutant lines with improved drought tolerance have higher stomatal densities than wild type Seng Cu and controls in normal condition. When plants of these mutant lines experienced the drought treatment, they showed a significant decrease in total stomata per leaf square unit. If the distance between the blue line (control) and red line (drought treatment) is wider, it means that the decreasing number of stomata per FOV is more significant. If the distance between two lines is narrow, the decreasing number of stomata per FOV is less significant. The rice line with highest reduction (average 86.90 stomata per FOV) is mutant line 68(3) having average 90.90 stomata per FOV in control conditions, average 4.00 stomata per FOV in drought treatment (Fig. 3). The negative control, DT18, showed the least change (reduced average of 14.6 stomata per FOV); the

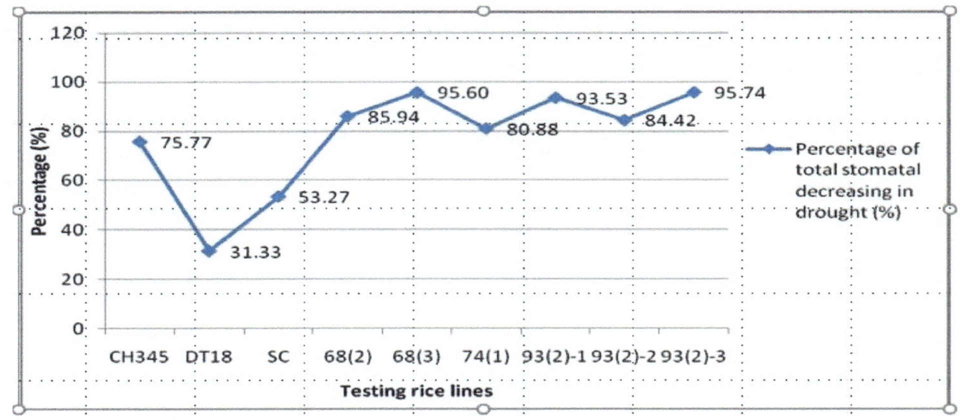

FIG. 4. Ratio of decreasing of stomata of testing rice lines in the drought stress compared with control conditions.

average 46.60 stomata per FOV in control condition and average 32.00 stomata per FOV in drought treatment. The reductions of total stomata per unit leaf square (FOV) are calculated and charted as a percentage of changes in drought compared with control conditions to evaluate the change level of each rice line (Fig. 4). There were variations in the level of stomatal density reduction among rice lines. The negative control DT18 showed the lowest level of reduction of 31.33%. With the wild type, SC, the reduction was 53.27% and all mutant lines had decreased significantly, by more than 80%. In particular, the reductions in mutant lines 68(3) and 93(2)-3 were more than 95%. The stomatal density of line CH345, the positive control, had decreased by 75.77%, not the highest reduction value due to its stomatal density; it was also not the highest in normal conditions.

3. CONCLUSIONS

The results showed that stomata density plays a key role in the tolerance ability of a plant under drought stress. Drought tolerance depends on the plant's ability to reduce the transpiration through stomata decrease under drought stress. Several studies have also demonstrated that rapid responses on stomatal characteristics were proposed to lead to drought tolerance in plants [17, 23, 24, 26].

REFERENCES TO CHAPTER 3–2

[1]   PILLITTERI, L.J., DONG, J., Stomatal development in Arabidopsis, Arabidopsis Book **6** 11 (2013); 11:e0162. doi: 10.1199/tab.0162

[2]   FRANKS, P.J., FARQUHAR, G.D., The mechanical diversity of stomata and its significance in gas-exchange control, Plant Physiol. **143** (2007) 78–87.

[3]   RAISSIG, M.T., et al., Mobile MUTE specifies subsidiary cells to build physiologically improved grass stomata, Science **355** (2017) 1215–1218; doi: 10.1126/science.aal3254.

[4]   STEBBINS, S.S., SHAH, Developmental studies of cell differentiation in the epidermis of monocotyledons: II. Cytological features of stomatal development in the Gramineae, Develop. Biol. **2** (2020) 477–00

[5]   RUDALL, P.J., CHEN, E.D., CULLEN, E., Evolution and development of monocot stomata, Am. J. Bot. **104** (2017) 1122–1141; doi: 10.3732/ajb.1700086.

[6]   HUGHES, J., et al., "Reducing stomatal density in barley improves drought tolerance without impacting on yield", Plant Physiology, (2017); DOI: https://doi.org/10.1104/pp.16.01844.

[7]   CAI, S., PAPANATSIOU, M., BLATT, M.R., CHEN, Z.H., Speedy grass stomata: Emerging molecular and evolutionary features. Mol. Plant **10** (2017) 912–914; doi: 10.1016/j.molp.2017.06.002.

[8]   LAWSON, T., KRAMER, D.M., RAINES, C.A., Improving yield by exploiting mechanisms underlying natural variation of photosynthesis, Curr. Opin. Biotechnol. **23** (2012) 215–220.

[9]   CAINE, R.S., et al., Rice with reduced stomatal density conserves water and has improved drought tolerance under future climate conditions, New Phytol. **221** (2019) 371–384.

[10]   BERTOLINO, L.T., CAINE, R.S., GRAY, J.E., Impact of stomatal density and morphology on water-use efficiency in a changing world, Front. Plant Sci. **10** (2019) 225; doi: 10.3389/fpls.2019.00225.

[11]   CHEN, Z.H., et al., Molecular evolution of grass stomata, Trends Plant Sci. **22** 2 (2017) 124–139.

[12]   MÜLLER, H.M., et al., The desert plant *Phoenix dactylifera* closes stomata via nitrate-regulated SLAC1 anion channel, New Phytol. **216** (2017) 150–162; doi: 10.1111/nph.14672.

[13]   GRANTZ, D.A., ZEIGER, E., Stomatal responses to light and leaf-air water vapor pressure difference show similar kinetics in sugarcane and soybean, Plant Physiol. **81** 3 (1986) 865–868.

[14]   VICO, G., MANZONI, S., PALMROTH, S., KATUL, G., Effects of stomatal delays on the economics of leaf gas exchange under intermittent light regimes. *New Phytol.* 192, (2011)640–652. doi: 10.1111/j.1469-8137.2011.03847;

[15]   MERILO, E., JÕESAAR, I., BROSCHÉ, M., KOLLIST, H., To open or to close: species-specific stomatal responses to simultaneously applied opposing environmental factors, New Phytol. **202** (2014) 499–508; doi: 10.1111/nph.12667.

[16]   MCAUSLAND, L., et al., Effects of kinetics of light-induced stomatal responses on photosynthesis and water-use efficiency, New Phytol. **211** (2016) 1209–1220; doi: 10.1111/nph.14000.

[17]   HAWORTH, M., et al., Allocation of the epidermis to stomata relates to stomatal physiological control: stomatal factors involved in the evolutionary diversification of the angiosperms and development of amphistomaty, Environ. Exp. Bot. **151** (2018) 55–63; doi: 10.1016/j.envexpbot.2018.04.010.

[18]   HEPWORTH, C., CAINE, R.S., HARRISON , E.L., SLOAN, J., GRAY, J.E., Stomatal development: Focusing on the grasses, Curr. Opin. Plant Biol. **41** (2018) 1–7; doi: 10.1016/J.PBI.2017.07.009.

[19]   MCKOWN, K.H., BERGMANN, D.C., Grass stomata, Curr. Biol. **28** (2018) R814–R816; doi: 10.1016/j.cub.2018.05.074.

[20]   YAN, W., ZHENG, S., ZHONG, Y., SHANGGUAN, Z.,. Contrasting dynamics of leaf potential and gas exchange during progressive drought cycles and recovery in *Amorpha fruticosa* and *Robinia pseudoacacia,* Sci. Rep. **7** (2017) 4470; doi: 10.1038/s41598-017-04760-zG.L.

[21]   VERMA, K.K., Characteristics of leaf stomata and their relationship with photosynthesis in *Saccharum officinarum* under drought and silicon application, ACS Omega **5** 37 (2020) 24145–24153.

[22]   LAWSON, T., VIALET-CHABRAND, S., Speedy stomata, photosynthesis and plant water use efficiency, New Phytol. **221** (2019) 93–98.

[23]   LI, T., ANGELES, O., RADANIELSON, A., MARCAIDA, M., MANALO, E., Drought stress impacts of climate change on rainfed rice in South Asia, Climatic Change **133** 4 (2014) 709–720.

[24]   KIM, Y., et al., Root response to drought stress in rice (*Oryza sativa* L.), Int. J. Mol. Sci. **21** 4 (2020) 1513.

[25]   ABARSHAHR, M., RABIEI, B., LAHIGI, H.S., Assessing genetic diversity of rice varieties under drought stress conditions, Not. Sci. Biol. **3** 1 (2011) 114–123.

3–3. VALIDATION OF DROUGHT RESILIENT MUTANTS USING BIOCHEMICAL APPROACHES

S. BIBI[1], M. ASHRAF[2],
Z.U. QAMAR[3], F. SARSU[4]

[1]Soil and Environmental Science Division,
Nuclear Institute for Agriculture and Biology, Faisalabad, Pakistan

[2, 3] Plant Breeding and Genetics Division,
Nuclear Institute for Agriculture and Biology, Faisalabad, Pakistan

[4]Joint FAO/IAEA Centre of Nuclear Techniques in Food and Agriculture,
Department of Nuclear Science and Applications,
International Atomic Energy Agency, Vienna

Abstract

Drought stress is a common occurrence that has a negative impact on rice production around the world. The objective of this study was to broaden the genetic variation of coarse rice, evaluation and identification of promising mutant lines for drought tolerance. The experiment was conducted at the Nuclear Institute for Agriculture and Biology, Faisalabad, during 2020 using twenty four promising rice mutants along with four standard varieties including Nagina22 (Tolerant variety), IR6, WAB-56-104 and BRRI-56, which were grown in tunnel and field conditions. Significant variations in plant height, tiller number, growth habit and panicles features were observed under drought conditions and after recovery. The highest value of TPC was recorded in BARRI-Dhan-56 under drought conditions and after recovery the highest value was observed in WAB56-104. The highest percentage cell membrane stability was 85.7 % in mutant 1011-1-1 while after recovery, the maximum cell membrane thermostability (%CMS) was recorded in mutant 233-2-3, mutant 582-1-1 and WAB56-104. Irrigation treatment significantly increased plant methylglyyoxal activity in all 28 rice genotypes as compared to drought and recovery. The level of proline was triggered to maximum under drought conditions in mutant 2711-1-1. Peroxidase was highest in mutant 54-1-1 under water stress condition as well as after recovery followed by mutant 149-2-2 as compared to check varieties. The highest value of catalase was recorded in mutant 1430-1-1 under drought conditions whereas it was highest in mutant 2-3-2 after recovery as compared to check varieties. Among all genotypes, Ascorbate peroxidase activity was highest in mutant 130-2-1 under drought conditions and the maximum activity was observed in mutant 2-3-2 after recovery. Among all genotypes, TSS in drought condition was highest in mutant 1227-1-1; during recovery, the maximum was observed in mutant 2-3-2. These findings showed that the genotypes that had higher performance in the respective attributes may be used for varietal development as well as in the future rice cross breeding programme for the improvement of traits of interest.

Key words: Drought, coarse rice, antioxidant enzymes, cell membrane thermostability, proline, methylglyyoxal activity.

1. INTRODUCTION

One of the most frequently cultivated cereal crops, rice is grown all over the world in various geographical, ecological and climatic environments. China, India, Indonesia and Pakistan all produce rice and Pakistan is fourth in the world in terms of Basmati rice output. Agricultural value added and GDP are 2.7% and 0.6%, respectively for rice. Punjab (59.46% of the harvest) and Sindh (32.13% of the harvest) are the two provinces where the majority of the rice crop is grown [1]. Rice production has decreased due to a reduction in cultivated area, effects of monsoon, late receding of water period and variable temperature in rice fields. Drought stress is a common occurrence that has a negative impact on rice production around the world [2]. Drought resistance is an efficient technique to combat the problem of yield loss induced by drought stress in rice [3, 4]. Drought tolerant lines are being developed and their drought tolerant properties are being studied by breeders and biologists [5]. Therefore, elucidating the mechanisms of rice adaptation and response to drought stress is essential to meet the growing global demand for food.

Drought stress in plants can also be caused by excessive water demand from leaves, where evapotranspiration exceeds the rate of water absorption despite adequate groundwater availability [6]. The visual impact of drought stress on plants can be noticed in the morphology of the plants. Leaf senescence [7] is a visual performance that is similarly influenced by water stress. This is linked to the increase in reactive oxygen species that causes leaf senescence [8], implying that osmoregulation is the most important mechanism in plants, and that a decrease in turgor leads to the accumulation of osmo-protectants. Under conditions of water scarcity, the accumulation of several osmolytes, such as proline, soluble sugar, phenolic and methylglyoxal increases and plays a significant role in plants for drought tolerance [9].

Proline is a five-carbon amino acid that is proteinogenic and its accumulation is linked to drought tolerance due to its role in maintaining leaf turgor and increases in stomatal conductance [10, 11]. When compared to well-watered settings, proline accumulation increases in all rice cultivars during drought [11]. Many physiological processes, including photosynthesis and mitochondrial respiration, soluble carbohydrates are essential for maintaining the balance [12]. Drought triggers the accumulation of soluble sugars, and protects the plants to some extent in unfavourable conditions [10, 13]. Drought can also induce an imbalance in the formation of reactive oxygen species (ROS) and their quenching in rice, resulting in oxidative damage and negatively influencing the life cycle of plants [12].

Therefore, the most effective strategy to enhance drought tolerance in rice is to increase antioxidants and non-antioxidant activity in rice organs. Plants have antioxidant defence mechanisms that protect them from oxidative damage [12]. The enzymatic antioxidants are superoxide dismutase (SOD), catalase (CAT), peroxidase (POD), ascorbate peroxidase (APX), etc. The effect of drought on these characteristics may help to explain characteristics of tolerance in a genotype. However, the key phase in plants when a stress occurs determines a significant part of this impact.

2. MATERIALS

The experiments were conducted at the Nuclear Institute for Agriculture and Biology (NIAB), Faisalabad, Pakistan during 2020. The plant material consisted of 24 rice mutants named as 2-3-1, 2-3-2, 54-1-1, 130-2-1, 149-2-2, 233-2-3, 582-1-1, 928-1-1, 1011-1-1, 1227-1-1, 1304-2-1, 1343-

1-1, 1351-1-1, 1424-1-2, 1430-1-1, 2331-1-2, 2332-1-1, 2331-2-2, 2334-1-1, 2529-1-1, 2530-1-1, 2588-2-1, 2606-2-3, 2711-1-1, along with four standard varieties, including Nagina22 (heat tolerant variety), IR6, WAB-56-104 and BRRI-56. All the mutants and varieties were grown in tunnel and field conditions at the experimental farm of NIAB, Faisalabad. For the analysis, leaf samples of equal fresh weight were taken from the plants after washing.

3.   METHODS

Evaluation of cell membrane thermostability from the stressed, normal and recovered plants was calculated, as explained by Barr and Weatherley [14]. Assessment of enzymatic and non-enzymatic antioxidants in 28 rice genotypes were used and samples were labelled and stored at -20°C in MAB LAB-11 (Marker Assisted Breeding), PBG (Plant Breeding and Genetics Division) for further biochemical analysis. Leaf samples (200 mg were used to extract the sample in 2 mL (100mM at pH7.4) of potassium phosphate buffer. Each sample was vertex homogenized for 5 min, then centrifuged at 10 000 rev./min for 10 min. at 4°C. After centrifugation supernatants were collected into labelled Eppendorf tubes and stored in a -20°C refrigerator for the future. Data were noted in triplicate for biochemical attributes.

The measurement of ascorbate peroxides (APX) activity was done by using the method of Dixit et al. [15]. The reagent was prepared by mixing 2.7 mL of phosphate buffer (100mM phosphate buffer having pH7.2), 100 μL of ascorbic acid (7.5mM), 100 μL of antioxidant enzyme extracted sample and lastly, 100 μL of H_2O_2 (300mM). The APX value was measured at the absorbance of 290 nm including blank with time scan (1 min) every 30 s, which showed a decrease in absorbance.

For the analysis of catalase (CAT) activity [16], the assay solution consisted of 2 .8 mL of phosphate buffer (100mM at PH 7.2), 100 μl of antioxidant enzyme extracted sample and 100 μL of H_2O_2 (300mM). The CAT value was measured at the absorbance of 240 nm, including blank with time scan (1 min) of every 30 s, which showed a decrease in absorbance.

For the analysis of peroxidae (POD) activity [17], the assay solution prepared was 2.7 mL (100mM at pH7.2) of phosphate buffer, 100 μL of antioxidant enzyme extracted sample, and 100 μL of guaiacol (3.4mM of guaiacol). An amount of 100 μL of H_2O_2 was added in each test tube before taking its absorbance value. The POD value was measured at the absorbance of 470 nm, including blank with time scan (1 min) of every 30 s, which showed an increase in absorbance.

The total phenolic content (TPC) was estimated by using the procedure in Ref. [18]. For the assessment, 0.2 g leaf samples were ground into 500 μL of ice cold methanol (95%) by using a mortar and pestle. Samples were placed in the dark for 48 h at room temperature and then centrifuged at 10 000 rev./min for 10 min. The supernatant of each sample was collected into new fresh tubes and stored at $-20\ ^\circ$C for TPC measurement. For assay, 100 μL of supernatant with 100 μL of 10% (vol./vol.) F-C reagent and after mixing 800 μL of 700mM Na_2CO_3 was added. Samples were then incubated at 25°C until a black colour was developed. The content was determined to have an absorbance of 735 nm by using a spectrophotometer, along with a blank. A linear regression equation was derived using a standard curve made with varied concentrations of gallic acid. The TPC concentration was calculated by applying the formula as TPC (mg/g FW) = (CxV/V$_t$ x W).

Proline was estimated in flag leaves according to the method in Ref. [19]. Fresh leaf samples (0.5 g) were ground in sulfosalicylic acid solution (3 %) and then centrifuged at 10 000 rev./min for 10 min.. Acid ninhydrin and glacial acetic acid (1:1) was added into the supernatant.The contents were then incubated at 100°C for 1 h. Toluene was added in a test tube with vigorous shaking for further extraction. The aqueous phase was incubated at room temperature and read at 515 nm absorbance using a spectrophotometer. For the preparation of the standard curve, L. Proline (Sigma) was employed. The amount of proline in the samples was measured in milligrams of proline per gram of dry matter.

Methylglyoxal (MG) was estimated basically in fresh leaves according to the method of Yadav et al. [20]. For the MG assay, neutralized supernatant was used in a 1 mL Eppendorf tube in which 250 µL of 1, 2-diaminobenzene, 100 µL of perchloric acid and 650 µL of the neutralized supernatant were added. A spectrophotometer was used to measure the absorption at 335 nm after 25 min. For the preparation of the standard curve, MG (Sigma) was employed. The amount of MG in the samples was measured from the standard curve in µmol/g of fresh weight. Cell membrane thermostability (CMT) was estimated; fully expended leaves were cut into small pieces in a test tube and washed three times with ddH_2O. The tubes were submerged for 24 h at 10 °C and after normalizing the sample at room temperature, an initial conductance was noted from drought (T_1) and control (C_1). Then, all glass tubes were kept in an autoclave with pressure of 0.10 MPa for 15 min at 121°C. The contents were then mixed at 25°C and the final conductance of drought (T_2) and control (C_2) samples were measured. Cell membrane stability was expressed as per cent relative injury (RI %) using the following equations:

$$\text{Relative injury \%} = [1-(1-T1/T2)/(C_1/C_2)] \times 100$$

Total soluble sugar (TSS) was assessed by the Nelson-Somogyi method [21], with little modification. Brown rice flour samples of 100 mg were homogenized in 2 mL of 80% ethanol after which the mixture was subjected to a vortex for 1 h at room temperature then centrifuged at 10 000 rev./min for 10 min. Afterwards, supernatants of each sample were collected into new labelled Eppendorf tubes. The pellets were washed twice with 80% ethanol and the previous steps repeated. Finally the supernatants of each sample were pooled and stored at −20°C till further use. For TSS reaction, a 200 µL sample was added in 3 mL of 0.15 g anthrone and prepared for incubation in a water bath at 97^{0}C for 15 min. Samples were taken out from the water bath and immediately placed in ice. Sample absorbance was measured at 625 nm, including blanks. The TSS was determined using the equation:

$$\text{Total soluble sugars} = OD625 \times \text{correction factor (38.64)} \times \text{Dilution factor (37.5)}.$$

4.   STATISTICAL ANALYSES

All recorded data were statistically analysed for each characteristic according to Steel et al. [21] to assess the gaps between p value (P = 0.05) and all traits.

5. RESULTS

5.1. Variability in biochemical attributes

Analysis of variance for biochemical attributes per plant is presented in Table A–1. The results indicated a highly significant difference among mutants/varieties, which allow further analysis. According to Table A–1, the interaction effects between treatment and genotype (T × G) are highly significant which is an indicator of the huge impact of drought stress on biochemical attributes per plant in rice mutants/ varieties.

5.2. Correlation coefficient of biochemical attributes

Correlation values estimated a strong relationship to evaluate attributes Table A–2. TPC results are extremely significant and have a positive relationship with CAT (0.464), POD (0.427), APX (0.559) and proline (0.279). POD was highly significant and positively correlated with CAT (0.3354), APX (0.5616), TPC (0.2124) and proline (0.1225). CAT showed positive correlation with APX (0.6934), POD (0.3354), TPC (0.4642) and proline (0.4118). APX revealed positive correlation with CAT (0.6934), POD (0.5616), proline (0.4778) and with TPC (0.4275). Proline is positively correlated with APX (0.4778), CAT (0.4118), POD (0.1225) and with TPC (0.279). Methylglyoxal is negatively correlated with POD (–0.0091), APX (–0.3918), CAT (–0.2663), TPC (–0.2134) and proline (–0.5370). TSS showed negative correlation with APX (–0.3823), CAT (– 0.3352), POD (–0.3908), Proline (–0.0007), TPC (–0.0855), and MG (–0.0498).

5.3. Mean performance of biochemical attributes

Irrigation treatment significantly increased plant methylglyoxal (MG) activity in all 28 rice genotypes as compared to drought and recovery as shown in Fig. A–1. However, a significant decrease in MG activity was observed in Nagina-22 and mutant 2-3-1, while it was retained in all other genotypes compared to that of standard varieties under drought conditions. Methylglyoxal activity, in the case of after watering, was significantly increased whereas a significant decrease was found in mutant 54-1-1, mutant 233-2-3, mutant 582-1-1, mutant 1351-1-1 and BARI-DHAN-56. The highest value of MG was recorded at 110.43µmole of MG/g of fresh weight in mutant 54- 1-1, followed by mutant 2331-1-2 and mutant 1351-1-1 under water stress conditions. After recovery, the maximum value for MG was recorded in mutant 2-3-1 followed by mutant 2332-2-2, mutant 1343-1-1 and IR6.

Proline showed significant variation in 28 rice genotypes (Fig. A–2). It ranged from 1.38 to 12.31µmole of proline/ g of fresh weight. The level of proline was highest in mutant 1304-2-1, mutant 2530-1-1 and Nagina-22, while drought condition was triggered by the proline content to maximum; the highest was showed in mutant 2711-1-1, followed by mutant 2529-1-1, mutant 2334-1-1, mutant 2530-1-1, mutant 2588-2-1, mutant 1227-1-1, mutant 130-2-1, Ng-22 and IR6. Good recovery was observed in mutants 54-1-1, 130-2-1, 149-2-2, 1227-1-1, 1430-1-1, 2332-1-1,2332-2-2, 2334-1-1 and WAB56-104.

The data revealed that the genotypes were highly significant (P = 0.05) with respect to total phenolic content (Fig. A–3). This attribute ranged from 1178.9 to 17052.2µmole of TPC/g of fresh weight. It was observed that all mutant genotypes significantly increased the TPC values under drought conditions as compared to normal irrigation and after watering. The highest value of total phenolic content (TPC) was recorded in BARRI-Dhan-56 (17 052.2μmole of TPC/g) followed by Nagina-22 (14 047.8μmole of TPC/g) and the minimum was recorded in mutant 1304-2-1 (4465.6μmole of TPC/g) under drought condition. After recovery, highest values were observed in WAB56-104 (10 225μmole of TPC/g) followed by Nagina-22 and mutant 1424-1-23.

It was observed that cell membrane thermostability (CMS) was minimum under normal growing conditions as compared to drought and recovery as given in Fig. A–4. The highest percentage of CMS was recorded at 85.7% in mutant 1011-1-1 followed by 130-2-1 (83%) and 1227-1-1 (81.8%), while the minimum was noted in IR-6 and mutant 2711-1-1. After recovery, the CMS percentage was high in varieties as compared to stressed plants and the maximum was recorded in mutant 233-2-3 (105.0%), mutant 582-1-1 and WAB56-104.

Peroxidase (POD) analysis showed significant variation in 28 rice genotypes presented in Fig. A–5. It ranges from 7.63 to 116.18mM of POD/g of fresh weight. All the treated populations showed significance in variance ($p < 0.5$) for peroxides and among all genotypes POD was highest in mutant 54-1-1, followed by mutants 2-3-1 and 2-3-2 under water stress conditions. Mutants 149-2-2 followed by 1351-1-1, 1430-1-1 and 1304-2-1 were highest under normal conditions whereas mutants 149-2-2, 1343-1-1 and 54-1-1 showed the maximum recovery as compared to the check varieties.

Data revealed that the genotypes were highly significant (P = 0.05) with respect to catalase (Fig. A–6). This attribute ranged from 1.78 to 11.9mM of CAT/g of fresh weight. Highest value of catalase was recorded at 5.32mM of CAT/g of fresh weight in Nagina-22 followed by 1011-1-1 (5.02) under normal conditions, whereas under drought conditions, the highest was recorded in mutant 1430-1-1 (11.9mM of CAT/g). followed by mutants 1011-1-1 (11.08mM of CAT/g), Nagina-22, mutant 2331-1-2 and mutant 2332-2-2. Among all genotypes, catalase was highest in mutant 2-3-2 followed by mutants 2-3-1 and 2332-2-2 after recovery as compared to check varieties.

Ascorbate peroxidase (APX) showed significant variation ($p < 0.5$) in 28 rice genotypes as shown in Fig. A–7. The APX ranged from 15.7 to 167.4mM of APX/g of fresh weight. Among all genotypes, APX activity was more or less similar in 28 genotypes under normal growing conditions, whereas the highest was found in mutant 130-2-1, followed by mutants 2588-2-1, 928-1-1, 2-3-2 and 2-3-1 under drought conditions. Highest recovery was observed in mutants 2-3-2 and 2-3-1.

Total soluble sugars (TSS) showed significant variation ($p < 0.5$) in 28 rice genotypes (Fig. A–8). They ranged from 1701.5 to 3133.1μg of TSS/g of fresh weight. Among all genotypes, TSS was highest in Nagina-22, WAB56-104, mutants 2530-1-1, 1424-1-2, and 130-2- under normal growing condition, whereas in drought conditions the highest was observed in mutant 1227-1-1, followed by 233-2-3, mutant 130-2-1, Nagina-22 and WAB56-104. Highest recovery was observed in mutant 2-3-2, mutants 2-3-1, 2331-1-2, 2334-1-1, 1424-1-2 and Nagina-22 and WAB56-104.

5.4. Two way principal component analysis

The principal component analysis (PCA) is formed by plotting the first principal component (PC1) and second principal component (PC2) that account for the largest variability of the observed parameters and also grouped the rice genotypes in the loading plot. The tested genotypes' eigenvalues from PCA were used to rank them for their drought tolerance. The first two PCs with eigenvalues more than explain the 64.9 % of total variation (Table A–3). PC1 accounted for 41.5%, while (PC2) accounted for 23.4 % of the total variation among the biochemical related traits under drought stress conditions. The two factor PCA plots showed higher impact of observed parameters under water stress as compared with normal conditions. The results showed that on the basis of the first and second components, PCA was used to evaluate the relationship between variables in rice mutants (Figs 9 and 10) of plot analysis. It was shown that positive components on both positive axes consist of highly tolerant mutants, i.e. 2-3-1, 2-3-2, 54-1-1,582-1-1, 928-1-1, 1304-2-1, 1343-1-1, 1351-1-1, 1430-1-1, 2331-1-2, 2588-2-1, Nagina, WAB56-104, BRRI Dhan56. A small number of genotypes from components A and C could be employed in a cross-breeding programme to enhance the desirable traits. Principal component analysis was used to create a bi-plot and showed that variables are very well distributed on the plot as a path. The distance between each variable and its corresponding component demonstrated how those variables contributed to the distinctness of mutants and their parents, as shown in Table A–3. Hence, the first component was most correlated to TPC, CAT, POD, APX and proline having a positive effect and MG, TSS and CMS having a negative effect under stress conditions. The second component exhibited positive effects for CAT, POD, APX and MG, but showed negative correlation for TPC, proline, TSS and CMS. The third component is positively related to POD, APX, proline and CMS and has a negative impact with TPC, CAT, MG and TSS. The fourth PC presented the positive results with TPC, CAT, MG, and CMS and showed the negative correlation with those attributes that play a significant role in drought stress conditions.

6.   DISCUSSION

Drought is a highly significant abiotic factor for rice yield stability among the various abiotic stresses. Rice cultivation with less water is a global problem that negatively affects crop development, crop efficiency and yield among rice growing countries, including Pakistan [22]. The development and popularization of upland rice, especially aerobic rice, is the best answer in this regard. Practically every country that produces rice has been developing aerobic rice lines for use in upland rice fields [23]. All of the promising rice mutants studied in this study were developed under drought conditions with no irrigation and rely primarily on rainwater. Due to a variety of morphological, physiological, and biochemical adaptations among rice, mutants were able to maintain their growth and performance under water scarcity, according to research.

Antioxidant, non-antioxidant enzyme activity and MG detoxification were investigated in the current study. The results revealed that MG increased in normal irrigation conditions, but decreased during drought stress and after recovery except in a few cultivars, i.e. mutant 2332-2-2 (Fig. A–1). Hoque et al. [24] reported that under stress, the toxic molecule MG accumulates and detoxification may be a mechanism for conferring tolerance to various abiotic stimuli. Rice mutants showed highest activity of the antioxidant enzymes as compared to standards during drought stress conditions which may help to detoxify the MG under water stress conditions. A similar trend was observed by El-Shabrawi et al. [25] that Pokkali showed greater activity of both necessary for the detoxification of MG when compared to IR64 under salt stress conditions.

According to Hernandez et al. [26], Total soluble sugars are a very effective attribute of drought stress, because they alter the release of carbohydrates from the source to sink organs in rice plants. Previously, it was shown that soluble sugars act as an osmoprotectant in the presence of various stressors, regulating osmotic adjustment, protecting membranes, and scavenging damaging reactive oxygen species [27]. Results showed that TSS increases under drought stress conditions more or less similar to normal sowing conditions. After watering, the rice plants increased the content of TSS (Fig. A–8). Similar results were found, where the levels of sugars decreased in response to drought in rice plants [28]. However, several studies have found that soluble sugar levels rise in drought conditions [26, 29–31].

Total phenolic contents are a type of secondary metabolite found in plants that has redox characteristics and is a component of non-antioxidant enzyme [32]. Jogawat et al. reported that plants produce more polyphenols in response to abiotic stress including flavonoids and phenolic acids, which help the plants cope with environmental conditions [33]. Results revealed that all mutant genotypes significantly increased the TPC values under drought conditions as compared to normal irrigation and after watering (Fig. A–3). These rice genotypes may have a lot of potential for growing in water stressed environments. Several reports [34–36] have also found an increase in phenolic content in rice plants growing under drought and salinity stress, which is consistent with our findings. According to Ahmed et al., salinity stress, on the other hand, causes a decrease in TPC in Tibetan wild barley plants [37].

The results revealed that proline is increased most rapidly under severe drought conditions as shown in (Fig. A–-2). Dien et al. [29] found the same results, that proline increased under drought stress environments, especially DA8 and Thierno Bande which increased proline more than other varieties. The proline content of the rice varieties rapidly reduced after re-watering, reaching levels near or equivalent to those in the control condition (Fig. A–2). Proline dehydrogenase, the first enzyme in the proline breakdown pathway, converts proline into 1-pyrroline-5-carboxylate (P_5C) due to osmotic stress. The enzyme P5C dehydrogenase then reverses into glutamate]. Several studies showed that proline is enhanced during drought stress and reverses when the stress is withdrawn [38–40].

Crop growth and development may be hampered by abiotic stresses, but the reproductive phase is most susceptible to stress conditions [41, 42]. Plants are subjected to injury caused by a variety of biotic and abiotic stressors on a regular basis, which can result in the loss of growth of the plants. Plants activate self-defence systems in response to injury in order to repair damaged tissues or protect against infections. An antioxidant enzyme has shown to increase in mRNA levels in rice plants when mechanical injury occurs. Specialized enzyme antioxidants, for instance CAT, APX and POD are activated in response to abiotic stressors and serve as the first line of defence in the process of detoxifying the negative effects of reactive oxygen species (ROS) [43–45]. Increased CAT and APX activities under stress were linked to increased ROS generation in the current study, as shown in Figs A–6 and A–7. Hasanuzzaman et al. [46] reported that an accumulation of ROS decreased the biosynthesis of CAT and impaired stability of the membrane. CAT is one of the best enzymes for removing H_2O_2 according to Hasanuzzaman. APX is an essential enzymatic antioxidant that detoxifies ROS in stressful situations and has a higher affinity for scavenging H_2O_2 than CAT [47]. In this study, APX activity increased in drought condition among all rice mutants with standard varieties (Fig. A–7). Similar results were reported by Kamarudin et al. [45] that catalase, ascorbate peroxidase, guaiacol peroxidase enzyme activities were increased in rice cultivars under drought stress. Another group of non-chloroplastic enzymes that detoxify H_2O_2 in the cell cytosol is peroxidases. However, the present study showed that a clear upregulation of POD was exhibited in 2-3-1, 2-3-2, 54-1-1, 130-2-1, 582-1-1, 928-1-1, 1227-1-1, 1424-1-2, 2332-1-1,

2332-2-2, 2334-1-1 and 2529-1-1 with standard varieties, i.e. Nagina-22, IR6, WAB-56 and BARI-Dhan 56 and mutants 149-2-2, 233-2-3, 928-1-1, 1011-1-1, 1304-2-1, 1343-1-1, 1351-1-1, 1430-1-1, 2331-1-2, 2530-1-1, 2588-2-1, 2606-2-3 and 2711-1-1 showed downregulation under drought conditions as shown in Fig. A–5. This could help to ensure efficient H_2O_2 removal in cell compartments other than the chloroplast. An imbalance between the formation of ROS and the ability of antioxidants to quench them has been hypothesized as a cause of decreased enzyme function. Similar findings were found in rice tolerant genotype increased antioxidant activities being subjected to drought stress [39, 43, 45].

CMTS is a good indication of plant stress tolerance. The osmotic potential in leaf tissues is hypothesized to affect CMS as determined by the PEG test. The higher the drought stress, the higher the solute concentrations in cell sap and leaf tissues. The main contributors to osmotic potential were sugars, total phenolic content, proline, MG and antioxidants. In this study, rice cultivars differ in their relative cell membrane stability (CMS) (Fig. A–8). As a result, the level of injury to plant cells and organelles is determined by cell membrane stability. According to Naveed et al. [3], genotypes with higher CMTS can be classified as having better heat tolerance than those with lower CMTS. Therefore, at the seedling stage of rice, CMTS can be used as a consistent physiological marker for heat tolerance. In this study, eight indicators of biochemical traits were chosen during the rice reproductive stage, when drought damage was most likely to occur. To discover differences in drought tolerance among the 28 cultivars under study, these eight individual parameters were combined into eight separate comprehensive indices using PCA. Two way PCA showed the drought tolerant cultivations on the basis of these eight attribute, i.e. 2-3-1, 2-3-2, 54-1-1, 582-1-1, 928-1-1, 1304-2-1, 1343-1-1, 1351-1-1, 1430-1-1, 2331-1-2, 2588-2-1, Nagina, WAB56-104 and BRRI Dhan56 (Figs A-9, A-10).

7. CONCLUSION

In this study, all of the aforementioned attributes are sought to be included in order to choose suitable mutants for the rice breeding programme. Among the 28, only 12 mutants (54-1-1, 130-2-1, 1011-1-1, 2332-2-3, 582-1-1, 1351-1-1, 2711-1-1, 2529-1-1, 2530-1-1, 2588-1-1, 1227-1-1) performed better in both physiological and biochemical terms under induced drought throughout the entire trial. Mutants 1424-1-23, 1227-1-1, Nagina-22 and web-56 showed good recovery after watering.

ACKNOWLEDGEMENTS

The authors are thankful to the IAEA for providing technical and financial support through IAEA Coordinated Research Project Pak-22224, Breeding High Yielding Rice for Limited-Water Conditions through Mutagenesis and Related Approaches. The authors also thank the Pakistan Atomic Energy Commission (PAEC) and Nuclear Institute for Agriculture and Biology (NIAB) Faisalabad for their active cooperation and encouragement throughout the project.

TABLE A–1. ANALYSIS OF VARIANCE OF 28 COARSE RICE MUTANTS/VARIETIES EVALUATED FOR BIOCHEMICAL ATTRIBUTES

Traits	SOV	DF	SS	MS	F	P
APX	Replication (R)	2	477	238		
	Treatment (T)	2	244810	122405	630.90	0.0000
	Genotypes (G)	27	36249	1343	43.66	0.0000
	TxG	54	51447	953	30.98	0.0000
	Error R x T	4	811	203		
	Error R x T x G	162	4982	31		
	Total	251	338775			
	Grand mean	74.905				
	CV (RxT)	19.01				
	CV(RxTxG)	7.40				
CAT	Replication (R)	2	11.51	5.755		
	Treatment (T)	2	806.37	403.186	186.32	0.0001
	Genotypes (G)	27	146.42	5.423	16.59	0.0000
	TxG	54	243.89	4.516	13.82	0.0000
	Error RxT	4	8.66	2.164		
	Error RxTxG	162	52.96	0.327		
	Total	251	1269.80			
	Grand mean	5.4063				
	CV (RxT)	27.21				
	CV(RxTxG)	10.58				
POD	Replication (R)	2	492	246.0		
	Treatment (T)	2	43778	21889.0	54.63	0.0012
	Genotypes (G)	27	64147	2375.8	34.56	0.0000
	TxG	54	50779	940.3	13.68	0.0000
	Error RxT	4	1603	400.7		
	Error RxTxG	162	11136	68.7		
	Total	251	171934			
	Grand mean	43.810				
	CV (RxT)	45.69				
	CV(RxTxG)	18.93				
TPC	Replication (R)	2	2.840E+07	1.420E+07		
	Treatment (T)	2	7.802E+08	3.901E+08	8.90	0.0337
	Genotypes (G)	27	4.634E+08	1.716E+07	3.73	0.0000
	TxG	54	6.037E+08	1.118E+07	2.43	0.0000
	Error RxT	4	1.753E+08	4.384E+07		
	Error RxTxG	162	7.452E+08	4600204		
	Total	251	2.796E+09			
	Grand mean	43.810	7564.7			
	CV (RxT)	45.69	87.52			
	CV(RxTxG)	18.93	28.35			
MG	Replication (R)	2	5639	2819.3		
	Treatment (T)	2	60910	30454.9	41.36	0.0021

	Genotypes (G)	27	8706	322.5	2.02	0.0041
	TxG	54	18632	345.0	2.16	0.0001
	Error RxT	4	2945	736.3		
	Error RxTxG	162	25902	159.9		
	Total	251	122734			
	Grand mean	87.737				
	CV (RxT)	30.93				
	CV(RxTxG)	14.41				
Proline	Replication (R)	2	1.71	0.855		
	Treatment (T)	2	1759.12	879.562	207.41	0.0001
	Genotypes (G)	27	179.27	6.640	6.52	0.0000
	TxG	54	324.02	6.000	5.89	0.0000
	Error RxT	4	16.96	4.241		
	Error RxTxG	162	165.00	1.019		
	Total	251	2446.09			
	Grand mean	6.1302				
	CV (RxT)	33.59				
	CV(RxTxG)	16.46				
TSS	Replication (R)	2	9290959	4645479		
	Treatment (T)	2	2.915E−07	1.457E+07	10.64	0.0250
	Genotypes (G)	27	1.171E−07	434003	4.44	0.0000
	TxG	54	6302483	116713	1.19	0.2006
	Error RxT	4	5479093	1369775		
	Error RxTxG	162	1.585E−07	97849.5		
	Total	251	7.780E−07			
	Grand mean	2438.7				
	CV (RxT)	47.99				
	CV(RxTxG)	12.83				
CMT	Replication (R)	2				
	Treatment (T)	2				
	Genotypes (G)	27				
	TxG	54				
	Error RxT	4				
	Error RxTxG	162				
	Total	251				
	Grand mean					
	CV (RxT)					
	CV(RxTxG)					

APX: ascorbate peroxides; CAT: catalase; POD: peroxidase; TPC: total phenolic content, MG: methylglyoxal, TSS: total soluble sugars, CMT: cell membrane thermostability.

TABLE A–2. PEARSON'S CORRELATION COEFFICIENTS ILLUSTRATING THE RELATIONSHIP OF ANTIOXIDANT AND NON-ANTIOXIDANT ENZYMES

	APX	CAT	POD	Proline	TPC	TSS	MG
APX	1.000						
CAT	0.693	1.000					
POD	0.562	0.335	1.000				
Proline	0.478	0.412	0.122	1.000			
TPC	0.428	0.464	0.212	0.279	1.000		
TSS	-0.382	0.335	-0.391	-0.001	-0.086	1.000	
MG	-0.392	0.266	-0.009	-0.537	-0.213	-0.050	1.000

APX: ascorbate peroxide;, CAT: catalase; POD: peroxidase; TPC: total phenolic content;
MG: methylglyoxal; TSS: total soluble sugars; CMT: cell membrane thermostability.

TABLE A–3. PRINCIPAL COMPONENT ANALYSIS OF COARSE RICE MUTANTS/VARIETIES ON THE BASIS OF ANTIOXIDANT AND NON-ANTIOXIDANT ENZYMES

Variable	PC1	PC2	PC3	PC4	PC5	PC6	PC7	PC8
Eigenvalue	3.3204	1.8710	0.9975	0.5753	0.4854	0.3433	0.2286	0.1784
Proportion	0.415	0.234	0.125	0.072	0.061	0.043	0.029	0.022
Cumulative	0.415	0.649	0.774	0.846	0.906	0.949	0.978	1.000

Rotated factor loadings and communalities (Varimax rotation)					
Traits	Factor1	Factor2	Factor3	Factor4	Communality
TPC	0.733	-0.002	-0.150	0.601	0.922
APX	0.886	0.172	0.152	-0.205	0.879
CAT	0.850	0.134	-0.082	0.040	0.748
POD	0.352	0.654	0.511	-0.172	0.843
MG	-0.686	0.563	-0.034	0.232	0.843
Proline	0.759	-0.514	0.031	-0.054	0.844
TSS	-0.237	-0.810	-0.016	-0.112	0.725
CMS	-0.224	-0.397	0.826	0.267	0.960
Variance	3.3204	1.8710	0.9975	0.5753	6.7643
% Variance	0.415	0.234	0.125	0.072	0.846

APX: ascorbate peroxides; CAT: catalase; POD: peroxidase; TPC: total phenolic content;
MG: methylglyoxal; TSS: total soluble sugars; CMT: cell membrane thermostability.

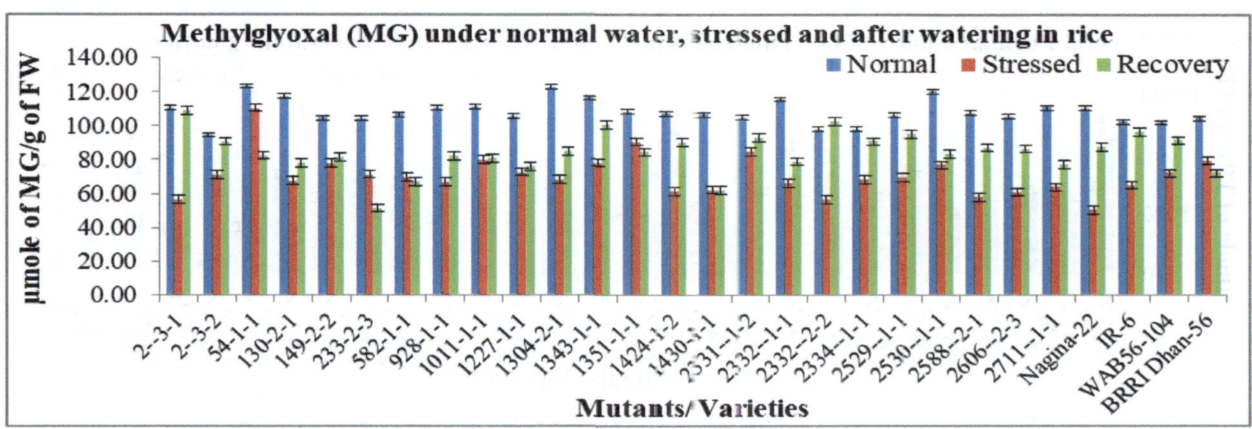

FIG. A–1. Methylglyoxal estimation, 1–24 mutants; 24–28 varieties; series 1 is normal; series 2 is stressed; and series 3 is after recovery.

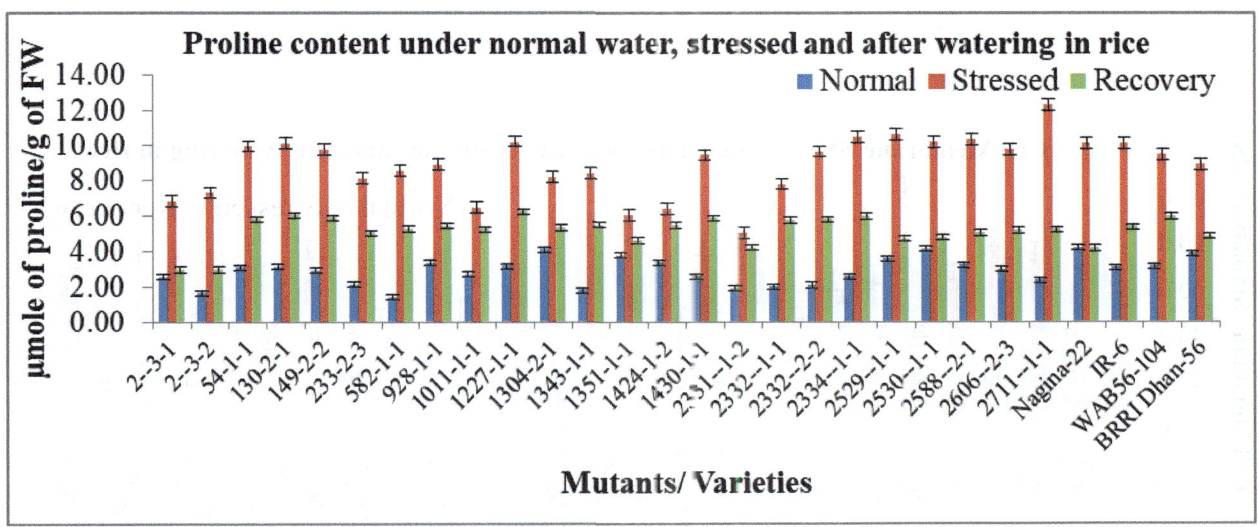

FIG. A–2. Proline estimation, 1–24 mutants; 24–28 varieties; series 1 is normal; series 2 is stressed; and series 3 is after recovery.

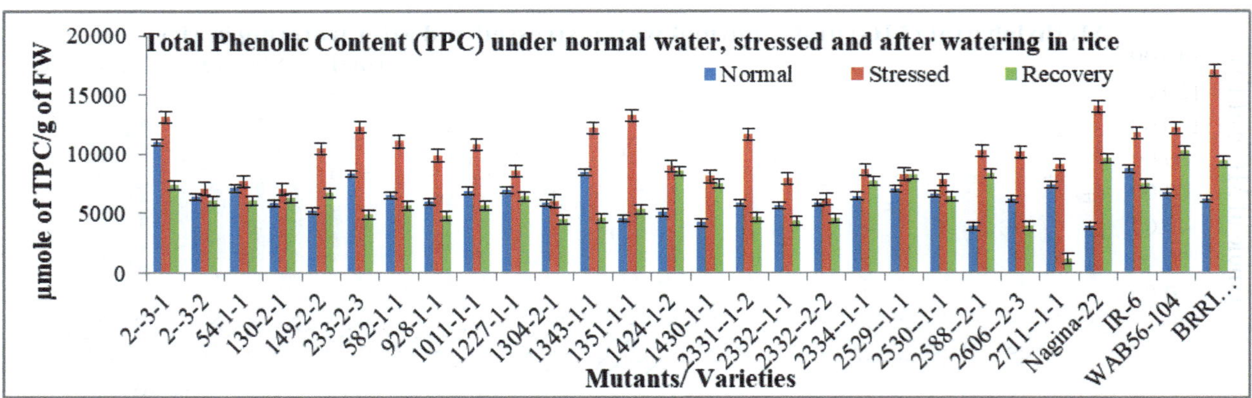

FIG. A–3. *Total phenolic content estimation, 1–24 mutants; 24–28 varieties; series 1 is normal; series 2 is stressed; and series 3 is after recovery.*

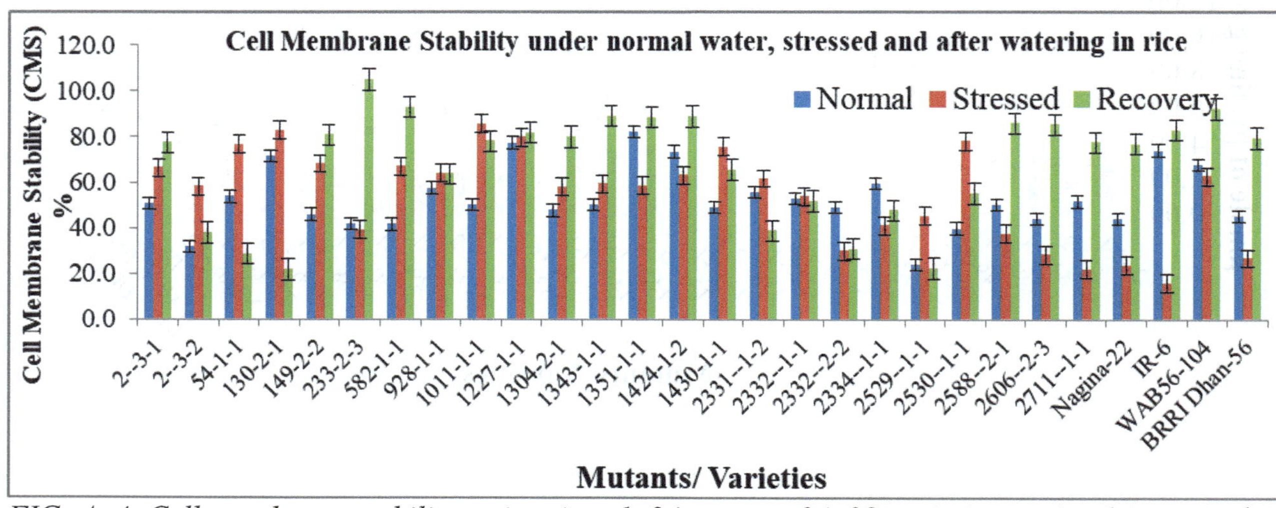

FIG. A–4. *Cell membrane stability estimation, 1–24 mutant; 24–28 varieties; series 1 is normal; series 2 is stressed; and series 3 is after recovery.*

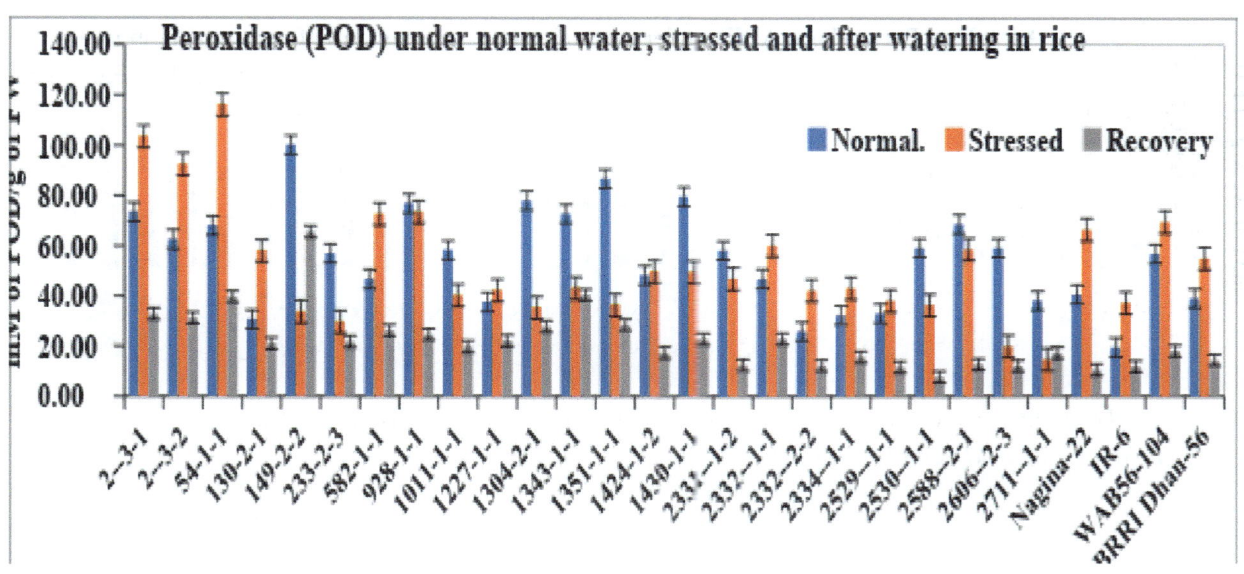

FIG. A–5. Peroxidase estimation, 1–24 mutant; 24–28 varieties; series 1 is normal; series 2 is stressed; and series 3 is after recovery.

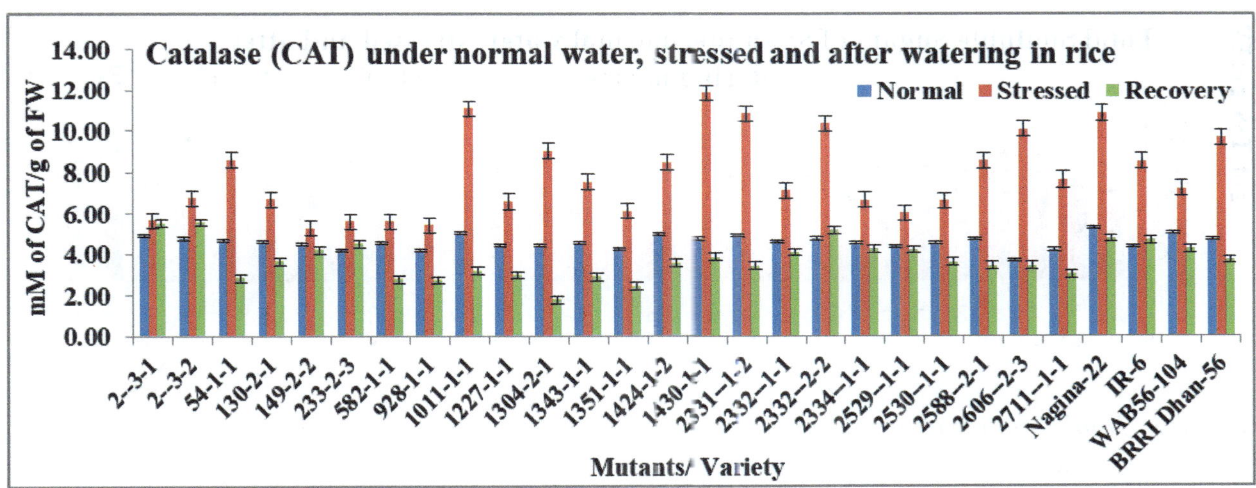

FIG. A–6. Catalase estimation, 1–24 mutant; 24–28 varieties; series 1 is normal series 2 is stressed; and series 3 is after recovery.

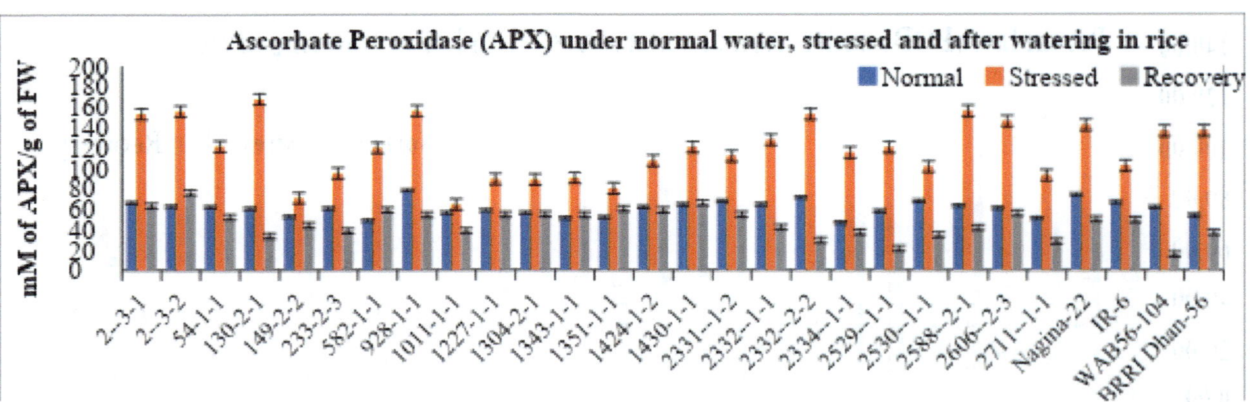

FIG. A–7. Ascorbate peroxidase estimation, 1–24 mutant; 24–28 varieties; series 1 is normal; series 2 is stressed; and series 3 is after recovery.

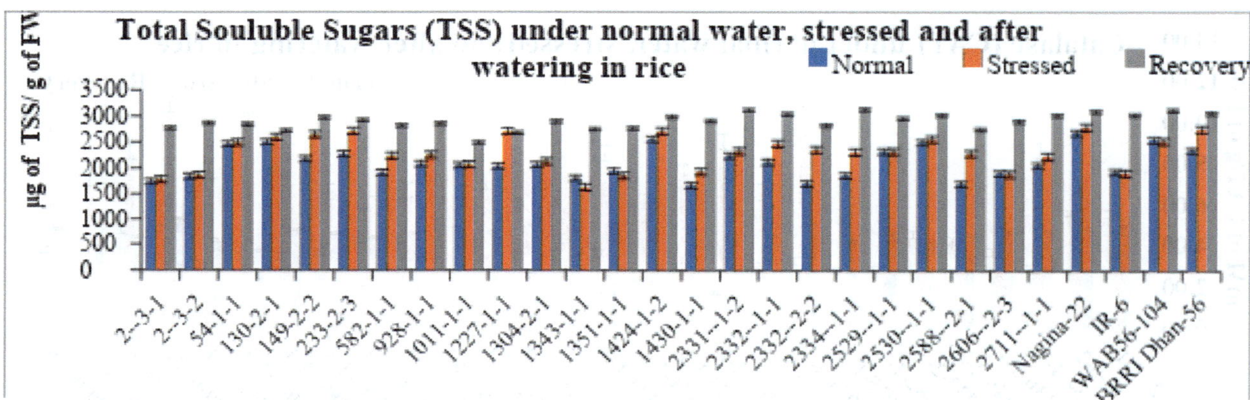

FIG. A–8. Total soluble sugar estimation, 1–24 mutant; 24–28 varieties; series 1 is normal; series 2 is stressed; and series 3 is after recovery.

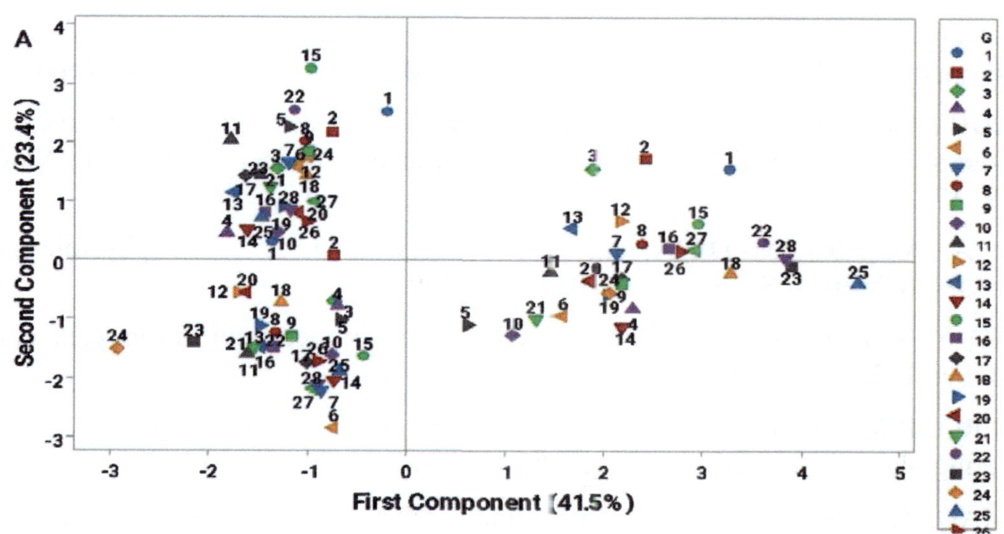

FIG. A–9. Two way PCA display of the rice mutants with standard named as 1=2-3-1, 2=2-3-2, 3=54-1-1, 4=130-2-1, 5=149-2-2, 6=233-2-3, 7=582-1-1, 9=928-1-1, 9=1011-1-1, 10=1227-1-1, 11=1304-2-1, 12=1343-1-1, 13=1351-1-1, 14=1424-1-2, 15=1430-1-1, 16=2331-1-2, 17=2332-1-1, 18=2331-2-2, 19= 2334-1-1, 20=2529-1-1, 21=2530-1-1, 22=2588-2-1, 23=2606-2-3, 24=2711-1-1, 25=Nagina-22, 26=IR-6, 27=WAB56-104, 28=BRRI Dhan56.

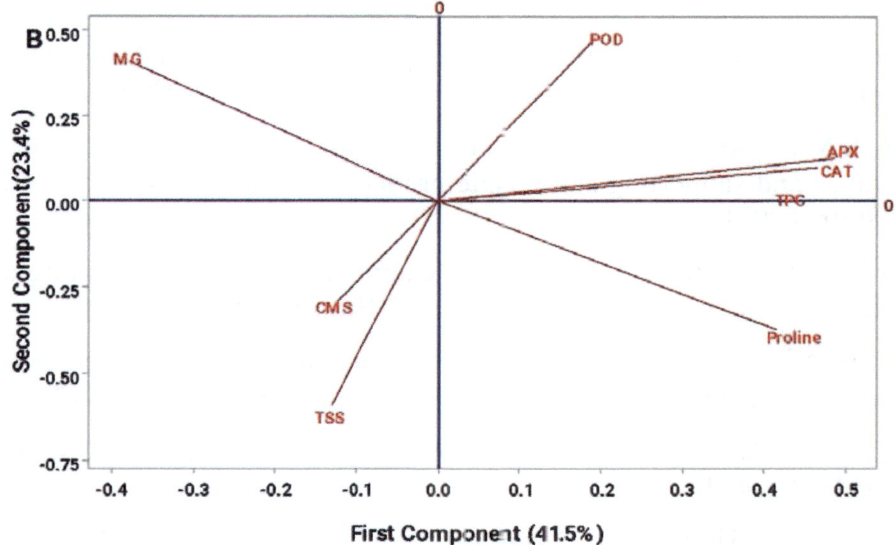

FIG. A–10. Two way PCA impacted by measurements of variables as Catalase (CAT); peroxidase (POD); (APX); total phenolic content (TPC); proline; methylglyoxal (MG); total soluble sugars (TSS) for control and drought stressed plants as parameters reflecting drought tolerance.

REFERENCES TO CHAPTER 3–3

[1]   FAS-UDA report, https://ipad.fas.usda.gov/pdfs/Pakistan/ Pakistan_December2009_MonthlyReport.pdf.

[2]   UPADHYAYA, H., PANDA, S.K., Drought stress responses and its management in rice. In Advances in rice research for abiotic stress tolerance, Woodhead Publishing, Sawston, UK (2019) 177–200.

[3]   GUPTA, A., RICO-MEDINA, A., CAÑO-DELGADO, A.I. The physiology of plant responses to drought, Science **368** (2020) 266–269.

[4]   ZHAO, Y., ABA receptor PYL9 promotes drought resistance and leaf senescence, Proc. Natl Acad. Sci. **113** 7 (2016) 1949–1954.

[5]   RIVERO, R.M., et al., Delayed leaf senescence induces extreme drought tolerance in a flowering plant, Proc. Natl Acad. Sci. **104** 49 (2007) 19631–19636.

[6]   AHMAD, Z., et al., Growth, physiology, and biochemical activities of plant responses with foliar potassium application under drought stress: A review, J. Plant Nutr. **41** 13 (2018) 1734– 1743.

[7]   SUPRASANNA, P., NIKALJE, G.C., RAI, A.N., "Osmolyte accumulation and implications in plant abiotic stress tolerance", Osmolytes and Plants Acclimation to Changing Environment: Emerging Omics Technologies (IQBAL, N., NAZAR, R., A. KHAN, N., Eds), Springer, New Delhi (2016) 1–12. https://doi.org/10.1007/978-81-322-2616-1_1.

[8]   MISHRA, M., WUNGRAMPHA, S., KUMAR, G., SINGLA-PAREEK, S.L., PAREEK, A., How do rice seedlings of landrace Pokkali survive in saline fields after transplantation? Physiol. Biochem. Photosynthesis Res. **150** 1 (2021) 117–135.

[9]   GILL, S.S., TUTEJA, N., Reactive oxygen species and antioxidant machinery in abiotic stress tolerance in crop plants, Plant Physiol. Biochem. **48** 12 (2010) 909–930.

[10]   ZARGAR, S.M., et al., Physiological and multi-omics approaches for explaining drought stress tolerance and supporting sustainable production of rice, Front. Plant Sci. **12** (2022); doi: 10.3389/fpls.2021.803603.

[11]   BARRS, H.D., WEATHERLEY, P.E., A re-examination of the relative turgidity technique for estimating water deficits in leaves, Austral. J. Biol. Sci. **15** 3 (1962) 413–428.

[12]   DIXIT, V., PANDEY, V., SHYAM, R., Differential antioxidative responses to cadmium in roots and leaves of pea (*Pisum sativum* L. cv. Azad), J. Exp. Bot. **52** 358 (2001) 1101–1109.

[13]   MAEHLY, A.C., CHANCE, B., The assay of catalases and peroxidases, Methods Biochem. Anal. **1** (1954) 357–424.

[14]   CHANCE, B., MAEHLY, A.C., Assay of catalase and peroxidase, Methods Enzymol. **2** (1955) 764–775.

[15]   AINSWORTH, E.A., GILLESPIE, K.M., Estimation of total phenolic content and other oxidation substrates in plant tissues using Folin–Ciocalteu reagent, Nature Prot. **2** 4 (2007) 875–877.

[16] BATES, L.S., WALDREN, R.P., TEARE, I.D., Rapid determination of free proline for water-stress studies, Plant Soil **39** 1 (1973) 205–207.

[17] YADAV, S.K., SINGLA-PAREEK, S.L., REDDY, M.K., SOPORY, S.K.,. Transgenic tobacco plants overexpressing glyoxalase enzymes resist an increase in methylglyoxal and maintain higher reduced glutathione levels under salinity stress, FEBS Lett. **579** 27 (2005) 6265–6271.

[18] OSER, B.L., Hawks Physiological Chemistry, McGraw-Hill, New York (1979).

[19] BITA, C., GERATS, T., Plant tolerance to high temperature in a changing environment: Scientific fundamentals and production of heat stress-tolerant crops, Front. Plant Sci. **4** (2013) 273.

[20] LIU, H., ZHAN, J., HUSSAIN, S. NIE, LIXIAO, Grain yield and resource use efficiencies of upland and lowland rice:(Cultivars under aerobic cultivation, Agronomy **9** 10 (2019) 591.

[21] HOQUE, T.S., et al., Methylglyoxal: An emerging signaling molecule in plant abiotic stress responses and tolerance, Front. Plant Sci. **7** (2016) 1341.

[22] EL-SHABRAWI H, et al., Redox homeostasis, antioxidant defense, and methylglyoxal detoxification as markers for salt tolerance in Pokkali rice, Protoplasma **245** (2010) 85–96; doi: 10.1007/s00709-010-0144-6.

[23] HERNANDEZ, J.O., et al., Morpho-anatomical traits and soluble sugar concentration largely explain the responses of three deciduous tree species to progressive water stress, Front. Plant Sci. **12** (2021) 738301-738301.

[24] PARVAIZ, A., SATYAWATI, S., Salt stress and phyto-biochemical responses of plants: A review, Plant Soil Environ. **54** 3 (2008) 89.

[25] XU WEI, et al., Drought stress condition increases root to shoot ratio via alteration of carbohydrate partitioning and enzymatic activity in rice seedlings, Acta Physiol. Plantarum **37** 2 (2015) 9.

[26] DIEN, D.C., MOCHIZUKI, T., YAMAKAWA, T., Effect of various drought stresses and subsequent recovery on proline, total soluble sugar and starch metabolisms in rice (*Oryza sativa* L.) varieties, Plant Prod. Sci. **22** 4 (2019) 530–545.

[27] MOOLPHUERK, N., PATTANAGUL, W., Pretreatment with different molecular weight chitosans encourages drought tolerance in rice (*Oryza sativa* L.) seedling, Not. Bot. Hort. Agrobotanici Cluj-Napoca **48** 4 (2020) 2072–2084.

[28] GUJJAR, R.S., ROYTRAKUL, S., CHUEKONG, W., SUPAIBULWATANA, K., A synthetic cytokinin influences the accumulation of leaf soluble sugars and sugar transporters, and enhances the drought adaptability in rice, Biotech. **11** 8 (2021) 1–14.

[29] SASI, M., et al., Plant growth regulator induced mitigation of oxidative burst helps in the management of drought stress in rice (*Oryza sativa* L.), Environ. Exp. Bot. **185** (2021) 104413.

[30] CHUNTHABUREE, S., SANITCHON, J., PATTANAGUL, W., THEERAKULPISUT, P., Effects of salt stress after late booting stage on yield and antioxidant capacity in pigmented rice grains and alleviation of the salt-induced yield reduction by exogenous spermidine, Plant Prod. Sci. **18** 1 (2015) 32–42.

[31] QUAN, N.T., et al., 2016. Involvement of secondary metabolites in response to drought stress of rice (*Oryza sativa* L.), Agriculture **6** 2 (2016) 23.

[32] MINH, L.T., et al., Effects of salinity stress on growth and phenolics of rice (*Oryza sativa* L.), Int. Lett. Nat. Sci. **57** (2016).

[33] JOGAWAT, A., et al., Crosstalk between phytohormones and secondary metabolites in the drought stress tolerance of crop plants: A review, Physiol. Plant. **172** 2 (2021) 1106–1132.

[34] Ahmed, I.M., et al., Secondary metabolism and antioxidants are involved in the tolerance to drought and salinity, separately and combined, in Tibetan wild barley, Environ. Exp. Bot. **111** (2015) 1–12.

[35] NASRIN, S., SAHA, S., BEGUM, H.H., SAMAD, R.,. Impacts of drought stress on growth, protein, proline, pigment content and antioxidant enzyme activities in rice (*Oryza sativa* L. var. BRRI dhan-24), Dhaka Univ. J. Biol. Sci. **29** 1 (2020) 117–123.

[36] SWAPNA, S., SHYLARAJ, K.S.,. Screening for osmotic stress responses in rice varieties under drought condition, Rice Sci. **24** 5 (2017) 253–263.

[37] LUM, M.S., HANAFI, M.M., RAFII, Y.M., AKMAR, A.S.N., Effect of drought stress on growth, proline and antioxidant enzyme activities of upland rice, J. Animal Plant Sci. **24** 5 (2014).

[38] EDREIRA, J.I.R., MAYER, L.I., OTEGUI, M.E., Heat stress in temperate and tropical maize hybrids: Kernel growth, water relations and assimilate availability for grain filling, Field Crops Res. **166** (2014) 162–172.

[39] FAHAD, S., et al., Crop production under drought and heat stress: Plant responses and management options. Front. Plant Sci. (2017) 1147.

[40] WANG, X., et al., Differential activity of the antioxidant defence system and alterations in the accumulation of osmolyte and reactive oxygen species under drought stress and recovery in rice (*Oryza sativa* L.) tillering, Sci. Rep. **9** 1 (2019) 1–11.

[41] MELANDRI, G., et al., Biomarkers for grain yield stability in rice under drought stress, J. Exp. Bot. **71** 2 (2020) 669–683.

[42] KAMARUDIN, Z.S., et al., Growth performance and antioxidant enzyme activities of advanced mutant rice genotypes under drought stress condition, Agronomy **8** 12 (2018) 279.

[43] HASANUZZAMAN, M., HOSSAIN, M.A., SILVA, J.A., FUJITA, M., "Plant response and tolerance to abiotic oxidative stress: Antioxidant defense is a key factor", Crop Stress and its Management: Perspectives and Strategies, Springer, Dordrecht (2012) 261–315.

[44] HASANUZZAMAN, M., et al., Reactive oxygen species and antioxidant defense in plants under abiotic stress: Revisiting the crucial role of a universal defense regulator, Antioxidants **9** 8 (2020) 681.

[45] RAJPUT, V.D., Recent developments in enzymatic antioxidant defence mechanism in plants with special reference to abiotic stress, Biology **10** 4 (2021) 267.

[46] HOSSEN, M.S., et al., Comparative physiology of Indica and Japonica rice under salinity and drought stress: An intrinsic study on osmotic adjustment, oxidative stress, antioxidant defense and methylglyoxal detoxification, Stresses **2** 2 (2022) 156–178.

[47] NAVEED, M., AHSAN, M., AKRAM, H.M., ASLAM, M., AHMED, N., Measurement of cell membrane thermo-stability and leaf temperature for heat tolerance in maize (*Zea mays* L): Genotypic variability and inheritance pattern, Maydica **61** 2 (2018) 7.

PART 4

MOLECULAR MARKER-BASED SELECTION STRATEGIES FOR DROUGHT RESILIENT MUTANTS

4–1. SELECTION FOR DROUGHT TOLERANCE USING MOLECULAR MARKERS IN GAMMA RAY INDUCED MUTANTS OF RICE

M.M. ISLAM[1], Md.A. HOSSAIN[2,3], M.A. HOSSAIN[3], F. SARSU[4],
MST.S.R. KHANOM[1], A.C. SHARMA[3], U. SAHA[5], S.N. BEGUM[1]

[1]Plant Breeding Division, Bangladesh Institute of Nuclear Agriculture,
BAU Campus, Mymensingh, Bangladesh

[2]Department of Biotechnology and Genetic Engineering,
Noakhali Science and Technology University, Noakhali, Bangladesh

[3]Department of Genetics and Plant Breeding,
Bangladesh Agricultural University, Mymensingh, Bangladesh

[4]Joint FAO/IAEA Centre of Nuclear Techniques in Food and Agriculture,
International Atomic Energy Agency, Vienna

[5]Department of Biotechnology,
Bangladesh Agricultural University, Mymensingh, Bangladesh

Abstract

Drought stress is now posing a serious threat to food security in the rainfed ecosystem of Bangladesh. The main objective of the study was to evaluate the rice mutant lines for drought tolerance using SSR markers. Seventeen rice genotypes, including ten M_5 rice mutants with their three parents (Binadhan-17, Galon and NERICA-4), two tolerant (Binadhan-19 and BRRI dhan71) and two susceptible (Binadhan-11 and IR- 64) check varieties were used for the experiment. DNA extraction of leaf samples was performed by the CTAB method followed by PCR amplification with five SSR markers, i.e. RM279, RM152, RM315, RM234 and RM324. The amplicon size for each marker allele of all 17 genotypes was measured and a total of 42 alleles were detected. Furthermore, genetic diversity values for all SSR markers varied from 0.727 to 0.872 with an average value of 0.799. In addition, the PIC values ranged from 0.860 to 0.702 with an average value of 0.778. The findings of genetic diversity analysis put forward that the SSR markers for NERICA-4/M_5/P-2, Galon/M_5/P-1, Binadhan-17/M_5/P-3, Binadhan-17/M_5/F-4 and Binadhan- 17/M_5/P-5 mutants were considered to be drought tolerant as tolerant checks from the UPGMA dendrogram. Based on the molecular results, the mutants could be utilized for exploring the genetic variability for developing better drought tolerant rice varieties.

Key words: Rice, mutation breeding, drought tolerance, SSR markers.

1. INTRODUCTION

Rice ensures food security for more than half of the world's population and has grown in more than hundred countries, with 90% of the total global production from Asia [1, 2]. Plant growth as well as productivity of rice is significantly affected by numerous biotic and abiotic stresses as reported by Dixit et al. [3]. With the changing climate pattern, drought is becoming one of the significant constraints among other stresses which would affect rice production severely, especially in the rainfed ecosystem. Drought is defined as water stress, mainly due to lack of rain during the crop growing period. Shortage of water is the main obstacle for rice production in rainfed ecosystems since most of the rice varieties are susceptible to water stress [4]. About 1 000 000 ha of land in

Bangladesh is affected due to drought stress, particularly in Barind Tract and other northern districts [5]. For the drought-prone areas, development of drought tolerant rice varieties is needed to maintain food security. Several tools have been developed to enhance plant breeding through advances in molecular genetics and biotechnology. Researchers used SSR markers at the molecular level to find out the genetic relationship of drought tolerance in rice [6–8], enabling precise classification of the studied genotypes. The development of SSR markers and their application to the genetic dissection of agronomically significant characteristics has shown to be an effective method [9]. Traditional breeding approaches for improving rice drought tolerance are delayed because of regional variances and seasonal fluctuations in drought timing and intensity, along with the complexity of drought tolerance characteristics and the difficulties in selecting trait combinations [10]. Diversity analysis based on phenotypic traits may not provide an accurate assessment of genetic variations because the features are impacted by environmental influences [11]. As a result, SSR markers have been widely used to determine genetic divergence among genotypes, since it is unaffected by environmental influences. Use of SSR markers to identify accessions with genes and genomic areas that influence target attributes can accelerate efforts in breeding drought tolerant rice due to the fact that SSR markers are transmitted from generation to generation [8]. The genetic improvement of drought tolerance is a big challenge because it is a complex trait controlled by several quantitative trait loci (QTLs) [12]. The association of QTLs controlling multiple drought tolerance features in rice chromosomes has been reported using SSR markers [13]. The multi-allelic nature of SSR markers, as well as their high polymorphism, reliability, reproducibility, cost effectiveness, mono-locus and ease of analysis make it possible to establish a link between individuals with a small number of markers [10, 14]. SSR markers are also valuable for precisely selecting complicated breeding characteristics, reducing labour costs and allowing the assembling of many desired features into a single cultivar [15]. SSR markers also successfully utilize mutation breeding of rice for drought tolerance [16].

For developing improved drought tolerant rice varieties with high yield and good qualitative characters, knowledge of genetic diversity and relationship among the genotypes of the crop plays a significant role in plant breeding. The purpose of this experiment was to identify the SSR markers associated with drought tolerance in rice mutants. Using SSR markers to find QTLs influencing drought tolerance related characteristics has the potential to speed up breeding programmes that can help to alleviate the problem of food security. The results of this study are likely to help future evaluation and yield trials for drought tolerance in rice production under current climate change circumstances.

2.   METHODOLOGY

2.1.   Plant materials

The experiment was conducted at the molecular laboratory of the Plant Breeding Division of the Bangladesh Institute of Nuclear Agriculture (BINA) from October 2020 to December 2020. The experiment was carried out by using 17 rice genotypes, including ten promising M_5 generation rice mutants with their parents and checking varieties at the seedling stage (Table 1). Of these, two drought tolerant varieties and two drought susceptible varieties were considered as positive and negative checks.

TABLE 1. LIST OF 17 RICE GENOTYPES USED FOR THE EXPERIMENT

Name of the genotypes	Types	Origin
Binadhan-17/M$_5$/P-3	Mutant	BINA
Binadhan-17/M$_5$/P-4	Mutant	BINA
Binadhan-17/M$_5$/P-5	Mutant	BINA
NERICA-4/M$_5$/P-2	Mutant	BINA
NERICA-4/M$_5$/P-3	Mutant	BINA
NERICA-4/M$_5$/P-5	Mutant	BINA
NERICA-4/M$_5$/P-6	Mutant	BINA
Galon/M$_5$/P-1	Mutant	BINA
Galon/M$_5$/P-3	Mutant	BINA
Galon/M$_5$/P-6	Mutant	BINA
Binadhan-17	Parent (P) – high yielding	BINA
NERICA-4	Parent (P) – drought tolerant	Africa
Galon	Parent (P) – drought tolerant landrace	Bangladesh
Binadhan-11	Drought susceptible check	BINA
IR-64	Drought susceptible check	IRRI
BRRI dhan71	Drought tolerant check	BRRI
Binadhan-19	Drought tolerant check	BINA

IRRI: International Rice Research Institute; BRRI Bangladesh Rice Research Institute.

2.2. Methods

Fresh, robust leaves from 25 day old seedlings were used for DNA extraction as per the CTAB method. Twenty SSR markers associated with drought tolerance in rice were used to survey the parents, Binadhan-17, NERICA-4 and Galon, for polymorphism. Five markers were found to be polymorphic and these were used for genotyping of the mutants. PCR analysis was performed in a 10 µL reaction sample containing 2 µL of template DNA (50 ng/µL), 5 µL of master mix, 2 µL of nuclear free water, 1 µL each of 10µM forward and reverse primers using a thermal cycler with a single 96-well. After initial denaturation for 4 min at 94°C, each cycle comprised 45 s denaturation at 94°C, 45 s annealing at 55°C and extension for 2 min at 72°C with a final extension for 10 min at 72°C at the end of 35 cycles and then incubated at 4°C. Then the PCR products were run for separating electrophoretically on an 8.0% polyacrylamide gel, which was placed in an electrophoresis tank containing 1.0X TBE buffer. Ethidium bromide (0.5 µg mL^{-1}) was used to soak the gel for 20 min and then destained with distilled water for a few minutes with gentle shaking. A photograph was taken of the resolved DNA bands using a gel documentation system. The size of the amplified DNA fragments was determined by comparing the migration distance of the molecular weight of the 100 bp DNA Ladder used.

2.3.   Statistical analysis

Software was used to determine the molecular weight of each SSR marker allele and to analyse alleles of the markers. This software also was used to evaluate summary data such as the number of alleles per locus, the main allele frequency, gene diversity and polymorphism information content (PIC) values. The Nei's genetic distance coefficient was used to form a dendrogram representing the genetic relationships between genotypes based on the unweighted pair group method with arithmetic mean (UPGMA),.

3.   RESULTS

Seventeen rice genotypes, including ten promising M_5 generation rice mutants with their parents and check varieties were successfully amplified with the five microsatellite markers where primer pairs were referred to as loci and DNA bands as alleles (Table 2) (Figs 1 and 2). The amplicon size for each marker allele of all 17 genotypes was measured and a total of 42 alleles were detected. The number of alleles ranged from 7 to 11 per locus, with an average of 8.4 alleles per loci. The highest number of alleles (11) was found for RM279 followed by RM152 (10), RM315 (8), RM234 (7), and RM324 produced the lowest number of alleles (6). The gene diversity value ranged from 0.727 (RM234) to 0.872 (RM152 and RM279). The highest PIC value was found for RM279 (0.860) followed by RM152 (0.859), RM324 (0.757).

TABLE 2. ALLELE NUMBER, ALLELE FREQUENCY, GENETIC DIVERSITY AND PIC VALUE OF 17 RICE GENOTYPES FOR FIVE SSR MARKERS.

Locus name	No. of alleles	Allele frequency (%)	Gene diversity	PIC
RM152	10	0.235	0.872	0.859
RM234	7	0.471	0.727	0.702
RM279	11	0.235	0.872	0.860
RM324	6	0.294	0.789	0.757
RM315	8	0.471	0.734	0.712
Mean	8.4	0.341	0.799	0.778

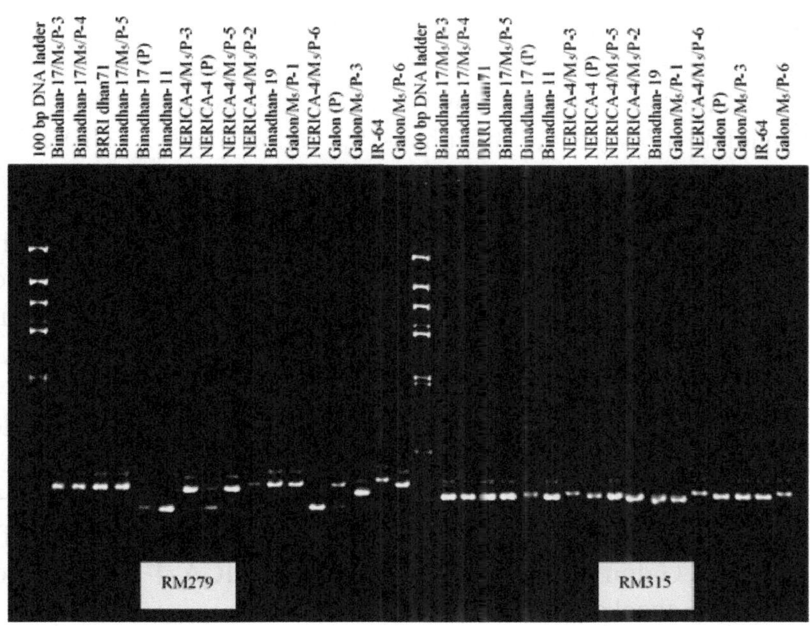

FIG. 1. DNA profile of 17 rice genotypes with RM152 and RM234 markers.

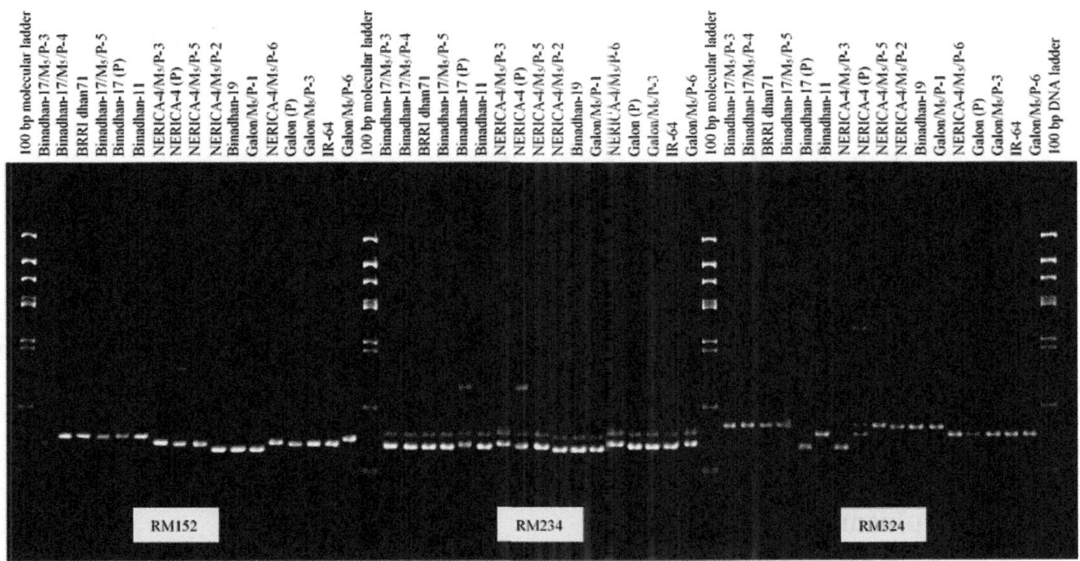

FIG. 2. DNA profile of 17 rice genotypes with RM279, RM324 and RM315 markers.

3.1. Genetic distance based on UPGMA dendrogram

Among the mutants, less genetic distance based on pairwise comparisons was observed for Binadhan-17/M₅/P-3, Binadhan-17/M₅/P-4 and Binadhan-17/M₅/P-5, which were genetically more similar mutants to BRRI dhan71, but more dissimilar to Binadhan-19 and Binadhan-17, which were developed from mutation of parent Binadhan-17. The lowest genetic distance was found in Binadhan-17 (P) with BRRI dhan71, Galon/M₅/P-5, NERICA-4 (P) and NERICA-4/M₅/P-6, whereas more dissimilarity was found in Binadhan-11 with Binadhan-17 (P), Binadhan-19, Galon/M$_5$/P-1, NERICA-4/M$_5$/P-2, NERICA-4/M$_5$/P-3 and NERICA-4/M$_5$/P-6. Besides, IR-64 was mostly similar to Galon/M$_5$/P-3 followed by Galon (P). The greatest dissimilarity of NERICA-4 (P) was with Binadhan-17/M$_5$/P-4, Binadhan-19, Galon/M$_5$/P-1, NERICA-4/M$_5$/P-2 and NERICA-4/M$_5$/P-3.

Cluster analysis was performed using the UPGMA (unweighted pair group method with arithmetic mean) method to group the studied genotypes based on similarity coefficient. All the rice genotypes (mutants, parents and checks) were separated into five main distinct clusters/groups (I, II, III, IV and V). Drought tolerant mutants and checks were divided into four clusters (II, III and V) and cluster IV had susceptible and moderately tolerant genotypes (Fig. 3). The most diverse genotypes in cluster V were Galon/M$_5$/P-6 and NERICA-4/M$_5$/P-6 with Binadhan-19, followed by BRRI dhan71. Cluster II had two mutants, Galon/M$_5$/P-1 and NERICA-4/M$_5$/P-2, which are tolerant for drought stress due to similarity with Binadhan-19. Cluster III had all mutants of Binadhan-17 (P) similarity with BRRI dhan71 and dissimilar relationship with the genotypes of Cluster IV. In Cluster IV, Galon/M$_5$/P-3, Galon (P), NERICA-4 (P), NERICA-4/M$_5$/P-5 showed a similar relationship with susceptible checks IR-64 and Binadhan-11.

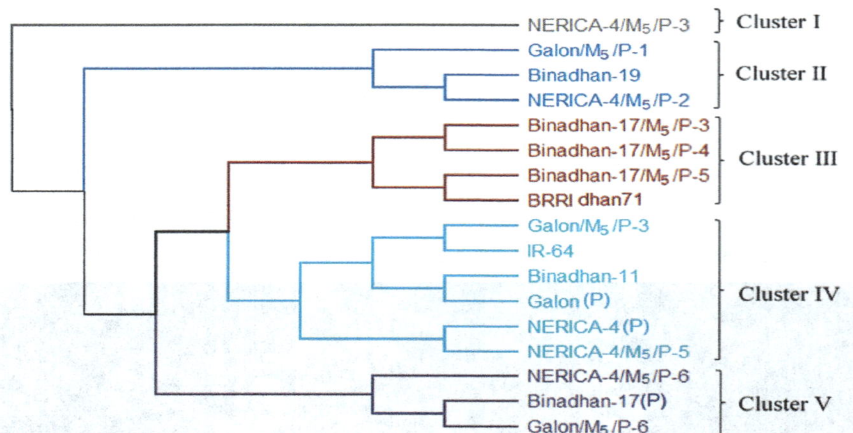

FIG. 3. Dendrogram showing the genetic relationship between 17 rice genotypes based on the alleles detected by five SSR markers clustered using the UPGMA using Nei's genetic distance.

4. DISCUSSION

4.1. SSR polymorphism analysis

Identifying genetic relationships and the divergence of genetic resources is an important step in genotype selection with desired traits. It contributes to reducing the use of closely related genotypes in breeding operations, which would otherwise result in genetic depression and less genetic variation [17]. According to Anupam et al. [18], genetic diversity analysis solely based on phenotypic traits may not be a reliable measure of genetic differences as they are influenced by environmental factors. Genetic diversity among the genotypes is not only to distinguish the genotypes by phylogenetic tree, it also indicates a chance of finding new and useful genes, as the genotypes with most distinct DNA profiles are likely to contain a greater number of novel alleles [19].

The current study was thus carried out to determine genetic diversity and relationship among the chosen rice genotypes in order to find the presence of markers in relation to similar banding patterns with checks. Genetic diversity is the most widely used parameter for assessing genetic variation within the genotypes [20]. The assessment of genetic diversity of genotypes is one of the potential approaches for variety development [21]. Because of their capacity to detect a large number of distinct alleles correctly and efficiently, SSRs are regarded as excellent for assessing genetic diversity, fingerprinting for varietal identification, and assessing seed purity [22, 23]. Rice genotypes with high levels of genetic diversity discovered in this study might be useful resources for widening the genetic base and attaining quick advancements in rice breeding. To evaluate the genetic relatedness among the selected 17 rice genotypes, five polymorphic SSR markers were used related to drought stress tolerance.

Changes in the number of alleles per locus may have been related to differences in rice genotypes, diversity of SSR markers used, and differences in methods of detection and evaluation [24]. According to Jain et al. [25], the number of alleles studied varied from 7 to 11, with a mean value of 8.4 per locus. The results indicated that there is favourable allelic diversity, which is necessary for assessing genetic diversity. The use of different genotypes might explain the variation in the number of alleles discovered per locus [24, 25]. Like the findings of Wang et al. [26], the mean expected gene diversity in our study was 0.799, which was also comparable in a microsatellite based study in rice. It could be attributed to the high rate of exchange of genetic materials by the recorded higher mean gene diversity in the present study among the genotypes [10]. It was reported that the number of alleles per locus ranged from 2 to 8 with average 3.8 using BRRI released varieties [27]. In the crop development programme, genetic relationship analysis among different genotypes is an essential component and plays a key role in their useful utilization [8].

To discriminate genotypes depending on their genetic links, PIC values represent a marker's relative allelic polymorphism [24]. The fact that the chosen microsatellites were highly polymorphic with a mean PIC value shows that they were very informative in differentiating the studied genotypes. The PIC value is an excellent indication of a marker's effectiveness for linkage analysis since it represents the chance of the marker being identified in the genotypes [28]. It also reflects genotypic allelic diversity. Average PIC values observed in this study (0.778) were comparable to 0.8912 reported by Yesmin et al. [29]. The differences in PIC values may be linked to the selection of different markers and the diversity of the studied genotypes. The PIC values varied among loci and ranged from 0.702 to 0.860, with an average of 0.778; comparable to previous estimates of microsatellite analysis in rice i.e. 0.76–0.95 with an average of 0.855 [30], and 0.239–0.765 with an average of 0.508 [31]. Ngangkham et al. [32] revealed that a PIC value of 0.5 or above for a microsatellite marker is deemed highly informative, confirming that SSR markers are employed for genetic investigations and identifying the polymorphism rate of a marker at a given locus.

4.2. Cluster analysis based on Nei's genetic distance

The clustering pattern in the present study indicated the existence of variability among the rice genotypes. Vanlalsanga et al. [33] used SSR markers to categorize 50 rice genotypes into three groups. It can be revealed that Binadhan-17/M$_5$/P-3, Binadhan-17/M$_5$/P-4, Binadhan-17/M$_5$/P-5, Galon/M$_5$/P-1, NERICA-4/M$_5$/P-2 and NERICA-4/M$_5$/P-5 were tolerant on the basis of their genotyping compared with their checks. To a considerable extent, groupings generated from genetic diversity and UPGMA clustering were incongruent, showing actual genetic differences among the genotypes under investigation at the DNA level and the perfect nature of the clustered genotypes. The findings are also useful for molecular fingerprinting and more effective genotype selection for crossing and improvement of rice breeding techniques for drought tolerance, which will help to sustain genetic development.

5. CONCLUSIONS

It is important to study the genetic diversity of the rice mutants in comparison to their parents and check varieties. This will not only provide information on their phylogenetic relationship, but will also indicate a chance of finding new and useful genes, as the accessions with more distinct DNA profiles are likely to contain a greater number of novel alleles. The findings of genetic diversity analysis revealed that the SSR markers for NERICA-4/M_5/P-2, Galon/M_5/P-1, Binadhan-17/M_5/P-3, Binadhan-17/M_5/P-4 and Binadhan-17/M_5/P-5 rice mutants appeared to be drought tolerant like the check (BRRI dhan71, NERICA-4 Binadhan-19) genotypes from the UPGMA dendrogram. The findings are useful for molecular fingerprinting and more effective genotype selection for future field trials as well as improving rice breeding strategies for drought tolerance, which will maintain the steady state of genetic improvement.

REFERENCES TO CHAPTER 4–1

[1] FUKAGAWA, N.K., ZISKA, L.H., Rice: Importance for global nutrition, J. Nutr. Sci. Vitamin. **65** (2019) S2–S3.

[2] KRUPA, K.N., SHASHIDHAR, H.E., DALAWAI, N., REDDY, M., SWAMY, H.V., Molecular marker based genetic diversity analysis in rice genotypes (*Oryza sativa* L.) using SSR markers, Int. J. Pure Appl. Biosci. **5** (2017) 668–674.

[3] DIXIT, S., et al., Marker assisted forward breeding to combine multiple biotic-abiotic stress resistance/tolerance in rice, Rice **13** 1 (2020) 1–15.

[4] MOSTAJERAN, A., RAHIMI-EICHI, V., Effects of drought stress on growth and yield of rice (*Oryza sativa* L.) cultivars and accumulation of proline and soluble sugars in sheath and blades of their different age leaves, Am.-Eur. J. Agric. Environ. Sci. **5** 2 (2009) 264–272.

[5] KADER, M.A., et al., Development of drought tolerant rice variety BRRI dhan66 for the rainfed lowland ecosystem of Bangladesh, Bangladesh Rice J. **23** 1 (2019) 45–55.

[6] MASUD, M.A., et al., Screening for drought tolerance and diversity analysis of Bangladeshi rice germplasms using morphophysiology and molecular markers, Biologia **77** 1 (2021) 21–37.

[7] DONDE, R., et al., Assessment of genetic diversity of drought tolerant and susceptible rice genotypes using microsatellite markers, Rice Sci. **26** 4 (2019) 239–247.

[8] VERMA, H., BORAH, J.L., SARMA, R.N., Variability assessment for root and drought tolerance traits and genetic diversity analysis of rice germplasm using SSR markers, Sci. Rep. **9** 1 (2019) 1–19.

[9] MIAH, G., et al., A review of microsatellite markers and their applications in rice breeding programs to improve blast disease resistance, Int. J. Mol. Sci. **14** 11 (2013) 22499–22528.

[10] GABALLAH, M.M., et al., Identification of genetic diversity among some promising lines of rice under drought stress using SSR markers, J. Taibah Univ. Sci. **15** 1 (2021) 468–478.

[11] ANUPAM, A., et al., Genetic diversity analysis of rice germplasm in Tripura State of Northeast India using drought and blast linked markers, Rice Sci. **24** 1 (2017) 10–20.

[12] SAHEBI, M., et al., Improvement of drought tolerance in rice (*Oryza sativa* L.): Genetics, genomic tools, and the WRKY gene family, BioMed Res. Int. (2018) 1–20.

[13] LANG, N.T., NHA, C.T., HA, P.T.T., BUU, B C., Quantitative trait loci (QTLs) associated with drought tolerance in rice (*Oryza sativa* L.), SABRAO J. Breeding Genetics **45** 3 (2013) 409–421.

[14] SUVI, W.T., SHIMELIS, H., LAING, M., MATHEW, I., SHAYANOWAKO, A.I.T., Assessment of the genetic diversity and population structure of rice genotypes using SSR markers, Acta Agric. Scand. Section B. Soil Plant Sci. **70** 1 (2019) 76–86.

[15] COLLARD, B.C., MACKILL, D.J., Marker-assisted selection: An approach for precision plant breeding in the twenty-first century, Phil. Trans. R. Soc. B: Biol. Sci. **36** 3 (2008) 557–572.

[16] AHMADIKHAH, A., MARUFINIA, A., Effect of reduced plant height on drought tolerance in rice, Biotech **6** 2 (2016) 1–9.

[17] TABKHKAR, N., RABIEI, B., LAHIJI, H.S., CHALESHTORI, M.H., Genetic variation and association analysis of the SSR markers linked to the major drought-yield QTLs of rice, Biochem. Genetics **56** 4 (2018) 356–374.

[18] ANUPAM, A., et al., Genetic diversity analysis of rice germplasm in Tripura State of Northeast India using drought and blast linked markers, Rice Sci. **24** 1 (2017) 10–20.

[19] MESSMER, M.M., MELCHINGER, A.E., BOPPENMAIER, J., BRUNKLAUS-JUNG, E., HERRMANN, R.G., Relationships among early European maize inbreds: I. Genetic diversity among flint and dent lines revealed by RFLPs, Crop Sci. **32** 6 (1992) 1301–1309.

[20] NEI, M., TAJIMA, F., TATENO, Y., Accuracy of estimated phylogenetic trees from molecular data, J. Mol. Evol. **19** 2 (1983) 153–170.

[21] NACHIMUTHU, V.V., et al., Analysis of population structure and genetic diversity in rice germplasm using SSR markers: An initiative towards association mapping of agronomic traits in *Oryza sativa,* Rice **8** 1 (2015) 1–25.

[22] GANIE, S.A., et al., Assessment of genetic diversity of *Saltol* QTL among the rice (*Oryza sativa* L.) genotypes, Physiol. Mol. Biol. Plants **22** 1 (2016) 107–114.

[23] ALJUMAILI, S.J., et al., Genetic diversity of aromatic rice germplasm revealed by SSR markers, Biomed. Res. Int. (2018) 1–11.

[24] PATHAICHINDACHOTE, W., PANYAWUT, N., SIKAEWTUNG, K., PATARAPUWADOL, S., MUANGPROM, A., Genetic diversity and allelic frequency of selected Thai and exotic rice germplasm using SSR markers, Rice Sci. **26** 6 (2019) 393– 403.

[25] ASHRAF, H., et al., SSR based genetic diversity of pigmented and aromatic rice (*Oryza sativa* L.) genotypes of the western Himalayan region of India, Physiol. Mol. Biol. Plants **22** 4 (2016) 547–555.

[26] WANG, C.H., et al., Genetic diversity and classification of *Oryza sativa* with emphasis on Chinese rice germplasm, Heredity **112** 5 (2014) 489–496.

[27] ALI, M.S., et al., "Genetic diversity of BRRI released rice varieties using microsatellite markers", Proc. 3rd Int. Rice Congr. 2010, poster presentation abstract, p. 46.

[28] ISLAM, M.Z., et al., Variability assessment of aromatic rice germplasm by phyno-genetic traits and population structure analysis, Sci. Rep. **8** 1 (2018) 1–14.

[29] YESMIN, N., et al., Unique genotypic differences discovered among indigenous Bangladeshi rice landraces, Int. J. Genomics (2014) 1–11.

[30] AFIUKWA, C.A.A., et al., Screening of some rice varieties and landraces cultivated in Nigeria for drought tolerance based on phenotypic traits and their association with SSR polymorphisms, Afr. J. Agric. Res. **11** 29 (2016) 2599–2615.

[31] HOSSAIN, M.M., et al., Genetic diversity analysis of aromatic landraces of rice (*Oryza sativa* L.) by microsatellite markers, Genes, Genomes Genomics **6** (2012) 42–47.

[32] NGANGKHAM, U., et al., The potentiality of rice microsatellite markers in assessment of cross-species transferability and genetic diversity of rice and its wild relatives, Biotech **9** 6 (2019) 1–19.

[33] VANLALSANGA, SINGH, S.P., SINGH, Y.T , Rice of Northeast India harbor rich genetic diversity as measured by SSR markers and Zn/Fe content, BMC Genetics **20** (2019) 1–13.

4–2. CHARACTERIZATION AND GENE MAPPING OF A DROUGHT TOLERANCE AND DECREASED TILLERING MUTANT IN RICE *Oryza Sativa* L.

L. REN[1], J. ZHANG[1], T. LAN[1], T. LAN[1, 2,]

[1]Key Laboratory of Genetics, Breeding and Multiple Utilization of Crops,
Ministry of Education, Fujian Agriculture and Forestry University, Fuzhou, China

[2]Fujian Provincial Key Laboratory of Crop Breeding by Design,
Fujian Agriculture and Forestry University, Fuzhou China

Abstract

The characteristics and gene mapping of drought tolerance and decreased tillering mutants in rice lay a foundation for the study of the mechanism of rice drought tolerance, tiller development and morphological changes to adapt to arid environments. Drought tolerance identification, physiological indexes and related gene expression analysis, leaf microscopic observation and agronomic character investigation of rice drought tolerance and decreased tillering mutant *dtdt* were carried out, and wild type Huanghuazhan was used as control. Based on the *dtdt* × Zhonghua11 hybrid F_2 population, genetic analysis was performed, and mutants were mapped using BSA-seq and molecular marker methods. After drought stress and rehydration, the survival rate of the mutant seedlings was 100% and that of the wild type seedlings was 6%, and the mutant seedlings were shorter than the wild type seedlings. Using 20% PEG6000 to simulate drought stress, the mutants showed increased DHAR activity and CAT activity, and decreased H_2O_2 content and MDA content compared with the wild type. Microscopic observation of leaves showed that compared with the wild type, the stomatal density and volume of the mutant were reduced, but there was no significant difference in stomatal opening. There was no significant difference in the stomatal opening of the mutants before and after stress, but the number of closed stomata in the wild type increased significantly after stress. The investigation of agronomic characters showed that compared with the wild type, the tiller number and plant height of the mutant were significantly reduced. Genetic analysis showed that the mutant trait of *dtdt* is controlled by a pair of recessive genes. The mutant gene was mapped between the two markers RM19410 and ID3341597 with a physical distance of 282.5 kb. The mutant *dtdt* exhibited the characteristics of drought tolerance, reduced tillering, and its drought tolerance may be regulated by stomatal density reduction, stomatal volume reduction and ROS pathway. *DTDT* may be a new gene regulating tillering in rice.

Key words: rice; drought tolerance; tillering; stomata; gene mapping.

1. INTRODUCTION

At present, the shortage water is a global problem that restricts agricultural production. Up to 43% of China's area is arid or semi-arid, and there is a great imbalance in time and space distribution, which makes the contradiction between supply and demand of water resources in China more acute and is one of the biggest crises faced by China's agricultural production [1]. Rice is an important food crop, with more than half of the world's population relying on it as the main food source [2]. Developing drought resistant varieties not only can save water resources, but is also conducive to stable production, can save energy and reduce environmental pollution. Thus, the drought resistance capability of rice is becoming more and more important [3, 4], among them, research on rice drought resistance gene excavation and function, to understand the mechanism of rice drought tolerance.

Plants respond to drought through multiple physiological and biochemical regulatory mechanisms, such as different signal transduction and osmotic pressure regulation [4]. Stomata are the main channel for gas exchange between plant leaves and the outside world, and play an important role in regulating plant photosynthesis, respiration and transpiration [5]. When plants are stressed by adversity, reactive oxygen species (ROS) molecules will be produced, which continuously accumulate beyond the normal level, causing membrane lipid peroxidation and malondialdehyde (MDA) generation, damaging cell structure and causing irreversible damage [6, 7]. ROS scavenging system includes antioxidant protective enzymes, such as superoxide dismutase (SOD), peroxidase (POD), catalase (CAT) and non-enzyme substances such as reduced glutathione (GSH) and ascorbic acid (AsA). In order to maintain the homeostasis of plants, ROS scavenging systems will continuously remove ROS generated by various metabolic reactions in coordination with each other to ensure the healthy growth and development of plants [8, 9]. At present, some rice drought tolerance genes have been cloned by mutants and QTL mapping. *DST* and *OsSRO1* regulate stomatal closure and increase drought tolerance of rice by regulating the accumulation of H_2O_2 in rice [10, 11]. *DSM1* encodes a mitogen-activated protein kinase that regulates the response to drought stress by scavenging ROS in plants. *DS8* encodes a NAP1 protein, which plays an important role in actin filament activity and affects sensitivity to drought through ABA-mediated stomatal closure [12]. DHS is a key control factor for the synthesis of epidermal wax in rice, and DHS negatively regulates the biosynthesis of epidermal wax by promoting ROC4 degradation to improve drought tolerance [13]. *DRO1* improves drought resistance of rice by controlling root growth angle and changing root structure [14].

The Tiller number has a significant impact on rice yield [15]. Studies have shown that tillering production and later development are regulated by a complex network of genetic, hormonal and environmental factors. To further improve the potential of rice production, the International Rice Research Institute (IRRI) proposed the concept of cultivating ideal plant type (IPA), which is characterized by fewer ineffective tillers, more grains per panicle, and thick and strong stems [16]. IPA1 is a transcription factor *OsSPL14* containing the SBP domain, and its function is regulated by *OsmiR156*. The *OsSPL14* mutation has relieved OsmiR156's inhibition on IPA1, thus producing phenotypes such as reduced tillering, lodging resistance and high yield, which is known as the ideal plant type [17]. Rice *MOC1* gene positively regulates the formation of tiller buds, and *moc1*mutants show a single-rod phenotype [18]. *MOC2* encodes a fructose-1, 6-bisphosphatase, whose mutation inhibits the growth of tiller buds [19]. After the mutation of *MOC3*, the formation of tillering buds was also blocked [20]. A series of rice mutants *D3, D10, D14, D17, D27* and *D53* all showed short stems and multiple tillers, which were mostly related to strigolactone [21–26].

In this study, a drought tolerant and decreased tillering mutant (*dtdt*) was selected from a population of radiation induced progenies of *indica* rice cultivar Huanghua Zhan (HHZ). Its agronomic characters, leaf microstructure, drought resistant physiological indexes and related gene expression quantity characteristics were analysed, based on *dtdt* × ZH11 hybrid building population of F₂ generation, genetic analysis. At the same time, the method of bulked segregant analysis and sequencing (BSA-seq) and molecular marker method were used, with gene mapping, for further research on the function of the mutant *dtdt*.

2. METHODOLOGY

The *indica* rice variety Huanghuazhan (HHZ), the drought tolerance and decreased tillering mutant *dtdt* derived from the radiation induced HHZ, and the *japonica* rice variety Zhonghua 11 (ZH11) were used. Wild type HHZ and mutant *dtdt* were simultaneously sown in paddy soil in rectangular plastic plates (36 cm×28 cm×4.5 cm), and drought stress was carried out at the two-leaf stage for 14–16 days. After re-watering (1.5 L/plate), the survival rate of seedlings was calculated.

2.1. Physiological index measurement

Wild type HHZ and mutant *dtdt* were seeded in 96-well hydroponic boxes using kimura B formula and replaced every seven days. The seedlings were treated with 20% PEG6000 when they grew to the third-leaf stage, and the H_2O_2 content of seedlings on day 0 and day 3 of treatment was measured. Three biological replicates were performed, and shoots of five seedlings were taken from each replicate. At the same time, nitrogen blue tetrazole (NBT) staining analysis was performed on seedlings at day 0 and day 3 of treatment. Five biological replicates were selected, and a single third leaf was taken from each replicate.

Wild type HHZ and mutant *dtdt* were seeded in 96-well hydroponic boxes using kimura B formula and replaced every seven days. The seedlings were treated with 20% PEG6000 when they were grown to the third-leaf stage. CAT activity (UV absorption method), DHAR activity and MDA content (thiobarbituric acid method) of 0–5 day seedlings were measured in three biological replicates, and shoots of five seedlings were taken from each replicate.

2.2. Expression analysis of *CAT* and *DHAR* related genes

Wild type HHZ and mutant *dtdt* were seeded in 96-well hydroponic boxes using kimura B formula and replaced every seven days. When they grew to the third-leaf stage, the shoots were taken, frozen with liquid nitrogen and stored in an ultra-low temperature refrigerator. At the same time, the shoots were treated with 20% PEG6000 for 24 h and frozen with liquid nitrogen. RNA was extracted using a plant RNA small amount extraction kit, a reverse transcription kit and a real-time fluorescence quantitative PCR kit. The *Actin* gene was used as an internal reference to calculate the relative gene expression level. Each sample has three biological replicates, and each biological replicate includes three technical replicates. Quantitative primers for selected genes are shown in Table 1.

TABLE 1. QRT-PCR PRIMER SEQUENCES OF CAT AND DHAR RELATED GENES

Primer name	Forward primer（5'-3'）	Reverse primer（5'-3'）
Actin	TGGCATCTCTCAGCACATTCC	TGCACAATGGATGGGTCAGA
OsCATA	AGGAGGCAGAAGGCGACGATACA	TCTTCACATGCTTGGCTTCACGTT
OsCATB	GGCTGTCGGGAAAAGTGTGTCATTG	TTTCAGGTTGAGACGTGAAGCCAGC
OsCATC	TCAAGAGATGGATCGACGCACTCTC	GAAGCAGATTGCAACGCTGATCG
OsDHAR1	ATGGGCGTGGAGGTGTGCGTCAAGG	CCTTGCTCTTCAAGAACGTTGTGAAGC

2.3. Microscopic observation of leaves

Wild type HHZ and mutant *dtdt* were seeded in 96-well hydroponic boxes using kimura B formula and replaced every seven days. The seedlings were treated with 20% PEG6000 when they grew to the third-leaf stage. The middle part of the third leaf, about 0.5 cm, was taken after 0 and 3 days of treatment, respectively, fixed by glutaraldehyde, washed by buffer solution, ethanol gradient dehydration, tert-butanol replacement, and freeze dried. The samples were observed and photographed using a scanning electron microscope. Stomatal number and stomatal opening and closing characteristics were observed in nine different visual fields.

2.4. Investigation and statistical analysis of agronomic traits

Wild type HHZ and mutant *dtdt* were planted in the university experimental field in summer of 2021, and the fertilizer, water and pesticide management were the same as that of the general field. In the mature stage of rice, plant height, tiller number, effective panicle number, panicle length, full grain number and seed-setting rate were investigated, and the data were analysed statistically.

2.5. Genetic analysis and gene mapping

F_1 generations were obtained by crossing *dtdt* as female parent and ZH11 as male parent. The F_1 generation was self-crossbred to obtain the F_2 generation, and an F_2 small population was sown in the university experimental field. After sowing, the phenotype of the population was tracked and observed. About 60 days later, phenotype identification was carried out on the F_2 generation population, and the numbers of wild type and mutant of the F_2 population were counted, and the data were statistically analysed.

A large F_2 population (4000 seeds) was sown in the university experimental field. About 60 days after sowing, 193 individual plants consistent with the mutant phenotype were selected from the F_2 generation population, and they were mixed to a mutant pool. NGS sequencing and analysis of the parent and mutant DNA pools were performed using a bulked segregant analysis and sequencing (BSA-seq) method. Linkage analysis was also performed using molecular markers.

3. RESULTS

3.1. Seedling drought tolerance of *dtdt*

To identify the drought tolerance of *dtdt* at the seedling stage, wild type HHZ and mutant *dtdt* seedlings were cultured to two-leaf stage (about 14 days) and subjected to drought stress, while the control group was watered normally. The results showed that under normal conditions, the plant height of the mutant *dtdt* at the seedling stage was significantly lower than that of the wild type HHZ (Figs 1 (A), (C), (E). After 10 days of water-break, the mutant *dtdt* showed leaf curling, while the wild type HHZ showed leaf curling, wilting and plant bending (Fig. 1 (D). After re-watering for three days, the leaves of the mutant *dtdt* were fully expanded and the seedling survival rate was 100%. Only three plants of wild type HHZ recovered, and the seedling survival rate was 6% (Fig. 1 (F). It can be seen that the drought tolerance of the mutant *dtdt* at the seedling stage was significantly stronger than that of the wild type HHZ.

FIG. 1. Drought tolerance phenotype of dtdt *at the seedling stage. A, C, E: normal watering control group; B, D, F: drought stress treatment group; A, B are before the water-break; C, D are ten days after the water-break; E, F are three days after re-watering. HHZ: wild type;* dtdt: *mutant.*

3.2. Physiological indices of drought tolerance and expression characteristics of related genes of *dtdt*

To investigate the drought tolerance related physiological indexes of mutant *dtdt*, the mutant *dtdt* and wild type HHZ were seeded in 96-well hydroponic boxes and treated with 20% PEG6000 at the third-leaf stage. Hydrogen peroxide content was determined and nitrogen blue tetrazole (NBT) staining was performed at days 0 and 3 after drought treatment. The results showed that the hydrogen peroxide content of mutant *dtdt* was significantly lower than that of wild type HHZ in both normal and drought conditions (Fig. 2). After stress, the hydrogen peroxide content of wild type HHZ was significantly increased, while the hydrogen peroxide content of mutant *dtdt* was significantly decreased (Fig. 2).

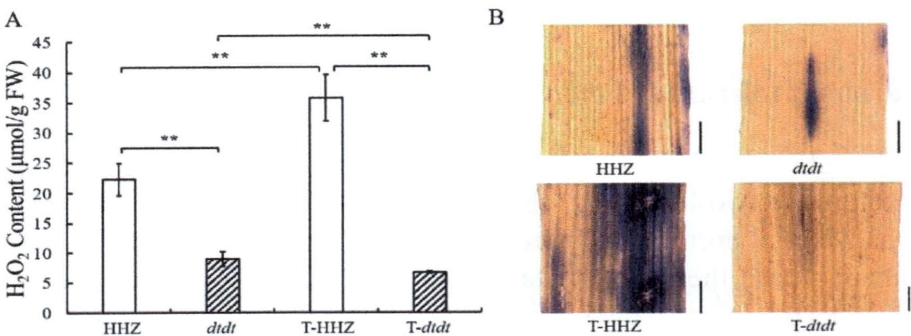

FIG. 2. H₂O₂ content and NBT staining of dtdt. *A: hydrogen peroxide content; B: NBT staining. HHZ; wild type:* dtdt; *mutant; T-HHZ: wild type under stress;* T-dtdt: *mutant under stress; Bar = 200 μm: **: significant difference (P < 0.01).*

The rice seedlings were treated with 20% PEG6000 at the third-leaf stage, and CAT activity was measured from 0 to 5 days of drought stress. The data were investigated for general statistical analysis. The results showed that the CAT activity of both mutant *dtdt* and wild type HHZ increased first and then decreased with the increase of drought stress time, and the CAT activity of mutant *dtdt* was almost always higher than that of wild type HHZ (Fig. 3 (A)). Real time PCR (qRT-PCR) analysis of three CAT isozymes in mutant *dtdt* and wild type HHZ showed that the relative expression levels of *OsCATA* and *OsCATC* in mutant *dtdt* were significantly higher than those in wild type HHZ under normal growth conditions. After 24 h of drought stress, the relative expression level of *OsCATA* in mutant *dtdt* was significantly higher than that of wild type HHZ (Fig.2.3B) (Fig. 2 (B)), which was consistent with the change of CAT activity.

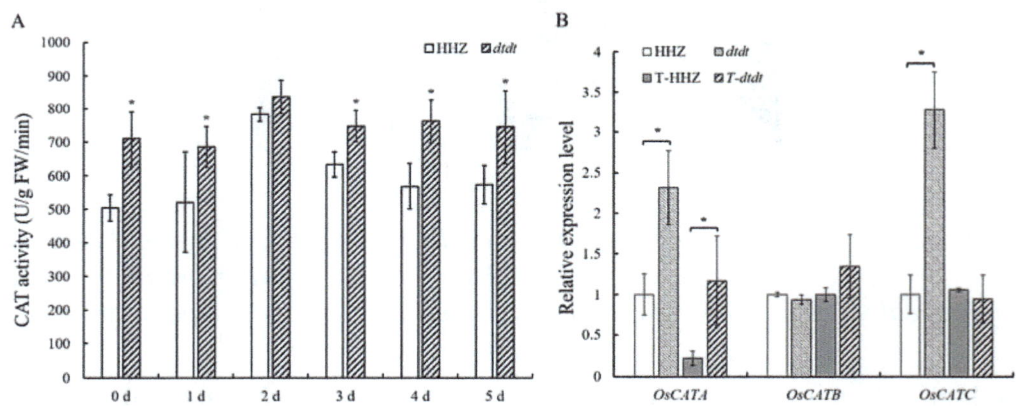

FIG. 3. CAT activity and related gene expression of dtdt. *A: CAT activity; the abscissa represents the days of drought stress; B: the expression level of CAT-related genes; HHZ: wild type;* dtdt: *mutant; T-HHZ: wild type under stress;* T-dtdt; *mutant under stress; *: represents the significant difference between mutant and wild type under the same treatment (P < 0.05).*

The rice seedlings were treated with 20% PEG6000 at the third-leaf stage, and the activity of DHAR was measured from 0 to 5 days of drought treatment. The data were investigated for general statistical analysis. The results showed that the DHAR activity of both mutant *dtdt* and wild type HHZ showed a trend of first increasing and then decreasing, and the DHAR activity of mutant *dtdt* was higher than that of wild type HHZ most of the time (Fig. 4 (A)). The expression analysis of *DHAR*-related genes in the mutant *dtdt* and wild type HHZ showed that the expression of *OsDHAR1* in both the mutant *dtdt* and wild type HHZ increased after 24 h of drought stress, and the expression of *OsDHAR1* in the mutant *dtdt* was always lower than that in the wild type HHZ (Fig. 4 (B)). This is different from DHAR activity.

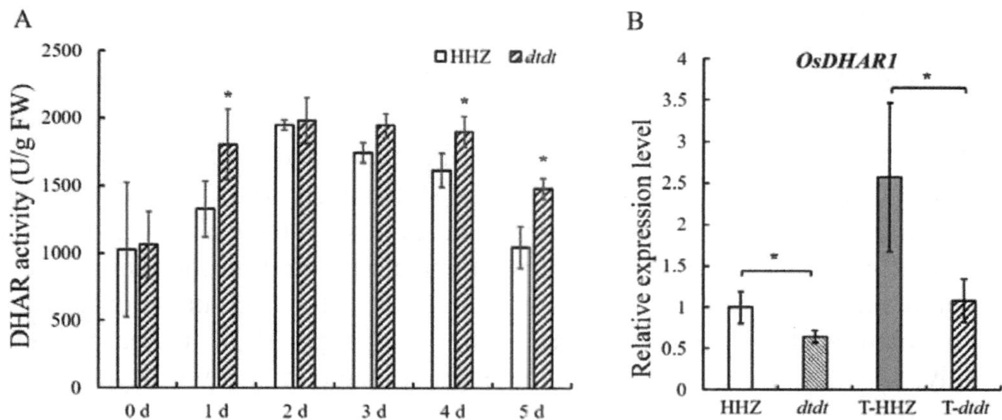

FIG. 4. Dehydroascorbate reductase activity and related gene expression of dtdt. A: dehydroascorbate reductase activity; the abscissa represents the days of drought stress; B: expression level of DHAR-related genes; HHZ, wild type; dtdt: mutant; T-HHZ: wild type under stress; T-dtdt: mutant under stress; *: significant difference between mutant and wild type under the same treatment (P < 0.05).

In addition, mutant *dtdt* and wild type HHZ were cultured in soil, and the rice seedlings were subjected to drought treatment at the third-leaf stage. The MDA content was measured from 0 to 5 days of drought treatment, and the data were investigated for general statistical analysis. The results showed that MDA content was always lower in mutant *dtdt* in comparison to wild type HHZ, and the difference was most significant on day 5 of drought stress (Fig. 5). The characteristics of MDA content were consistent with the drought tolerance phenotype of *dtdt*.

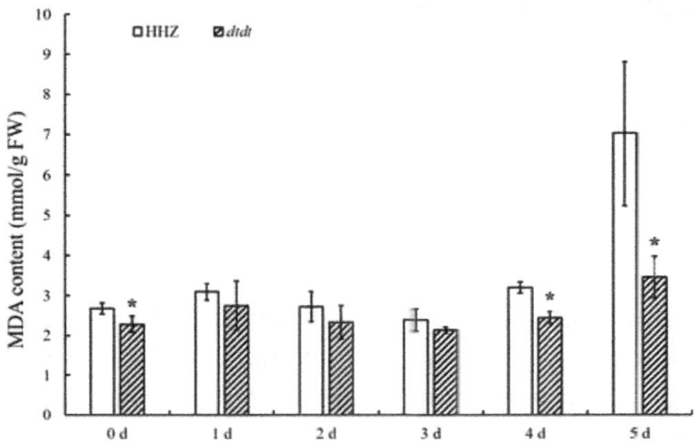

FIG. 5. Malondialdehyde content of dtdt. MDA: Malondialdehyde content; HHZ: wild type; dtdt: mutant; abscissa represents days of drought stress; *: significant difference between mutant and wild type under the same treatment (P < 0.05).

3.3. Leaf stomatal characteristics of *dtdt*

To observe the mutant *dtdt* leaf stomata characteristics, the mutant *dtdt* and wild type HHZ were sown in 96-well hydroponic boxes, and grown to the third-leaf stage using 20% PEG6000 processing. The middle portion of the third leaf at days 0 and 3 of drought stress was fixed and dried, and the abaxial leaf stomata configuration of mutant *dtdt* and wild type HHZ were observed by using scanning electron microscopy. The results showed that the stomatal volume of the mutant *dtdt* was smaller than that of the wild type HHZ (Fig. 6 (A)), and the stomatal number of the mutant *dtdt* was significantly less than that of the wild type HHZ. There are 77 and 94 stomata in nine microscopic fields in mutant *dtdt* and wild type HHZ, respectively, under normal conditions, and 80 and 102 stomata in nine microscopic fields in mutant *dtdt* and wild type HHZ, respectively, after stress of 3 days. At the same time, the statistics of stomatal aperture showed that under normal conditions, the proportion of three types of stomatal states in the mutant *dtdt* were not significantly different compared with the wildtype HHZ. The stomatal apertures in the mutant *dtdt* were not obviously different under both normal and drought conditions. However, under drought conditions, the number of completely closed stomata in the wild type HHZ increased significantly. Correspondingly, the number of completely opened stomata in the wild type HHZ reduced remarkably (Fig. 6 (B)).

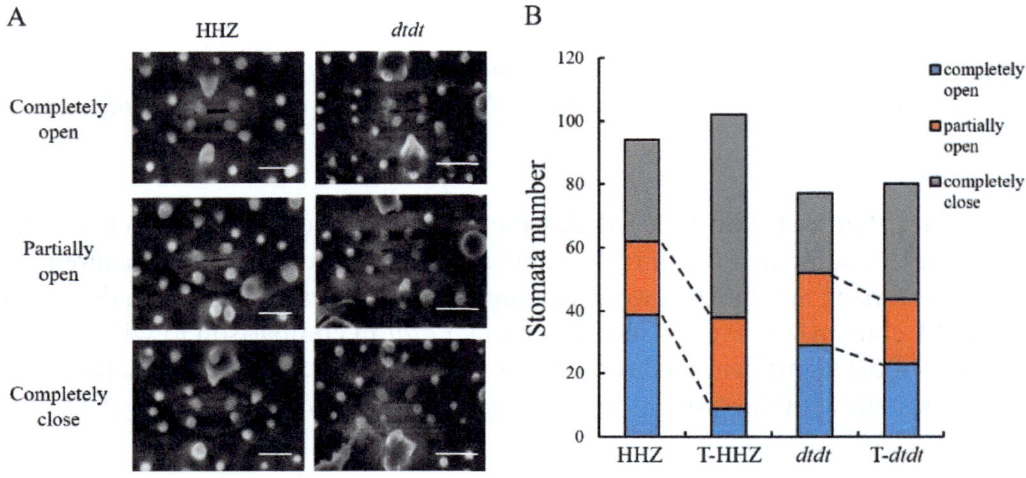

FIG. 6. Stomatal characteristics of dtdt. *A: three types of stomata in wild type and mutant; B: proportion of the three types of stomata; HHZ: wild type;* dtdt: *mutant; T-HHZ: wild type under stress;* T-dtdt: *mutant under stress; Bar = 10 μm.*

3.4. Agronomic traits of *dtdt*

To investigate the agronomic traits of mutants *dtdt* and wild type HHZ, they were cultivated in university experimental fields, and their agronomic traits were investigated when they grew to maturity. As shown in Fig. 7, the plant height, tiller number, effective panicle number and seed number per plant of the mutant *dtdt* were significantly reduced, as compared with wild type HHZ. Additionally, the panicle length and seed setting rate per plant were not obviously different between mutant *dtdt* and wild type HHZ. The plant height of mutant *dtdt* at maturity was significantly lower than that of wild type HHZ, which was like that at seedling stage. Meanwhile, the tiller number of the mutant *dtdt* was significantly lower than that of the wild type HHZ, resulting in a corresponding decrease in the number of effective panicles. Under the condition of similar panicle length and seed setting rate, the grain number per plant of the mutant *dtdt* was significantly lower than that of the wild type HHZ.

FIG. 7. Agronomic traits of dtdt. *A: plant height; B: tiller numbers; C: effective panicle numbers; D: panicle length; E: number of seeds per plant; F: seed setting rate per plant; G: mature plant; HHZ: wild type;* dtdt: *mutant (n = 25); Bar = 20 cm; *: significant difference (P < 0.05).*

3.5. Genetic analysis of *dtdt*

To confirm the few-tillering genetic characteristics of the mutant *dtdt*, F_1 generations were obtained by crossing *dtdt* as the female parent and *japonica* rice variety ZH11 as the male parent. All F_1 plants displayed normal tillering phenotype, indicating that the few-tillering phenotype of the mutant *dtdt* was recessive. F_1 generation was self-crossbred to obtain F_2 generation seeds, which were planted in the university experimental field. After 60 days of sowing, tiller traits of the F_2 population were investigated. The results showed that there were 245 rice plants in the F_2 population, of which 189 and 56 plants had the same tillering phenotype as wild type HHZ and mutant *dtdt*, respectively, with the ratio of 3.375:1 equal to the segregation ratio of 3:1 (χ^2=0.6). In conclusion, the few-tillering trait of the mutant *dtdt* was controlled by a pair of genes.

3.6. Gene mapping of *dtdt*

The F_2 segregation population was constructed by crossing the mutant *dtdt* with *japonica* rice ZH11, including 4000 plants. A total of 193 plants consistent with the mutant phenotype were selected from the F_2 population. The same amount of leaves were taken from each of 193 plants and DNA was extracted to construct the few-tillering pool. After BSA-seq, the total original data were 190.9G, sequencing data were Q20>97.25%, Q30>91.75%; GC content was 42.14-42.52%; and 97.289-98.181% of the sequencing data could be successfully matched to the reference genome. Therefore, the data volume of the samples reached the standard, the sequencing quality was qualified, the GC content distribution was normal, and the comparison between the sequencing data and the rice reference genome was normal, which could be used for subsequent mutation detection and gene mapping. Using the Nippon bare genome sequence as a reference, SNP and InDel markers in the genomes of *dtdt* and ZH11 were analysed, and the allelic frequency distribution of polymorphic markers in the few-tillering pool genome was further analysed. The results showed that there was an obvious peak on chromosome 6, and the allele frequency of the scatter reached 0 at about 3MB, which was homozygous (Fig. 8).

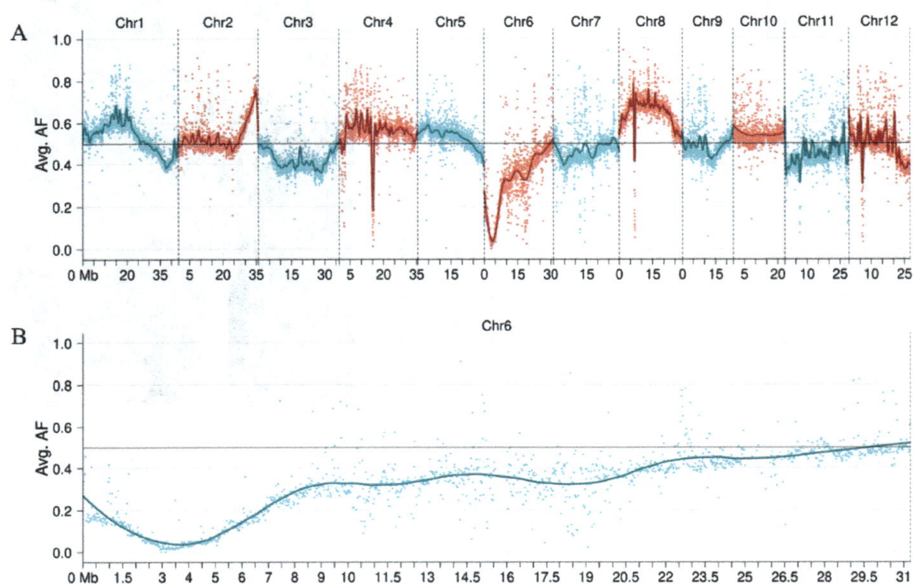

FIG. 8. Allele frequency distributions of mutant phenotypes extreme pool of F₂ population. A: genome-wide allele frequency distribution map; B: chromosome 6 magnification of allele frequencies.

In this study, 177 few-tillering individual plants in the F₂ population were used for linkage analysis using Indels markers obtained by BSA-seq and SSR markers on chromosome 6. The results showed that seven markers, including RM587 and ID5206262 on chromosome 6, were linked to *dtdt*. Through recombination analysis between markers and genes, *dtdt* was finally mapped between markers RM19410 and ID3341597, with a physical distance of 282.5 kb (Figs 9 and 10).

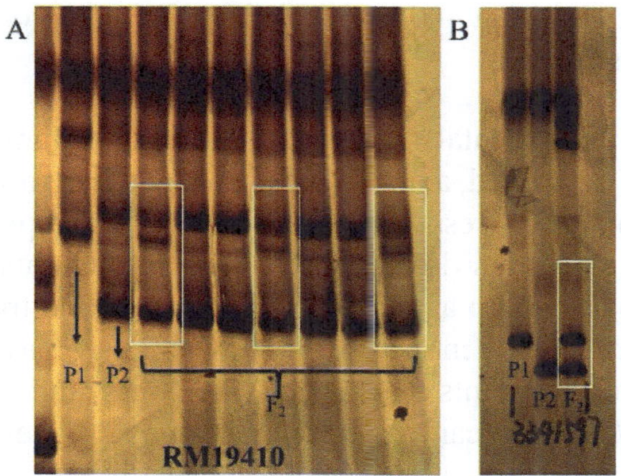

FIG. 9. Partial polyacrylamide gel electrophoresis of molecular markers. A: gel electrophoresis map of molecular marker RM19410; B: gel electrophoresis map of molecular marker ID3341597; P₁: Japonica rice ZH11; P₂: mutant dtdt; *white box indicates the recombinant plant.*

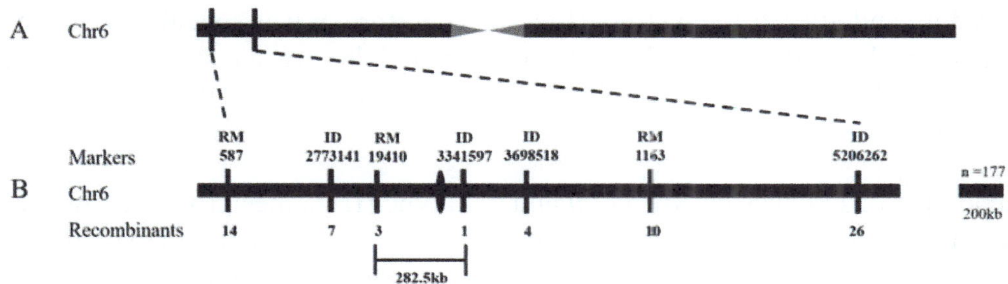

FIG. 10. Linkage map of drought tolerance and decreased tillering gene dtdt *on chromosome 6 of rice. A-B:* dtdt *is mapped between RM19410 and ID3341597 on chromosome 6, with the physical distance about 282.5 kb.*

4.   DISCUSSION

After radiation-induced mutagenesis of *indica* rice Huanghuazhan, a mutant *dtdt* with drought tolerance and reduced tillering at the seedling stage was screened from the M_3 generation (Figs 1 and 7). CAT can promote the decomposition of H_2O_2 into molecular oxygen and water and remove hydrogen peroxide in the body, thus preventing cells from being damaged by excessive H_2O_2. Therefore, CAT is one of the key enzymes in biological defence systems [4.60] [27]. DHAR is a key enzyme in the ascorbate-glutathione redox cycle. MDA is a product of membrane lipid peroxidation caused by peroxide accumulation in cells, and its degree indirectly reflects the strength of antioxidant capacity of plants. The decrease of MDA content indicates the enhancement of drought tolerance of plants [4.61] [28]. To explore the drought tolerance mechanism of mutant *dtdt*, the above physiological indicators of *dtdt* were detected in this study. The results showed that under drought stress, MDA content of mutant *dtdt* was lower than that of wild type HHZ, and DHAR activity and CAT activity were higher than that of wild type HHZ. Additionally, the expression levels of CAT related genes also increased (Figs 3–5), which was contrary to the fact that the H_2O_2 content of the mutant *dtdt* was significantly lower than that of the wild type HHZ (Fig. 2). The changing trend of physiological indicators of *dtdt* was consistent with its drought tolerance phenotype, suggesting that the mutant *dtdt* might improve its drought tolerance by regulating H_2O_2 homeostasis *in vivo*.

This study also found that the stomatal density and stomatal volume of the mutant *dtdt* on the abaxial surface of the leaves decreased, and there was no significant difference in stomatal aperture of the mutant *dtdt* before and after stress, while the number of completely closed stomata of wild type HHZ increased significantly after drought stress (Fig. 6). Stomata are known to play a critical role in gas metabolism such as carbon assimilation, respiration and transpiration. The decrease of stomatal density and stomatal volume in the leaves of mutant *dtdt* may greatly reduce transpiration rate, thus maintaining water in plants under drought stress and showing drought tolerance. Overexpression of *OsEPF1* in rice can affect stomatal development, leading to a decrease in stomatal density and stomatal index, as well as a decrease in stomatal volume, thus improving drought tolerance by limiting water loss [29]. The increase of H_2O_2 content in *DST* loss function mutant leads to increased stomatal closure and decreased stomatal density, thus enhancing drought tolerance and salt tolerance. Overexpression of *OsTF1L* promotes stomatal closure and enhances drought tolerance of rice [30]. In this study, the H_2O_2 content of the mutant *dtdt* was lower than that of the wild type HHZ (Fig.2), and the stomatal opening of the mutant *dtdt* showed no significant difference before and after drought stress (Fig. 6). Therefore, it was speculated that the mutant *dtdt* could improve its drought tolerance by reducing stomatal density and stomatal volume rather than regulating stomatal aperture by H_2O_2, which may be different from *DST* and *OsTF1L* in drought tolerance mechanism.

To investigate the phenotypic differences between *dtdt* and wild type HHZ, the agronomic traits of *dtdt* were investigated. The results showed that tiller number, effective panicle number, plant height and seed number per plant of *dtdt* were significantly reduced compared with HHZ (Fig. 7). This phenotype was not completely identical with. the *moc1* mutant, which did not have tillering with only one main stem. It was speculated that *dtdt* and *moc1* might belong to different types in regulating tillering. In this study, tiller number and plant height of the mutant *dtdt* decreased, showing a 'decrease' characteristic, suggesting that the growth and development of the mutant *dtdt* was affected.

Genetic analysis showed that *dtdt* mutation trait was controlled by a pair of recessive genes. By BSA-seq, it was found that the extreme pool allelic frequency was close to 0 on chromosome 6 (Fig. 8). Molecular marker linkage analysis showed that there were 3 and 1 recombinant plants in RM19410 and ID3341597, respectively. Therefore, *dtdt* was mapped between RM19410 and ID3341597, with a physical distance of 282.5 kb (Figs 9 and 10). At present, no tiller-related genes have been reported in this region, so *dtdt* may be a new gene regulating tillering in rice.

5. CONCLUSIONS

In this study, a drought tolerance and reduced tillering mutant *dtdt* was screened from the radiation induced offspring of Huanghuazhan, an *Indica* rice variety. Through the investigation of agronomic traits, microscopic observation, drought tolerance test at the seedling stage and analysis of physiological indexes related to drought tolerance, it was found that the stomatal density and stomatal volume of the mutant *dtdt* decreased, while the content of H_2O_2 and MDA decreased, the activity of DHAR and CAT increased, and the expression of *CAT*-related genes also increased. The combination of BSA-seq and molecular markers mapped *dtdt* between the markers RM19410 and ID3341597, with a physical distance of 282.5 kb. Therefore, it is speculated that the mutant *dtdt* may improve its drought tolerance by reducing stomatal density and stomatal volume, and regulate the cell damage caused by drought stress by regulating H_2O_2 content. The mechanism through which the mutant *dtdt* regulates stomata and H_2O_2 to regulate its drought tolerance needs to be further studied. It is also worth studying how the mutant *dtdt* affects tillering and plant development. *DTDT* may be a new gene regulating tillering in rice. However, whether *DTDT* regulates drought tolerance at the seedling stage and affects plant development the same gene needs further functional verification.

REFERENCES TO CHAPTER 4–2

[1]   WU, B, LU, Q., Features of desertification and significance of desertification control in China, China Pop. Resour. Environ. **1** (2002) 101–103 (in Chinese with English abstract).

[2]   TANAKA, T, et al., The rice annotation project database (RAP-DB): 2008 update, Nucleic Acids Res. **36** (2008) D1028–D1033.

[3]   LUO, L.J., ZHANG, Q.F., The status and strategy on drought resistance of rice (*Oryza sativa* L.), Chinese J. Rice Sci. **3** (2001) 50–55 (in Chinese with English abstract).

[4]   TCHINGA, J.B., Effects of drought stress on rice physiology and biochemistry, Open J. Nat. Sci. **8** 4 (2020) 220–226.

[5]   LIU, J., WANG, B.S., XIE, X.Z., Regulation of stomatal development in plants, Hereditas (Beijing), **33** 2 (2011) 131–137 (in Chinese with English abstract).

[6]   ZOROV, D.B., JUHASZOVA, M., SOLLOTT, S.J., Mitochondrial reactive oxygen species (ROS) and ROS-induced ROS release, Physiol. Rev. **94** 3 (2014) 909–950.

[7]   TIAN, M., RAO, L.B., LI, J.Y., Reactive oxygen species (ROS) and its physiological functions in plant cells, Plant Physiol. Commun. **2** (2005) 235–241 (in Chinese with English abstract).

[8]   LIN, Y. F., LI, W., DAI, L.Y., Research progress of antioxidant enzymes functioning in plant drought resistant process, Crop Res. **29** 3 (2015) 326–330 (in Chinese with English abstract).

[9]   WANG, Y.H., LI, J.Y., Molecular basis of plant architecture, Annu. Rev. Plant Biol. **59** (2008) 253–279.

[10]   HUANG, X.Y., et al., A previously unknown zinc finger protein, DST, regulates drought and salt tolerance in rice via stomatal aperture control, Genes Develop. **23** 15 (2009) 1805– 1817.

[11]   YOU, J., et al., The SNAC1-targeted gene OsSRO1c modulates stomatal closure and oxidative stress tolerance by regulating hydrogen peroxide in rice, J. Exper. Bot. **64** 2 (2013) 569–583.

[12]   HUANG, L.C., et al., A Nck-associated protein 1-like protein affects drought sensitivity by its involvement in leaf epidermal development and stomatal closure in rice, Plant J. **98** 5 (2019) 884–897.

[13]   WANG, Z.Y., et al., The E3 Ligase DROUGHT HYPERSENSITIVE negatively regulates cuticular wax biosynthesis by promoting the degradation of transcription factor ROC4 in Rice, Plant Cell **30** 1 (2018) 228–244.

[14]   UGA, Y., et al., Control of root system architecture by DEEPER ROOTING 1 increases rice yield under drought conditions, Nature Genetics **45** 9 (2013) 1097–1102.

[15]   MEHDY, M.C., Active oxygen species in plant defense against pathogens, Plant Physiol. **105** 2 (1994) 467–472.

[16]   XING, Y.Z., ZHANG, Q.F., Genetic and molecular bases of rice yield, Annu. Rev. Plant Biol. **61** (2010) 421–442.

[17]   JIAO, Y.Q,, et al., Regulation of OsSPL14 by OsmiR156 defines ideal plant architecture in rice, Nature Genetics **42** 6 (2010) 541–544.

[18]   LI, X.Y., et al., Control of tillering in rice, Nature **422** 6932 (2003) 618–621.

[19]   KOUMOTO, T., et al., Rice monoculm mutation *moc2*, which inhibits outgrowth of the second tillers, is ascribed to lack of a fructose-1,6-bisphosphatase, Plant Biotechnol. **30** 1 (2013) 47–56.

[20]   LU, Z.F, *MONOCULM 3*, an ortholog of *WUSCHEL* in rice, is required for tiller bud formation, J. Genetics Genomics **42** 2 (2015) 71–78.

[21]   ISHIKAWA, S., et al., Suppression of tiller bud activity in tillering dwarf mutants of rice Plant Cell Physiol. **46** 1 (2005) 79–86.

[22]   ARITE, T., et al., *DWARF10*, an *RMS1/MAX4/DAD1* ortholog, controls lateral bud outgrowth in rice, Plant J. **51** 6 (2007) 1019–1029.

[23]   ARITE, T., et al., *d14*, a strigolactone-insensitive mutant of rice, shows an accelerated outgrowth of tillers, Plant Cell Physiol. **50** 8 (2009) 1416–1424.

[24]   ZOU, J.H., The rice *HIGH-TILLERING DWARF1* encoding an ortholog of Arabidopsis MAX3 is required for negative regulation of the outgrowth of axillary buds, Plant J. **48** 5

[25]   LIN, H., DWARF27, an iron-containing protein required for the biosynthesis of strigolactones, regulates rice tiller bud outgrowth, Plant Cell **21** 5 (2009) 1512–1525.

[26]   JIANG, L., et al., DWARF53 acts as a repressor of strigolactone signalling in rice *Nature* **504** 7480 (2013) 401–405.

[27]   VIGHI, I.L., et al., Changes in gene expression and catalase activity in *Oryza sativa* L. under abiotic stress, Genetics Mol. Res. **15** 4 (2016) 10; 4238/gmr15048977.

[28]   MELANDRI, G., et al., Biomarkers for grain yield stability in rice under drought stress, J. Exper. Bot. **71** 2 (2020) 669–683.

[29]   CAINE, R.S., et al., Rice with reduced stomatal density conserves water and has improved drought tolerance under future climate conditions, New Phytol. **221** 1 (2019) 371–384.

[30]   BANG, S.W., et al., Overexpression of *OsTF1L*, a rice HD-Zip transcription factor, promotes lignin biosynthesis and stomatal closure that improves drought tolerance, Plant Biotechnol. J. **17** 1 (2019) 118–131.

4–3. IDENTIFICATION OF CAUSAL MUTATIONS THROUGH MUTMAP+ ANALYSIS OF GAMMA RAY INDUCED MUTANTS OF RICE

J. BAGRI[1], S.L. SINGLA-PAREEK[2], A. PAREEK[1, 3]

[1]School of Life Sciences, Jawaharlal Nehru University, New Delhi, India

[2]International Centre for Genetic Engineering and Biotechnology, New Delhi, India

[3]National Agri-Food Biotechnology Institute, Mohali, India

Abstract

Drought stress poses the most significant limitation on crop yield, impacting approximately one-third of the world's arable land and potentially exacerbating existing climate changes. To identify genes responsible for drought tolerance, mutation breeding proves to be a highly effective approach. Based on physiological, biochemical, and molecular profiles under drought stress, two of the potential rice mutant line, M3_1 were selected, and the detailed characterization was performed by the MutMap+ approach. MutMap+ has the advantage of both bulk segregant analysis and whole-genome re-sequencing approaches and enables the identification of candidate genes and causal SNPs that are linked to desired traits. In the present study, candidate genes (SNPs) were identified in the M3_1 rice line (genome region) that are known to be associated with drought stress tolerance. Studies also reported their involvement in various physiological processes, including signal transduction and abiotic stresses. To conclude this study, a number of potential causal SNPs and metabolic targets were identified for further in-depth investigations of genetic adaptation for tolerance against drought stress.

Key words: Rice, drought stress, causal SNPs, MutMap+, whole genome sequencing.

1. INTRODUCTION

Indexed mutant collections are extremely significant genetic resources for functional genomic studies in *Arabidopsis thaliana* and other model plant species [1, 2]. Multiple mutant collections have been established in rice in diverse genetic backgrounds, including Nipponbare, Dong Jin, Zhonghua 11, Hwayoung and Kitaake [3–5]. Rice mutants have been generated through various mechanisms, including DNA insertion [6], transposon/retrotransposon insertion [7], RNAi based gene silencing [8], TALEN-based genome editing [9], CRISPR/Cas9 genome editing, and mutation induction, using EMS and irradiation [4, 10]. Several databases have been established to facilitate the use of the mutant collections [10]. These approaches have advanced the characterization of approximately 2000 genes [11, 12]. The most common approach utilized is to screen a population of randomly mutagenized lines for a phenotype of interest and to identify the genes corresponding to these mutations subsequently. Generating transfer (T)-DNA or other insertional alleles simplifies subsequent cloning of the mutated genes, but limits the breadth of allelic diversity It is tedious to generate such populations in desired genetic backgrounds. Physical mutagens, such as gamma rays, which cause both point and larger base deletion mutations, make it easy to generate large mutagenized populations in almost any genetic background and generate a wide variety of alleles. However, cloning the causative genes from such alleles is more difficult. Diverse methods exist for these purposes, in which marker based mapping is gaining momentum. Map based cloning has been the primary strategy for isolating genes that impart agronomic traits for decades [13]. To this end, an outcross has been performed between a mutant line of interest with a second line, harbouring numerous genomic variants, most commonly small nucleotide polymorphisms (SNPs). The use of next generation, high throughput sequencing technologies and experimental approaches that leverage bulked segregant mapping has greatly facilitated the process of cloning genes corresponding to point mutations [14]. Variations of this include NGM [15], CloudMap [16], NIKS [17], SHOREmap [18], SIMPLE [19], MutMap [20], and MutMap+ [21]. Typically, these pipelines rely on SNP segregation derived from sequence variations of a distinct parental ecotype or the numerous random SNPs induced by mutagenesis. Comparison of SNP frequencies between homozygous mutant lines and reference lines may reveal regions of the genome linked to the causative mutation. However, the above technique requires a crossed with parental genotype, then further screened up to F2 progeny but it is time consuming and laborious, while the MutMap+ technique does not require to be crossed with the parental genotype [21].

The MutMap+ facilitates gene isolation and breeding of crops by reducing the time and labour required for identifying agronomically important genes [21]. As DNA sequencing is becoming easier and cheaper, the cost of identifying such genes could be markedly reduced. If a causal SNP cannot be identified, the SNPs flanking the regions harbouring causal mutations for the desired phenotypes (those with an SNP index of 1) can be used as DNA markers for marker-assisted selection by crossing the mutant to the wild type. Moreover, if mutagenesis is done in an elite crop cultivar, then mutants and associated SNP markers can be made available to breeders to generate new varieties.

In this study, we used MutMap+, a versatile extension of MutMap based on selfing of heterozygous plants showing a wild type phenotype, and identified gamma induced M_2 progeny segregating for wild type and a drought tolerant mutant phenotype of interest. In this study, we sequenced the M_3 progeny, which were grouped separately into bulks, and then underwent whole genome sequencing. Utilizing MutMap+, we bypassed the requirement for backcrossing to the wild type parent, enabling the identification of mutations in genes that do not necessitate crossing.

2. METHODOLOGY

2.1. Growth conditions and screening

The rice (*Oryza sativa* L.) gamma mutant was used in this study. In the M_2 mutant population created by gamma ray treatment of seeds from an elite indica cultivar, mega and high-yielding varieties were identified. Rice seedlings from each of the M_3 mutant plants were subjected to drought stress (see method in Section 3) and utilized to make the wild type (drought sensitive bulk) and mutant bulks (drought tolerant bulk) for whole genome sequencing. Figure 1 shows a flow chart of the overall screening strategy.

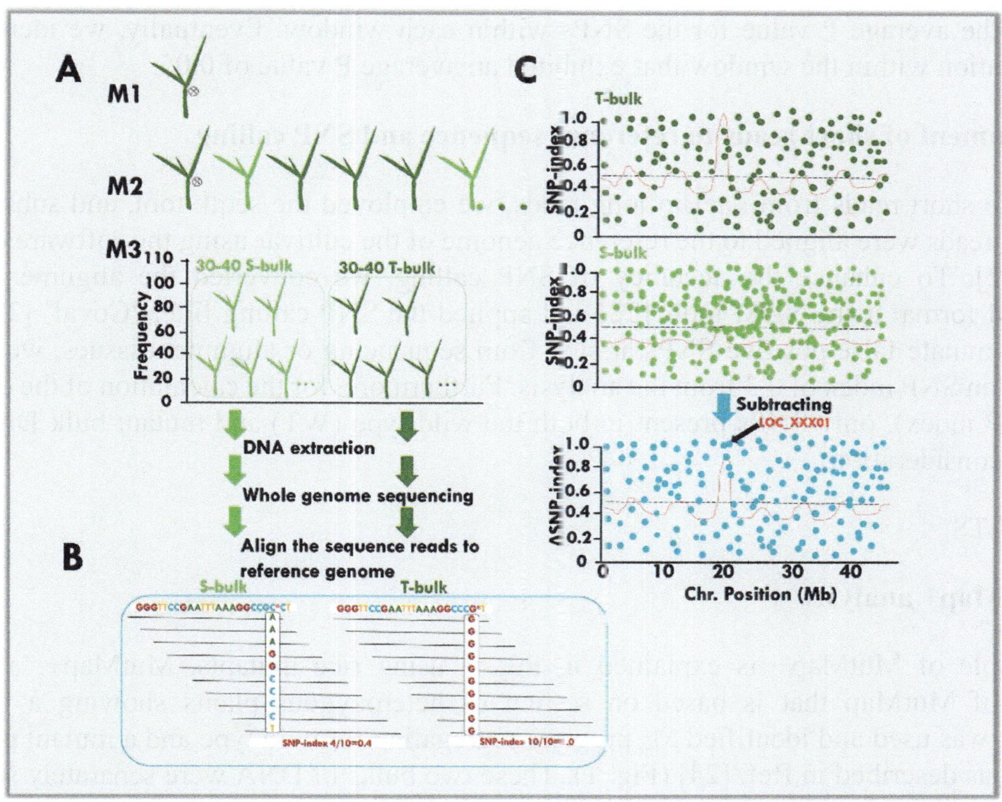

FIG. 1. Schematic representation of MutMap+ methodology. (A) Seeds harvested following gamma induced mutagenesis of rice (IR64) seeds used to establish M_1 generation.at this stage most of mutations incorporated by gamma-rays induced are in the heterozygous state. M_2 progeny obtained from a self-fertilized M_1 plant segregate for wild type (indicated by light green colour) and mutant (dark green colour) phenotypes. Here we focus on wild type heterozygous individuals. Genomic DNA from 30–40 M_3 mutant and wild type M_3 progeny are separately bulked and subjected to whole genome sequencing. The resulting short reads are aligned to reference sequence of the cultivar used for mutagenesis. (B) SNP index is calculated for each SNP, and plots relating SNP index and chromosome positions are obtained for both the mutant and wild type M_3 bulks separately. The two SNP index plots are compared to identify the genomic region with SNP index = 1 that is specific to the mutant bulk. (C) The Δ(SNP index) plot, which is derived by subtracting the SNP index value of the wild type bulk from that of the mutant bulk was analysed. The genomic region containing the causal mutation is expected to exhibit positive Δ(SNP index) values.

2.2. Whole genome sequence of bulked DNA

Genomic DNA was isolated from rice leaf samples from each M_3 plant and mixed in the same way to make a large amount of DNA that was used for sequencing. Following the manufacturer's instructions, the sequencing was done and a library made from genomic DNA.

2.3. Sliding window analysis for MutMap +

We performed sliding window analysis, utilizing a window size of 4 Mb and a 20–50 Kb increment. During this analysis, we calculated the average SNP index and conducted Fisher's exact test to determine the average P value for the SNPs within each window. Eventually, we identified the causal mutation within the window that exhibited an average P value of 0.05.

2.4. Alignment of short reads to reference sequence and SNP calling

To generate short reads from PacBio long reads, we employed the seqtk tool, and subsequently, these short reads were aligned to the reference genome of the cultivar using the software described in Ref. [22]. To enhance the accuracy of SNP calling, we converted the alignment files to SAM/BAM format using SAM tools [22] and applied the SNP calling filter 'Coval' [22, 23]. In order to eliminate false positive SNPs arising from sequencing or alignment issues, we removed SNPs with an SNP index of 0.3 from the analysis. Furthermore, for the calculation of the delta SNP Index (SNP index), only SNPs present in both the wild type (WT) and mutant bulk DNAs were taken into consideration.

3. RESULTS

3.1. MutMap+ analysis

The principle of MutMap+ is explained in Fig. 1 using rice mutants. MutMap+, a versatile extension of MutMap that is based on selfing of heterozygous plants showing a wild type phenotype, was used and identified M_2 progeny segregating for wild type and a mutant phenotype of interest, as described in Ref. [24] (Fig. 1). These two bulks of DNA were separately sequenced and aligned to the reference sequence of the parental cultivar. For each bulk, we generated the SNP index versus SNP genomic position graphs. In the majority of the genome, the ΔSNP index value is expected to be close to 0; however, it shows a notable positive deviation in the genomic region containing the causal mutation. To evaluate the statistical significance of these ΔSNP index values, we utilize Fisher's exact test.

In M3_1, we first identified a genomic region with a SNP index of 1.0 specific to the mutant bulk. We then closely examined the SNPs within this region on chromosomes. Among them, we found SNPs with a SNP index of 1.0 in the mutant bulk that belonged to various gene families, including protein tyrosine kinases (PTKs) and others. By conducting further analysis, we successfully determined the types of causal mutations in these genes, such as missense variants, three prime UTR variants, and five prime UTR variants. Subsequently, we carefully selected potential causal SNPs from these findings. These selected SNPs are being utilized with gene editing tools to further validate them.

4. DISCUSSION

Drought, salinity, high temperatures, and flooding are all believed to be threats to global food security as a result of climate change. In the background of WT, we selected the best (M3_1) gamma induced mutant rice lines, and whole genome sequencing was performed alongside that of the wild type. The genetic information related to the trait of drought tolerance, along with the allelic correlations of drought tolerance genes, has been previously documented in published studies by [24–26]. Moreover, the crucial QTLs (quantitative trait loci) associated with 'drought tolerance' were identified and finely mapped, as reported in Refs [27–30]. The development of climate resilient cultivars will make small scale farming more cost effective in the face of predicted climate change. If breeders combine it with the new 5Gs breeding strategy, it will also speed up genetic improvements and satisfy future food security needs [31]. In the post-genome sequence era, sequence based breeding has made it much easier to make cultivars that are resistant to disease [32, 33]. Upon annotating the candidate SNPs within the genome region on the chromosome for M3_1, we observed that these SNPs are situated within candidate genes associated with drought tolerance and developmental processes. Notably, SNPs belong to like protein tyrosine kinases (PTKs) and others have been previously linked to abiotic stress tolerance and are known to be involved in various physiological processes, including signal transduction and responses to abiotic stresses such as osmotic, drought and high salinity stress [34, 35].

5. CONCLUSIONS

MutMap+ offers the advantage of identifying candidate genes and causal SNPs associated with desired phenotypes through the combination of bulk segregant analysis and whole genome resequencing. In this study, we effectively applied the MutMap+ method to gamma induced mutations, uncovering candidate genes (SNPs) related to drought tolerance, as well as their involvement in various physiological and developmental processes. This study represents the first application of the MutMap+ method in this context.

REFERENCES TO CHAPTER 4–3

[1]   ALONSO, J.M., et al., Genome-wide insertional mutagenesis of *Arabidopsis thaliana*,. Science. **301** 5633 (2003) 653–657.

[2]   CHENG, X., et al., An efficient reverse genetics platform in the model legume *M. edicagotruncatula*,. New Phytol. **201** 3 (2014) 1065–1076.

[3]   WANG, L., et al., Construction of a genomewide RNA i mutant library in rice, Plant Biotechnol. J. **11** 8 (2013) 997–1005.

[4]   WEI, F.J., DROC, G., GUIDERDONI, E., YUE-IE, C.H., International consortium of rice mutagenesis: Resources and beyond, Rice **6** 1 (2013) 1–12.

[5]   LI, G., et al., The sequences of 1504 mutants in the model rice variety Kitaake facilitate rapid functional genomic studies, The Plant Cell. **29** 6 (2017) 1218–1231.

[6]   JEON, J.S., et al., T-DNA insertional mutagenesis for functional genomics in rice, Plant J. **22** 6 (2000) 561–570.

[7]   MIYAO, A., et al., Target site specificity of the Tos17 retrotransposon shows a preference for insertion within genes and against insertion in retrotransposon-rich regions of the genome, The Plant Cell, **15** 8 (2003) 1771–1780.

[8]   GHAG, S.B., ADKI, V.S., GANAPATHI, T.R., BAPAT, V.A., Plant platforms for efficient heterologous protein production, Biotechnol. Bioprocess. Eng. **26** (2021) 546–567; doi: 10.1007/s12257-020-0374-1.,

[9]   JIANG, W., et al., Demonstration of CRISPR/Cas9/sgRNA-mediated targeted gene modification in Arabidopsis, tobacco, sorghum and rice, Nucleic Acids Res. **41** 20 (2013) e188– e188.

[10]   VIANA, V.E., PEGORARO, C., BUSANELLO, C., COSTA DE OLIVEIRA, A., Mutagenesis in rice: The basis for breeding a new super plant, Front. Plant Sci. **10** (2019) 1326; doi: 10.3389/fpls.2019.01326.

[11]   DROC, G., et al., OryGenesDB: A database for rice reverse genetics, Nucleic Acids Res. **34** Suppl. 1 (2006) D736–D740.

[12]   YAMAMOTO, E., YONEMARU, J.I., YAMAMOTO, T., YANO, M., OGRO: The overview of functionally characterized genes in rice online database, Rice **5** 1 (2012) 1–10.

[13]   PETERS, J.L., CNUDDE, F., GERATS, T., Forward genetics and map-based cloning approaches, Trends Plant Sci. **8** 10 (2003) 484–491.

[14]   MICHELMORE, R.W., PARAN, I., KESSELI, R., Identification of markers linked to disease-resistance genes by bulked segregant analysis: A rapid method to detect markers in specific genomic regions by using segregating populations, Proc. Natl Acad. Sci. **88** 21 (1991) 9828–9832.

[15]   AUSTIN, R.S., Next-generation mapping of Arabidopsis genes, Plant J. **67** 4 (2011) 715– 725.

[16] MINEVICH, G., PARK, D.S., BLANKENBERG, D., POOLE, R.J., HOBERT, O., CloudMap: A cloud-based pipeline for analysis of mutant genome sequences, Genetics **192** 4 (2012) 1249–1269.

[17] NORDSTRÖM, K.J., et al., Mutation identification by direct comparison of whole-genome sequencing data from mutant and wild-type individuals using k-mers, Nature Biotechnol. **31** 4 (2013) 325–330

[18] SCHNEEBERGER, K., et al., SHOREmap: simultaneous mapping and mutation identification by deep sequencing, Nature Methods **6** 8 (2009) 550–551.

[19] WACHSMAN, G., MODLISZEWSKI, J.L., VALDES, M., BENFEY, P.N., A SIMPLE pipeline for mapping point mutations, Plant Physiol. **174** 3 (2017) 1307–1313.

[20] ABE, A., et al., Genome sequencing reveals agronomically important loci in rice using MutMap, Nature Biotechnol. **30** 2 (2012) 174–178.

[21] FEKIH, R., et al., MutMap+: genetic mapping and mutant identification without crossing in rice, PloS One **8** 7 (2013) e68529.

[22] LI, H., HANDSAKER, B., WYSOKER, A. FENNELL, T., RUAN, J.. The sequence alignment/map format and SAMtools, Bioinformatics **25** (2009) 2078–2079.

[23] NAKATA, M., et al., MutMapPlus identified novel mutant alleles of a rice starch branching enzyme IIb gene for fine-tuning of cooked rice texture, Plant Biotechnol. J. **16** 1 (2018) 111– 123.

[24] SOLIS, J., et al., Genetic mapping of quantitative trait loci for grain yield under drought in rice under controlled greenhouse conditions, Front. Chem. **5** (2018) 129.

[25] MUTHU, V., Pyramiding QTLs controlling tolerance against drought, salinity, and submergence in rice through marker assisted breeding, PloS One **15** 1 (2020) e0227421.

[26] RANJAN, R., et al., OsCPK29 interacts with MADS68 to regulate pollen development in rice, Plant Sci. **321** (2022) 111297.

[27] VIKRAM, P., et al., qDTY 1.1, a major QTL for rice grain yield under reproductive-stage drought stress with a consistent effect in multiple elite genetic backgrounds, BMC Genetics **12** (2011) 1–15.

[28] DIXIT, S., SINGH, A., KUMAR, A., Rice breeding for high grain yield under drought: A strategic solution to a complex problem, Int. J. Agron. **15** (2014).

[29] SANDHU, N., et al., Positive interactions of major-effect QTLs with genetic background that enhances rice yield under drought, Sci. Rep. **8** 1 (2018) 1626.

[30] CUI, Y., et al., Simultaneous improvement and genetic dissection of drought tolerance using selected breeding populations of rice, Front. Plant Sci. **9** (2018) 320.

[31] VARSHNEY, R.K., et al., 5Gs for crop genetic improvement, Curr. Opin. Plant Biol. **56** (2020) 190–196.

[32] VARSHNEY, R.K., SINGH, V.K., KUMAR, A.. POWELL, W., SORRELLS, M.E., Can genomics deliver climate-change ready crops? Current Opin. Plant Biol. **45** (2018) 205– 211.

[33] VARSHNEY, R.K., et al., Toward the sequence-based breeding in legumes in the post-genome sequencing era, Theor. Appl. Genetics **132** 3 (2019) 797–816.

[34] PAREEK, A., DHANKHER, O.P., FOYER, C.H., Mitigating the impact of climate change on plant productivity and ecosystem sustainability, J. Exper. Bot. **71** 2 (2020) 451–456.

[35] LIM, S.D., et al., Molecular dissection of a rice microtubule-associated RING finger protein and its potential role in salt tolerance in Arabidopsis, Plant Mol. Biol. **89** (2015) 365–384.

PART 5
SUMMARY PERSPECTIVE

5–1. INTEGRATING INDUCED MUTAGENESIS INTO THE PLANT BREEDING PIPELINE TO DEVELOP DROUGHT RESILIENCE IN CROP PLANTS

S. PENNA[1], S. FATMA[2], S. SIVASANKAR[2]

[1]Amity Center for Nuclear Biotechnology, Amity Institute of Biotechnology,
Amity University of Maharashtra Mumbai,
Mumbai, India

[2]Plant Breeding and Genetics Section,
Joint FAO/IAEA Centre of Nuclear Techniques in Food and Agriculture,
International Atomic Energy Agency, Vienna

Drought stress is the most critical abiotic stress affecting crop production over recent years and has become a research priority for plant breeders worldwide as it poses a serious challenge to food security. Erratic weather patterns, rising global temperatures and other environmental stresses further exacerbate crop productivity. Therefore it is necessary to find innovative and sustainable approaches to maintain or increase productivity. Current genetic resources might not be adequate to stabilize productivity under drought and breeders might need to explore wild germplasm, landraces and induced novel genetic variability to tap into unexplored diversity [1, 2]. Natural and induced genetic variation can produce phenotypes with tolerance through the generation of mutations at loci contributory to the trait, and these novel mutations can easily be introgressed into elite cultivars. Contrary to this, unlocking adaptive genetic variation hidden in related wild species and early landraces remains a major challenge for polygenic complex traits such as abiotic stress tolerance. In this context, induction of genetic variation and screening of mutant germplasm remain the most feasible approaches. Precise selection through statistically robust field experimental design and analysis is crucial to recover mutants that can be confidently identified as tolerant to drought. Such mutants further serve as important genomic resources for the establishment of genetic associations between secondary traits contributory to drought tolerance and the underlying genetic loci or gene(s). Modern genomic and analytical tools for high throughput screening can serve this purpose. These include genomic prediction, machine learning, and multi-trait gene editing, all of which can enable speed breeding and facilitate pre- breeding efforts for selection and characterization of drought adaptability and yield.

Complex traits such as drought require unravelling of the intricate mechanisms of genetic control of tolerance responses, which are essential for molecular breeding strategies. In breeding strategies, interactions between genetics and the environment (G × E) are also critical to understand the adaptation and deployment of drought tolerant genotypes to specific environments. Advanced genomics and genetic tools integrated with precise phenotyping and trait based breeding approaches are being developed to contribute to understand the genes and metabolic pathways underlying secondary traits contributory to drought tolerance in crops. Current advancements in integrating a gene-to-phenotype concept in crop improvement address secondary traits measured with phenotyping tools, such as NIR spectroscopy, canopy spectral reflectance, infrared thermography, magnetic resonance imaging and positron emission tomography, nuclear magnetic resonance and advanced imaging platforms, and correlate the phenotyping data with whole-genome sequencing information or candidate genomic regions [3]. The multiplicity and polygenic characteristics of the drought tolerance trait demand a multidisciplinary research approach integrating breeding with marker assisted selection, simulation modelling, physiology, and molecular genetics [4, 5]. Some of the important steps for the implementation of such an approach include:

(1) Accurate characterization of the major patterns of stress and their frequency of occurrence in the target environment.
(2) Evaluation of crop yield response to the major drought patterns (simulation modelling).
(3) Exploitation of crop phenology patterns for matching development stages (growth period, sowing, flowering, seed filling) with the most favourable period of soil moisture and climate regimes.
(4) Identification of plant traits that would maximize crop water productivity, precision of screening tools and protocols for consistent phenotypic description.
(5) Physiological dissection of crop water relations and water use efficiency.
(6) Marker based screening of breeding lines and populations.
(7) Use of comparative and functional genomics tools to elucidate drought tolerance traits and molecular/biochemical mechanisms.

The application of appropriate screening techniques for stress tolerance is one of the most crucial steps in the selection and/or development of stress tolerant varieties. Most crop species are sensitive to such stresses at all stages of plant development, mostly seed germination and pre- or post-flowering, as a consequence of which their growth and economic yield are substantially reduced under stress [6]. Thus, to identify germplasm with stress tolerance, it is necessary to develop screening methods that are simple and reproducible under the target environment conditions [7]. Applying uniform stress (drought nursery approach) through uniform soil water profile, uniform pre-stress crop growth, precise water application is also very important to differentiate among genotypes.

A major challenge for the rapid development of crop varieties, and the application of induced mutagenesis and mutation breeding where large populations are required, is the ability to quickly evaluate the germplasm. Hence reliable, simple, consistent and efficient screening/phenotyping methods and strategies are most needed [8]. It is also essential to generate genetic resources using mutational and biotechnological methods for predicting gene-to-phenotype associations and designing crop plants for climate resilience and enhanced food security. In this regard, interventions through genomic resources, germplasm sequencing, reverse genetic tools, mutant based functional genomics, sequencing based trait mapping, and genomics assisted breeding approaches will play a crucial role in developing drought stress tolerance in crop plants [5. 9]. Mutational genomics is an important tool to investigate the mutational events orchestrating genetic modification in mutant traits [10, 11]. Such mutational events can be characterized globally by using genomics technologies such as TILLING, TILLING by Sequencing and MutMap [12–14].

There has been steady progress and accomplishment under the IAEA Coordinated Research Programme (CRP) for the screening and selection of drought tolerant mutants with an aim to develop drought resilient mutant varieties in rice and sorghum. Mapping populations in these crops developed using the various mutants with drought tolerance identified as a result of research in this CRP can become a useful resource to determine relevant genetic associations. Further, the stable and well characterized rice and sorghum mutants developed can be deployed directly as new varieties or used indirectly in cross-breeding programmes for new varietal development.

REFERENCES TO CHAPTER 5–1

[1] CORTÉS, A.J., LÓPEZ-HERNÁNDEZ, F., BLAIR, M.W., Genome–environment associations, an innovative tool for studying heritable evolutionary adaptation in orphan crops and wild relatives, Front. Genet. **13** 910386 (2022); doi: 10.3389/fgene.2022.910386.

[2] KUMAR, M., RANI, K., PENNA, S., Induced mutagenesis for developing climate resilience in plants", Mutation Breeding for Sustainable Food Production and Climate Resilience (PENNA, S., JAIN, S.M., Eds), Springer, Singapore (2023) 171–203; https://doi.org/10.1007/978-981-16-9720-3_7.

[3] MIR, R.R., ZAMAN-ALLAH, M., SREENIVASULU, N., TRETHOWAN, R., VARSHNEY, R.K., Integrated genomics, physiology and breeding approaches for improving drought tolerance in crops, Theor. Appl. Genet. **125** 4 (2012) 25–45.

[4] SREEMAN, S.M., et al., Introgression of physiological traits for a comprehensive improvement of drought adaptation in crop plants, Front. Chem. **6** (2018); https://www.frontiersin.org/article/10.3389/fchem.2018.00092.

[5] FOOD AND AGRICULTURE ORGANIZATION OF THE UNITED NATIONS, Manual on Mutation Breeding, Third Edition (SPENCER-LOPES, M.M., FORSTER, B.P., JANKULOSKI, L., Eds), FAO, Rome (2018).

[6] SARSU, F., SPENCER-LOPES, M.M., SANGWAN, R., PENNA, S., "Specific techniques for increasing efficiency of mutation breeding: In vitro methods in plant mutation breeding". Manual on Mutation Breeding, Third Edition, ibid., pp. 205–226.

[7] SARSU, F., PENNA, S., NIKALJE, G.C., "Strategies for screening induced mutants for stress tolerance", Mutation Breeding for Sustainable Food Production and Climate Resilience (PENNA, S., JAIN, S.M., Eds), Springer, Singapore (2023); https://doi.org/10.1007/978-981-16-9720-3_6.

[8] REYNOLDS, M., et al., Breeder friendly phenotyping, Plant Sci. **295** (2020) 110396; https://doi.org/10.1016/j.plantsci.2019.110396.

[9] BOHRA, A., CHAND JHA, U., GODWIN, I.D., KUMAR VARSHNEY, R., Genomic interventions for sustainable agriculture, Plant Biotechnol. J. **18** 12 (2020) 2388–2405.

[10] PRASANNA, S., JAIN, S.M., Mutant resources and mutagenomics in crop plants, Emirates J. Food Agric. **29** 9 (2017) 651–657; doi:10.9755/ejfa.2017.v29.i9.86.

[11] JANKOWICZ-CIESLAK, J., MBA, C., TILL, B.J.,. "Mutagenesis for crop breeding and functional genomics", Biotechnologies for Plant Mutation Breeding (JANKOWICZ-CIESLAK, J., TAI, T., KUMLEHN, J., TILL, B., Eds), Springer, Cham (2017); https://doi.org/10.1007/978-3-319-45021-6_1.

[12] FEKIH, R., et al., MutMap+: Genetic mapping and mutant identification without crossing in rice, PLoS One **8** (2013); e68529; doi: 10.1371/journal.pone.0068529.

[13] TRAN, Q.H., BUI, N.H., KAPPEL, C., Mapping-by-sequencing via MutMap identifies a mutation in ZmCLE7 underlying fasciation in a newly developed EMS mutant population in an elite tropical maize inbred, Genes (Basel) **11** 3 (2020) 281; doi:10.3390/genes11030281

[14]   JIANG, C., et al., A reference-guided TILLING by amplicon-sequencing platform supports forward and reverse genetics in barley, Plant Commun. **11** 3(4) (2022) 100317; doi: 10.1016/j.xplc.2022.100317.

LIST OF CONTRIBUTORS

Ahmad, F.	Agrotechnology and Biosciences Division, Malaysian Nuclear Agency, Bangi, 43000 Kajang, Selangor, Malaysia
Ashraf, M.	Principal Scientist, Nuclear Institute for Agriculture and Biology, P.O. Box 128, Jhang Road, Faisalabad, Pakistan
Badigannavar, A.	Nuclear Agriculture and Biotechnology Division, Bhabha Atomic Research Centre, Mumbai, India
Bagri, J.	Jawaharlal Nehru University, New Delhi, India
Begum, S.N.	Plant Breeding Division, Bangladesh Institute of Nuclear Agriculture, Mymensingh, Bangladesh
Bibi, S.	Principal Scientist, Nuclear Institute for Agriculture and Biology, P.O. Box 128, Jhang Road, Faisalabad, Pakistan
Bimpong, I.K.	Joint FAO/IAEA Centre of Nuclear Techniques in Food and Agriculture, Department of Nuclear Sciences and Applications, International Atomic Energy Agency, 1400 Vienna, Austria
Carlos da Maia, L.	Plant Genomics and Breeding Center, Eliseu Maciel School of Agronomy, Federal University of Pelotas. Av. Eliseu Maciel s/n, Capão do Leão, RS, 96160-000, Brazil
Chatterjee, Y.	International Centre for Genetic Engineering and Biotechnology, New Delhi, India
Costa de Oliveira, A.	Plant Genomics and Breeding Center, Eliseu Maciel School of Agronomy, Federal University of Pelotas. Av. Eliseu Maciel s/n, Capão do Leão, RS, 96160-000, Brazil.
Ganapathi, T.R.	Nuclear Agriculture and Biotechnology Division, Bhabha Atomic Research Centre, Mumbai, India
Girish, G.	Agriculture Research Station, University of Agriculture Sciences Raichur, Hagari, India
Guo, Huijun	Engineering Lab for Crop Molecular Breeding; The National Key Facility for Crop Gene Resources and Genetic Improvement, Institute of Crop Science, Chinese Academy of Agricultural Sciences, Beijing 100081, China
Harun, A.R.	Agrotechnology and Biosciences Division, Malaysian Nuclear Agency, Bangi, 43000 Kajang, Selangor, Malaysia

Hasan, N.A.	School of Biology, Faculty of Applied Sciences, Universiti Teknologi MARA Cawangan Negeri Sembilan, Kampus Kuala Pilah, 72000 Negeri Sembilan, Malaysia
Hisham, S.N.	Department of Biological Sciences and Biotechnology, Faculty of Science and Technology, Universiti Kebangsaan Malaysia, 43600 Bangi, Selangor, Malaysia
Hossain, M.A.	Department of Genetics and Plant Breeding, Bangladesh Agricultural University, Mymensingh, Bangladesh
Hossain, Md.A.	Department of Biotechnology and Genetic Engineering, Noakhali Science and Technology University, Bangladesh; Department of Genetics and Plant Breeding, Bangladesh Agricultural University, Mymensingh, Bangladesh
Hussein, S.	Agrotechnology and Biosciences Division, Malaysian Nuclear Agency, Bangi, 43000 Kajang, Selangor, Malaysia
Indriatama, W.M.	Research Center for Radiation Process Technology, Gedung 720, KST B.J. Habibie Serpong, Research Organisation for Nuclear Technology, National Research and Innovation Agency, Indonesia
Islam, M.M.	Plant Breeding Division, Bangladesh Institute of Nuclear Agriculture, Mymensingh, Bangladesh
Ja'afar, M.A.N.	School of Biology, Faculty of Applied Sciences, Universiti Teknologi MARA Cawangan Negeri Sembilan, Kampus Kuala Pilah, 72000 Negeri Sembilan, Malaysia
Joseph, R.	Plant Genomics and Breeding Center, Eliseu Maciel School of Agronomy, Federal University of Pelotas. Av. Eliseu Maciel s/n, Capão do Leão, RS, 96160-000, Brazil
Kamarudin, Z.S.	Department of Crop Science, Faculty of Agriculture, Universiti Putra Malaysia, 43400 UPM, Serdang, Selangor, Malaysia
Kamissoko, S.	Institut d'Economie Rurale, BP 258, Rue Mohamed V, Bamako, Mali
Khan, A.	Senior Scientist, Nuclear Institute for Agriculture and Biology, P.O. Box 128, Jhang Road, Faisalabad, Pakistan
Khanom, Mst.S.R.	Plant Breeding Division, Bangladesh Institute of Nuclear Agriculture, Mymensingh, Bangladesh

Konaté, F.

Institut d'Economie Rurale, BP 258, Rue Mohamed V, Bamako, Mali

Kopp da Luz, V.

Plant Genomics and Breeding Center, Eliseu Maciel School of Agronomy, Federal University of Pelotas. Av. Eliseu Maciel s/n, Capão do Leão, RS, 96160-000, Brazil

Kouressy, M.

Institut d'Economie Rurale, BP 258, Rue Mohamed V, Bamako, Mali

Kumar, B.

International Centre for Genetic Engineering and Biotechnology, New Delhi, India

Kusuma Dewi, A.

Research Center for Radiation Process Technology, Gedung 720, KST B.J. Habibie Serpong, Research Organisation for Nuclear Technology, National Research and Innovation Agency, Indonesia

Lan, Tao

College of Agriculture, Fujian Agriculture and Forestry University, Fujian Province, China

Lan, Tong

College of Agriculture, Fujian Agriculture and Forestry University, Fujian Province, China

Le Duc, T.

Agricultural Genetics Institute, Hanoi, Viet Nam

Le Huy, H.

Agricultural Genetics Institute, Hanoi, Viet Nam; University of Science and Technology, Viet Nam National University, Hanoi

Li, Guiying

Engineering Lab for Crop Molecular Breeding; The National Key Facility for Crop Gene Resources and Genetic Improvement, Institute of Crop Science, Chinese Academy of Agricultural Sciences, Beijing 100081, China

Liu, Luxiang

Engineering Lab for Crop Molecular Breeding; The National Key Facility for Crop Gene Resources and Genetic Improvement, Institute of Crop Science, Chinese Academy of Agricultural Sciences, Beijing 100081, China

Maïga, A.

Institut d'Economie Rurale, BP 258, Rue Mohamed V, Bamako, Mali

Minh, N. Nguyen Thi

Agricultural Genetics Institute, Hanoi, Viet Nam

Minh, T. Vo Thi

Agricultural Genetics Institute, Hanoi, Viet Nam

Mishra, M.

International Centre for Genetic Engineering and Biotechnology. New Delhi, India

Nguyen, H. Thi — Agricultural Genetics Institute, Hanoi, Viet Nam

Oyebamiji, Y.O. — Department of Biological Sciences and Biotechnology, Faculty of Science and Technology, Universiti Kebangsaan Malaysia, 43600 Bangi, Selangor, Malaysia

Pareek, A. — Jawaharlal Nehru University, New Delhi, India; National Agri-Food Biotechnology Institute, Mohali, India

Pegoraro, C. — Plant Genomics and Breeding Center, Eliseu Maciel School of Agronomy, Federal University of Pelotas. Av. Eliseu Maciel s/n, Capão do Leão, RS, 96160-000, Brazil

Penna, S. — Amity Center for Nuclear Biotechnology, Amity Institute of Biotechnology, Amity University of Maharashtra Mumbai, Mumbai 410206, India

Pham, Xuan H. — Agricultural Genetics Institute, Hanoi, Viet Nam

Qamar, Z. — Principal Scientist, Nuclear Institute for Agriculture and Biology, P.O. Box 128, Jhang Road, Faisalabad, Pakistan

Rahman, S.A.A. — Agrotechnology and Biosciences Division, Malaysian Nuclear Agency, Bangi, 43000 Kajang, Selangor, Malaysia

Reflinur — Research Center for Genetic Engineering, Soekarno Science and Technology Area, Cibinong, Research Organization for Natural Sciences and Environment, National Research and Innovation Agency, Indonesia

Ren, Lijun — College of Agriculture, Fujian Agriculture and Forestry University, Fujian Province, China

Saha, U. — Department of Biotechnology, Bangladesh Agricultural University, Mymensingh, Bangladesh

Sahoo, R.N. — Jawaharlal Nehru University, New Delhi, India

Salleh, S. — Agrotechnology and Biosciences Division, Malaysian Nuclear Agency, Bangi, 43000 Kajang, Selangor, Malaysia

Sanogo, S. — Institut d'Economie Rurale, BP 258, Rue Mohamed V, Bamako, Mali

Sarsu, F. — Joint FAO/IAEA Centre of Nuclear Techniques in Food and Agriculture, Department of Nuclear Sciences and Applications, International Atomic Energy Agency, Vienna, Austria

Shamsudin, N.A.A.   Department of Biological Sciences and Biotechnology, Faculty of Science and Technology, Universiti Kebangsaan Malaysia, 43600 Bangi, Selangor, Malaysia

Sharma, A.C.   Department of Genetics and Plant Breeding, Bangladesh Agricultural University, Mymensingh, Bangladesh

Singla-Pareek, S.L.   International Centre for Genetic Engineering and Biotechnology, New Delhi, India

Sissoko, A.   Institut d'Economie Rurale, BP 258, Rue Mohamed V, Bamako, Mali

Sivasankar, S.   Joint FAO/IAEA Centre of Nuclear Techniques in Food and Agriculture, Department of Nuclear Sciences and Applications, International Atomic Energy Agency, 1400 Vienna, Austria

Tejeda, L.H.C.   Plant Genomics and Breeding Center, Eliseu Maciel School of Agronomy, Federal University of Pelotas. Av. Eliseu Maciel s/n, Capão do Leão, RS, 96160-000, Brazil

Teketé, M.L.   Institut d'Economie Rurale, BP 258, Rue Mohamed V, Bamako, Mali

Teme, N.   Institut d'Economie Rurale, BP 258, Rue Mohamed V, Bamako, Mali

Théra, K.   Institut d'Economie Rurale, BP 258, Rue Mohamed V, Bamako, Mali

Vaksmann, M.   CIRAD, UMR AGAP Institut, Montpellier F-34398, France

Venske, E.   Plant Genomics and Breeding Center, Eliseu Maciel School of Agronomy, Federal University of Pelotas. Av. Eliseu Maciel s/n, Capão do Leão, RS, 96160-000, Brazil

Wahid, A.N.A.   Agrotechnology and Biosciences Division, Malaysian Nuclear Agency, Bangi, 43000 Kajang, Selangor, Malaysia

Yebedié, A.,   Institut d'Economie Rurale, BP 258, Rue Mohamed V, Bamako, Mali

Yusof, S.N.   Department of Biological Sciences and Biotechnology, Faculty of Science and Technology, Universiti Kebangsaan Malaysia, 43600 Bangi, Selangor, Malaysia

Zhang, Jin   College of Agriculture, Fujian Agriculture and Forestry University, Fujian Province, China

Zhu, Li

Biotechnology Research Institute, Chinese Academy of Agricultural Sciences, Beijing 100081, China

IAEA
International Atomic Energy Agency

ORDERING LOCALLY

IAEA priced publications may be purchased from the sources listed below or from major local booksellers.

Orders for unpriced publications should be made directly to the IAEA. The contact details are given at the end of this list.

NORTH AMERICA

Bernan / Rowman & Littlefield

15250 NBN Way, Blue Ridge Summit, PA 17214, USA
Telephone: +1 800 462 6420 • Fax: +1 800 338 4550
Email: orders@rowman.com • Web site: www.rowman.com/bernan

REST OF WORLD

Please contact your preferred local supplier, or our lead distributor:

Eurospan

1 Bedford Row
London
WC1R 4BU
United Kingdom

Trade Orders and Enquiries:

Tel: +44 (0)1235 465576
Email: trade.orders@marston.co.uk

Individual Customers:

Tel: +44 (0)1235 465577
Email: direct.orders@marston.co.uk
www.eurospanbookstore.com/iaea

For further information:

Tel. +44 (0) 207 240 0856
Email: info@eurospan.co.uk
www.eurospan.co.uk

Orders for both priced and unpriced publications may be addressed directly to:

Marketing and Sales Unit
International Atomic Energy Agency
Vienna International Centre, PO Box 100, 1400 Vienna Austria
Telephone: +43 1 2600 22529 or 22530 • Fax: +43 1 26007 22529
Email: sales.publications@iaea.org • Web site: www.iaea.org/publications

Printed and bound by CPI Group (UK) Ltd, Croydon, CR0 4YY

19/05/2026

INSIDE
RED BULL RACING